Intended and Unintended Islanding of Distribution Grids

Other related titles:

You may also like
- PBPO094 | Salman K. Salman | Introduction to the Smart Grid: Concepts, technologies and evolution | 2017
- PBPO095 | S.M. Muyeen, Saifur Rahman | Communication, Control and Security Challenges for the Smart Grid | 2017
- PBPO097 | Dusmanta Kumar Mohanta, M. Jaya Bharata Reddy | Synchronized Phasor Measurements for Smart Grids | 2017
- PBPO073 | Vaccaro | Wide Area Monitoring, Protection and Control Systems: The enabler for smarter grids | 2016
- PBPO086 | Federico Milano | Advances in Power System Modelling, Control and Stability Analysis | 2016

We also publish a wide range of books on the following topics:
Computing and Networks
Control, Robotics and Sensors
Electrical Regulations
Electromagnetics and Radar
Energy Engineering
Healthcare Technologies
History and Management of Technology
IET Codes and Guidance
Materials, Circuits and Devices
Model Forms
Nanomaterials and Nanotechnologies
Optics, Photonics and Lasers
Production, Design and Manufacturing
Security
Telecommunications
Transportation

All books are available in print via https://shop.theiet.org or as eBooks via our Digital Library https://digital-library.theiet.org.

IET ENERGY ENGINEERING SERIES 231

Intended and Unintended Islanding of Distribution Grids

Edited by
Michael Finkel

The Institution of Engineering and Technology

About the IET

This book is published by the Institution of Engineering and Technology (The IET).

We inspire, inform and influence the global engineering community to engineer a better world. As a diverse home across engineering and technology, we share knowledge that helps make better sense of the world, to accelerate innovation and solve the global challenges that matter.

The IET is a not-for-profit organisation. The surplus we make from our books is used to support activities and products for the engineering community and promote the positive role of science, engineering and technology in the world. This includes education resources and outreach, scholarships and awards, events and courses, publications, professional development and mentoring, and advocacy to governments.

To discover more about the IET, please visit https://www.theiet.org/.

About IET Books

The IET publishes books across many engineering and technology disciplines. Our authors and editors offer fresh perspectives from universities and industry. Within our subject areas, we have several book series steered by editorial boards made up of leading subject experts.

We peer review each book at the proposal stage to ensure the quality and relevance of our publications.

Get involved

If you are interested in becoming an author, editor, series advisor, or peer reviewer please visit https://www.theiet.org/publishing/publishing-with-iet-books/ or contact author_support@ theiet.org.

Discovering our electronic content

All of our books are available online via the IET's Digital Library. Our Digital Library is the home of technical documents, eBooks, conference publications, real-life case studies and journal articles. To find out more, please visit https://digital-library.theiet.org.

In collaboration with the United Nations and the International Publishers Association, the IET is a Signatory member of the SDG Publishers Compact. The Compact aims to accelerate progress to achieve the Sustainable Development Goals (SDGs) by 2030. Signatories aspire to develop sustainable practices and act as champions of the SDGs during the Decade of Action (2020–2030), publishing books and journals that will help inform, develop, and inspire action in that direction.

In line with our sustainable goals, our UK printing partner has FSC accreditation, which is reducing our environmental impact to the planet. We use a print-on-demand model to further reduce our carbon footprint.

Published by The Institution of Engineering and Technology, London, United Kingdom

The Institution of Engineering and Technology (the "**Publisher**") is registered as a Charity in England & Wales (no. 211014) and Scotland (no. SC038698).

Copyright © The Institution of Engineering and Technology and its licensors 2024 First published 2024

All intellectual property rights (including copyright) in and to this publication are owned by the Publisher and/or its licensors. All such rights are hereby reserved by their owners and are protected under the Copyright, Designs and Patents Act 1988 ("**CDPA**"), the Berne Convention and the Universal Copyright Convention.

With the exception of:

   (i)   any use of the publication solely to the extent as permitted under:

       a. the CDPA (including fair dealing for the purposes of research, private study, criticism or review); or

       b. the terms of a licence granted by the Copyright Licensing Agency ("**CLA**") (only applicable where the publication is represented by the CLA); and/or

   (ii) any use of those parts of the publication which are identified within this publication as being reproduced by the Publisher under a Creative Commons licence, Open Government Licence or other open source licence (if any) in accordance with the terms of such licence,

no part of this publication, including any article, illustration, trade mark or other content whatsoever, may be used, reproduced, stored in a retrieval system, distributed or transmitted in any form or by any means (including electronically) without the prior permission in writing of the Publisher and/or its licensors (as applicable).

The commission of any unauthorised activity may give rise to civil or criminal liability.

Please visit https://digital-library.theiet.org/copyrights-and-permissions for information regarding seeking permission to reuse material from this and/or other publications published by the Publisher. Enquiries relating to the use, including any distribution, of this publication (or any part thereof) should be sent to the Publisher at the address below:

The Institution of Engineering and Technology
Futures Place,
Kings Way, Stevenage
Herts, SG1 2UA,
United Kingdom

www.theiet.org

While the Publisher and/or its licensors believe that the information and guidance given in this publication is correct, an individual must rely upon their own skill and judgement when performing any action or omitting to perform any action as a result of any statement, opinion or view expressed in the publication and neither the Publisher nor its licensors assume and hereby expressly disclaim any and all liability to anyone for any loss or damage caused by any action or omission of an action made in reliance on the publication and/or any error or omission in the publication, whether or not such an error or omission is the result of negligence or any other cause. Without limiting or otherwise affecting the generality of this statement and the disclaimer, whilst all URLs cited in the publication are correct at the time of press, the Publisher has no responsibility for the persistence or accuracy of URLs for external or third-party internet websites and does not guarantee that any content on such websites is, or will remain, accurate or appropriate.

Whilst every reasonable effort has been undertaken by the Publisher and its licensors to acknowledge copyright on material reproduced, if there has been an oversight, please contact the Publisher and we will endeavour to correct this upon a reprint.

Trade mark notice: Product or corporate names referred to within this publication may be trade marks or registered trade marks and are used only for identification and explanation without intent to infringe.

Where an author and/or contributor is identified in this publication by name, such author and/or contributor asserts their moral right under the CPDA to be identified as the author and/or contributor of this work.

British Library Cataloguing in Publication Data

A catalogue record for this product is available from the British Library

ISBN 978-1-83953-697-7 (hardback)
ISBN 978-1-83953-698-4 (PDF)

Typeset in India by MPS Limited

Image credits
Cover picture: St. Eustatius Hybridsystem – Overview by SMA Solar Technology AG

Part III:
Top left: João Peças Lopes, INESC TEC, Portugal
Top right: Thorsten Franzisi, LEW, Germany
Middle: Herwig Renner Graz University of Technology, Austria
Bottom left: SMA Solar Technology AG, Germany
Bottom right: Iberdrola Group, Bilbao, Spain

Contents

List of abbreviations	**xv**
Preface	**xix**
About the editor	**xxiii**
About the contributors	**xxv**

Part I	**Principles of island grid formation and their stability requirements**	**1**
1	**Introduction**	**3**
	Michael Finkel and Sebastian Palm	

1.1	Historic development	4
1.2	Energy transition changes the power grid	7
1.3	Definitions	12
	1.3.1 Disconnection of sub-grids	12
	1.3.2 Degrees of decoupling sub-grids	13
	1.3.3 Definition of electrical islands in the literature	14
	1.3.4 Definition of electrical islands in this book	15
1.4	Formation of sub-grids and islanded grids	16
1.5	Stable operation of sub-grids and islanded grids	20
	1.5.1 Power system strength, rotating inertia and short-circuit ratio	20
	1.5.2 Grid-forming and system-supporting properties of generation units	22
1.6	Opportunities and threats of islanded grids	23
	1.6.1 Intended islanding	23
	1.6.2 Unintended islanding	25
	References	26

2	**Basics of power system dynamics and stability**	**31**
	Herwig Renner, Petros Aristidou and Ziqian Zhang	

2.1	Dynamics in power systems	32
	2.1.1 Dynamic model basics	32
	2.1.2 Time/frequency range of concern	33
	2.1.3 Main components involved in power systems dynamics	33
	2.1.4 Synchronous generators	34
	2.1.5 Inverter-based generation	38
	2.1.6 Load	39
	2.1.7 High-voltage direct-current transmission	42
	2.1.8 Power system protection	43

viii *Intended and unintended islanding of distribution grids*

	2.1.9	Wide area monitoring and stability assessment	43
2.2	Definition and classification of power system stability		44
	2.2.1	General definition of power system stability	44
	2.2.2	Relevance of stability phenomena	45
2.3	Rotor angle stability		45
	2.3.1	Steady-state rotor angle stability	46
	2.3.2	Transient rotor angle stability	47
2.4	Frequency stability		49
2.5	Voltage stability		52
2.6	System strength		55
2.7	Impact of high penetration of inverter-based generation on power system stability		56
	2.7.1	Differences between grid-connected inverters and synchronous generators	56
	2.7.2	Small-signal stability of inverters	58
	2.7.3	Large-signal stability of inverters	58
2.8	Simulation and analysis models and method		60
	2.8.1	EMT-type models	60
	2.8.2	RMS-type dynamic models	61
	2.8.3	Power-flow models	62
	2.8.4	Modal analysis	63
	2.8.5	Comparison of models	63
References			64

3 Control of electric power systems   **69**
Michael Finkel, Herwig Renner and Ziqian Zhang

3.1	Frequency control and adjustment of generation to consumption		70
	3.1.1	System frequency	70
	3.1.2	Inertia management	71
	3.1.3	Types of control reserve and their provision	73
	3.1.4	Characteristics and control of synchronous generators	75
	3.1.5	Control behaviour of inverter-based resources	83
	3.1.6	Power sharing in islanded grids	85
3.2	Frequency characteristics of loads		86
	3.2.1	Underfrequency load shedding	87
	3.2.2	Rolling blackout	88
3.3	Control of reactive power and voltage		89
	3.3.1	Relationship between active and reactive powers and voltage	89
	3.3.2	Generator control (AVR)	91
	3.3.3	Voltage control in distribution grids	92
	3.3.4	Requirements for generation units to support the operation of the power system	97

| | | Contents | ix |

3.4	Further requirements for stable islanded grid operation	101
	3.4.1	Transition from interconnected to islanded grid operation and vice-versa	102
	3.4.2	Operation of the islanded grid	106
	References	107

Part II   Operational and planning issues   113

4   Behaviour at grid connection point   115
Michael Finkel, Georg Kerber and Herwig Renner

4.1	Introduction to grid codes	116
	4.1.1	Variable renewable energy impacts the way power systems are operated	116
	4.1.2	Tailoring grid connection code requirements to system context	117
	4.1.3	The role of grid codes in electricity system regulation	118
4.2	The EU network code on requirements for generators	120
	4.2.1	Frequency stability	122
	4.2.2	Voltage stability and robustness	127
	4.2.3	System restoration	132
4.3	Comparison of selected grid codes	134
	4.3.1	Frequency stability	135
	4.3.2	Robustness: FRT envelopes	138
	4.3.3	Black-start capability	139
4.4	Power quality aspects in islanded grids	140
	References	142

5   Power system restoration   149
Holger Becker and Christian Hachmann

5.1	Motivation for power system restoration and types of outages	149
	5.1.1	General	149
	5.1.2	Historical blackouts	149
	5.1.3	Restoration strategies	151
	5.1.4	Techno-economic trade-off	151
	5.1.5	Emergency backup supply	152
5.2	Restoration strategies	152
	5.2.1	Top-down	152
	5.2.2	Bottom-up	152
5.3	Ancillary services and secondary technology during power system restoration	153
	5.3.1	Blackstart	153
	5.3.2	Voltage control	154
	5.3.3	Frequency control	157
	5.3.4	Protection	160
	5.3.5	Information and communication technology	160

x *Intended and unintended islanding of distribution grids*

5.4	Different phases of power system restoration		161
	5.4.1	Planning phase	162
	5.4.2	Grid preparation phase	163
	5.4.3	Blackstart phase	163
	5.4.4	System and network restoration phase	164
	5.4.5	Load restoration phase	165
	5.4.6	Resume normal operation	165
5.5	Outlook: external impact due to renewable energy transition		166
	5.5.1	Decentralisation	166
	5.5.2	Weather dependency	166
	5.5.3	Changing load behaviour	166
	5.5.4	Start-up times	166
	5.5.5	Inverter-based generation and load	167
	5.5.6	High-voltage direct-current systems	167
	5.5.7	Decreasing distinction between system and load restoration	167
	5.5.8	Potential for distribution system islands	167
	5.5.9	Artificial intelligence in power system operation	168
References			168

6 Protection   **171**
Holger Kühn and Peter Schegner

6.1	Introduction		171
6.2	Short-circuit behaviour of inverters		172
	6.2.1	Switch-off mode in the event of a fault	173
	6.2.2	Zero-power mode	173
	6.2.3	Feed-in of reactive current in the positive sequence	174
	6.2.4	Feed-in of reactive current in the positive and negative sequence	175
6.3	Behaviour of inverters in different grid constellations		176
	6.3.1	Wind power plant – offshore grids	176
	6.3.2	Onshore grids	179
6.4	Development of short-circuit currents		181
6.5	Grid protection concepts and short-circuit contributions of inverters		183
	6.5.1	Extra-high-voltage grids and high-voltage grids	183
	6.5.2	Medium-voltage grids	183
6.6	Protection of inverter-based generators		185
6.7	Protection for island grids		187
References			189

7 Unintentional islanding detection   **191**
Sebastian Palm and Peter Schegner

7.1	Occurrence of unintentional electrical islands		191
	7.1.1	Phases of electrical islands	192
	7.1.2	Behaviour of distributed generation units	192
	7.1.3	Behaviour of electrical loads	194

	7.1.4	Approach for simple islanding scenarios	195
	7.1.5	Definition of the non-detection zone	195
	7.1.6	NDZ calculation of a simple arrangement	196
	7.1.7	Influence of P(f) and real load model	198
7.2	Purpose and principle of detection methods		199
	7.2.1	Islanding protection with voltage and frequency thresholds	200
	7.2.2	Additional detection methods	201
7.3	Description of selected islanding detection methods		202
	7.3.1	Voltage and frequency thresholds – PIDM	202
	7.3.2	Detection of voltage harmonics – PIDM	202
	7.3.3	Rate of change of frequency – PIDM	203
	7.3.4	Phase jump detection – PIDM	204
	7.3.5	Communication – PIDM	204
	7.3.6	Impedance measurement – AIDM	205
	7.3.7	Frequency shift – AIDM	205
	7.3.8	Phase shift – AIDM	207
	7.3.9	Q(f) control – AIDM	207
	7.3.10	Modulation of $\cos\varphi/\sin\varphi$ – AIDM	207
	7.3.11	Impedance insertion – AIDM	208
	7.3.12	Fault throwers – AIDM	208
7.4	Evaluation of detection methods		209
	7.4.1	Evaluation criteria	209
	7.4.2	Comparison of detection methods	210
References			212

8 Planning methods for secure islanding   **215**
Agnes M. Nakiganda and Petros Aristidou

8.1	Notations used in this chapter		216
8.2	Operational planning with static secure islanding constraints		217
	8.2.1	Day-ahead optimal planning model with static islanding constraints	218
	8.2.2	Network modelling for power flow constraints	221
8.3	Investment planning with static secure islanding constraints		224
	8.3.1	Changes in the objective	224
	8.3.2	Changes in the constraints	225
	8.3.3	Representative days	226
8.4	Incorporating transient islanding constraints in planning problems		227
	8.4.1	Incorporating transient frequency constraints in the investment planning problem	230
	8.4.2	Case study results	238
8.5	Concluding remarks		242
References			243

xii *Intended and unintended islanding of distribution grids*

9 Dynamic modelling for distribution grid analysis in the time domain   **247**
Christoph Brosinsky, Nayeemuddin Ahmed and Harald Weber

9.1	Introduction	247
	9.1.1 Modelling of power system dynamics	250
	9.1.2 Modelling of electromagnetic transient	251
	9.1.3 Modelling of electromechanical interactions (RMS)	253
	9.1.4 Model initialisation and numerical integration	254
	9.1.5 Transformation methods	255
9.2	Power system modelling	257
	9.2.1 The electric grid model	257
	9.2.2 Transmission and distribution line models	258
	9.2.3 Load modelling	259
	9.2.4 Modelling of conventional generation	260
	9.2.5 Thermal power plant	261
	9.2.6 Hydroelectric power plant	264
	9.2.7 Modelling of decentralised (renewable) generation	266
	9.2.8 Hydrogen storage power plant	267
9.3	Case study on the provision of ancillary services in an island network with high DER infeed	269
	9.3.1 Example island network	270
	9.3.2 Frequency regulation and HSPP performance evaluation	271
	9.3.3 Frequency regulation during high DER infeed	276
	9.3.4 Frequency regulation without HSPPs	278
9.4	Summary	280
	References	281

Part III   Islanded grids in practice   **285**

10 Insular power systems   **287**
Andreas Knobloch, Mohamed Mostafa, João Abel Peças Lopes,
Carlos Moreira, Leonel Carvalho, Hugo Morais, Lucas Pereira
and Michael Finkel

10.1	Caribbean island St. Eustatius	288
	10.1.1 Introduction	288
	10.1.2 Droop-based grid-forming control	288
	10.1.3 System operation without genset inertia	289
	10.1.4 Frequency stability at normal operation with large solar irradiance perturbations	290
	10.1.5 Frequency stability and uninterrupted power supply at sudden genset outage	291
	10.1.6 Fast fault clearing and voltage stability after short-circuit faults	292
	10.1.7 Fault current contribution	293
	10.1.8 Power quality and operation	293
10.2	Madeira Island	294
	10.2.1 About the Madeira Archipelago	294

Contents xiii

10.2.2	Sustainable energy action plan	296
10.2.3	Generation expansion plan for the Madeira electric power system	297
10.2.4	Grid expansion plan for the Madeira electric power system	298
10.2.5	Security of supply through reliability and generation adequacy assessment	299
10.2.6	System dynamic performance – the need of synchronous inertia and the role of battery energy storage	306
10.2.7	Madeira Grid Code – a short overview	310
10.2.8	Using storage at the secondary substation for voltage control	313
10.2.9	Conclusions	316
References		316

11 LINDA projects – droop-based practical examples   **319**
Christoph Steinhart, Tobias Lechner, Sebastian Seifried,
Dominik Storch, Michael Finkel and Georg Kerber

11.1	Motivation and aims	319
	11.1.1 Motivation	319
	11.1.2 Main objectives	320
11.2	Basic concept	320
	11.2.1 Basic requirements: SCR and inertia	320
	11.2.2 Leading power plant and load management	321
	11.2.3 Practical approach for the derivation of a frequency droop characteristic	324
11.3	Behaviour of DGs according to the droop concept	326
	11.3.1 Background	326
	11.3.2 Estimation of the frequency-dependent behaviour of PV systems	327
	11.3.3 Measured behaviour of a mixed PV system population	327
11.4	Interaction between load and frequency	329
	11.4.1 Influence of load behaviour on frequency stability	329
	11.4.2 Increase in frequency stability: load bank concept	330
	11.4.3 Challenge: RoCoF measurement	332
11.5	Practical examples of the LINDA concept	332
	11.5.1 First LINDA project – key issues and field test sight	332
	11.5.2 LINDA 2.0	333
	11.5.3 LINDA 4 H_2O: analysis of further application examples for the emergency power supply of critical infrastructures	334
11.6	Concluding remarks	336
References		336

12 Practical aspects of bottom-up grid restoration   **339**
Herwig Renner

12.1	Key issues for bottom-up grid restoration	340
12.2	Rules for grid restoration in Austria	343

xiv *Intended and unintended islanding of distribution grids*

12.3 Example of practical test of bottom-up grid restoration	343
12.3.1 Test schedule	343
12.3.2 Grid configuration during test	344
12.3.3 Black-start and provision of stable initial configuration	345
12.3.4 Start of storage pump in PP1	346
12.3.5 Synchronising island and fall-back with power imbalance	348
12.3.6 Synchronising with a second island	349
12.3.7 Grid protection during islanding tests	350
12.4 Conclusions	351
References	352

13 Unintended islanded grids   353
Enrique Romero-Cadaval, Eva González-Romera,
Francisco-José Pazos-Filgueira and Michael Finkel

13.1 Introduction	353
13.2 Operational experience on islanding events caused by large photovoltaic plants	354
13.2.1 Dangerous work conditions	355
13.2.2 Impossibility of network operation or maintenance	355
13.2.3 Failure of network automation	356
13.2.4 Damaged inverters	356
13.3 Technical reasons	356
13.3.1 Behaviour of photovoltaic inverters	356
13.3.2 Behaviour of an islanding network	357
13.4 Field test	357
13.5 Potential solutions	359
13.6 Inverter sensibility analysis	361
13.7 Modelling of a real network	364
13.8 Final remarks	367
References	368

14 Summary and outlook   369
Michael Finkel

14.1 Concluding remarks	369
14.2 Outlook and open questions	371
References	372

Glossary	**375**
Index	**383**

List of abbreviations

AC	alternating current
aFRR	automatic frequency restoration reserves
AHC	agglomerative hierarchical clustering
AIDM	active islanding detection method
AVC	automatic voltage controller
AVR	automatic voltage regulator
BESS	battery electric storage system
BEV	battery electric vehicle
BFM	branch flow model
BSU	black-start capable unit
CCGT	combined cycle gas turbine
CHP	combined heat and power
CIG	converter-interfaced generation
CLPU	cold load pick up
CoI	centre of inertia
CPP	conventional power plant
CSCR	composite short circuit ratio
DAE	differential-algebraic equation
DC	direct current
DCC	demand connection code
DER	distributed energy resources
DF	DistFlow
DG	distributed generation unit
DH	district heating
DFIG	doubly fed induction generator
DN	distribution networks
DSA	dynamic security assessment
DSO	distribution system operator
EESS	electrical energy storage system
EENS	expected energy not supplied
EHV	extra high voltage
EL	electrical load
EMT	electromagnetic transient
EnWG	Energy Industry Act
ESCR	equivalent circuit-based short circuit ratio
EU	European Union
EV	electric vehicle
FCR	frequency containment reserve
FFR	fast frequency response

FOC	fibre optic cable
FRT	fault ride-through
FSM	frequency-sensitive mode
GC	grid code
GFL	grid-following mode
GFM	grid-forming mode
HPP	hydroelectric power plant
HSPP	hydrogen storage power plant
HV	high voltage
HVDC	high-voltage direct-current
HVDC NC	HVDC network code
HVRT	high-voltage ride-through
IBR	inverter-based resources
ICT	information and communication technology
IDM	island detection method
IPS	interconnected power system
IR	instantaneous reserve
IVP	initial-value problem
LDC	line drop compensator
LFSM	limited frequency-sensitive mode
LFSM-O	limited frequency-sensitive mode (over-frequency)
LFSM-U	limited frequency-sensitive mode (under-frequency)
LOHC	liquid organic hydrogen carrier
LOLE	loss of load expectation
LPP	leading power plant
LV	low voltage
LVRT	low-voltage ride-through
MaStR	Marktstammdatenregister
mFRR	manual frequency restoration reserves
MG	microgrid
MILP	mixed-integer linear programming
MINLP	mixed-integer non-linear programming
MV	medium voltage
NDZ	non-detection zone
NER	national electricity rules
NC	network code
NLP	non-linear programming
NVD	neutral voltage displacement
OLTC	on-load tap changer
OP	operating point
P2H	power to heat
PCC	point of common coupling
PCR	primary control reserve
PF	power factor
POC	point of connection
PESSE	National Grid Safety Emergency Plan

List of abbreviations xvii

PI	proportional-integral
PIDM	passive islanding detection method
PLCC	power line carrier communications
PLL	phase-locked loop
PMU	phase measurement unit
PNSM	positive and negative sequence mode
PPM	power park modules
PSM	positive sequence mode
PSS	power system stabilizer
PV	photovoltaic system
RES	renewable energy sources
RfG	requirements for generators
RMS	root mean square
RoCoF	rate of change of frequency
SCR	short circuit ratio
SDLWindV	Verordnung zu Systemdienstleistungen durch Windenergieanlagen
SFS	Sandia frequency shift
SG	synchronous generator
SMCS	sequential Monte Carlo simulation
SOM	switch-off mode
SPMG	synchronous power-generating modules
SSCI	sub-synchronous control interactions
SSR	sub-synchronous resonance
STATCOM	static synchronous compensators
SVC	static VAR compensator
SysStabV	Systemstabilitätsverordnung
THD	total harmonic distortion
TPP	thermal power plant
TSO	transmission system operator
UCPTE	Union pour la Coordination de la Production et du transport de l'Electricit
UFLS	underfrequency load shedding
VDE	Verband der Elektrotechnik Elektronik und Informationstechnik e.V.
VRDT	voltage-regulated distribution transformer
VSC	voltage source converter
VSI	voltage source inverter
WAMS	wide area measurement system
WPP	wind power plant
WSCR	weighted short circuit ratio
ZPM	zero-power mode

Preface

Since the beginning of the electrical energy supply, it has always been necessary to clarify how a reliable, stable, and cost-effective electrical energy supply can best be realised under changing boundary conditions. While Thomas A. Edison pursued the decentralised approach with local DC grids, George Westinghouse, with the support of Nikola Tesla, prevailed in global competition with AC grids.

At the beginning of electrification, the AC grids were also constructed as isolated/islanded grids. Consequently, (almost) all power plants were equipped with frequency and voltage regulation. With the formation of interconnected electric power grids, the tasks of ancillary services (frequency and voltage control, system restoration, and operational management) were primarily assigned to the larger power plants connected to the transmission grids. Merging the local isolated grids into larger interconnected networks had technical and economic advantages, resulting in today's (transnational) electric power systems. Even if there have been changes in these power grids, e.g. shifts on the power plant side from coal to nuclear energy and gas-fired power plants, little has changed in the basic structure of the interconnected power systems.

However, the massive expansion of decentralised, renewable generation plants has led to a fundamental paradigm shift: Away from large centralised generation units to many smaller decentralised generation units (DGs). This has also been reflected in the operational management of the grids. Initially, the philosophy prevailed: If a problem occurs in the grid, the 'disturbing' DGs are switched off and the large power plants solve the problem. As the number of DGs increased, the issue of ancillary services also had to be reorganised, as switching off all DGs, e.g. at a certain frequency threshold, would have exceeded the control reserve available in the power system. As a result, more and more tasks to ensure frequency and voltage stability inevitably had to be assigned to DGs.

This initial situation makes it possible to recognise the relevance of the topic dealt with in this book: the intended and unintended islanding of distribution grids. Every interconnected system can split up into several subsystems due to faults, maloperation, etc., but also due to planned switching actions. This immediately raises the question of whether these subsystems can 'survive' and under which conditions these isolated/islanded subsystems can be operated in a stable manner. In the event of long-lasting power outages, e.g. due to natural disasters, the effects of war, etc., it is important to determine whether the DGs can be used for an emergency power supply or whether the grid can be rebuilt from the bottom up.

In the past, these questions have primarily been discussed in the transmission grid and in specially designed isolated grids, which could only be connected to the electric

xx *Intended and unintended islanding of distribution grids*

power system with great effort and expense. Furthermore, many renewable generation units were installed in the distribution grid. In addition, these units increasingly contribute to the ancillary services. As mentioned above, many renewable generation units have been installed in the distribution grid and are increasingly contributing to the stable and secure operation of electric power systems. This opens up the chance for local intended isolated/islanded grids but also increases the probability of unintended and undetected islanded grids, especially at medium- and low-voltage levels.

This brief introduction already shows very well the wide range of the topic: from cellular approaches or microgrids to interconnected grids, from grids with inverter-based resources (IBR) to grids with classic synchronous generators. This wide range makes it clear that a complete treatment of all aspects of islanded grid operation is not possible even in a comprehensive volume so that no exhaustive presentation can be expected from this book either. Recognising that there is no one expert for the entire field of intended and unintended islanding of distribution grids, an author team has come together to give the book the necessary depth through their specific expertise. The authors have selected what they consider to be the fundamental sub-areas and tailored their presentation to the specific aims of the book: To give young professionals in science and industry and master students in the field of electric power systems a quick and comprehensive introduction to the topic of intended and unintended islanded grids. With the help of the referred literature, the sub-areas can then be further deepened.

The structure of this book reflects the conscious endeavour to guide the reader through a logical sequence of topics, from the basic physical principles through operational and planning aspects to practical examples of intended and unintended islanding of distribution grids. The book can be divided into three main parts: Part I of the book, which comprises Chapters 1–3, is dedicated to the principles of island grid formation and their stability requirements. Chapter 1 introduces the topic and provides basic definitions and classifications. Chapter 2 summarises the basics of power systems dynamics and stability, while Chapter 3 focuses on the control of electric power systems.

Part II, compromising Chapters 4–9, is devoted to operational and planning issues. The operating behaviour of DGs at the grid connection point is defined to a large extent in national standards. Chapter 4, therefore, presents regulations concerning frequency and voltage stability, robustness, and system restoration. Power system restoration is a special form of an islanded grid and is described in Chapter 5. Even though power system restoration usually takes place in the transmission grid, many operational aspects are also relevant for islanded grid operation in the distribution grid. Chapter 6 is then dedicated to the topic of protection, and Chapter 7 presents and compares methods for detecting (unintended) electrical islands. The last two chapters of this section then focus on planning methods (Chapter 8) and the time-domain modelling of islanded grids (Chapter 9).

Part III, comprising Chapters 10–14, considers real projects in which islanded grids have been realised. First, the insular power systems on the Caribbean island of St Eustatius and on the island of Madeira are presented. The LINDA projects in

Chapter 11 present a concept for an emergency power supply with DGs. Chapter 12 is then dedicated to the practical aspects of bottom-up grid restoration. The example of an unintended isolated grid in the medium-voltage grid of the Spanish company Iberdrola S.A., which is often described in the literature, is presented in Chapter 13. The final Chapter 14 concludes the book, by briefly summarising key aspects and providing an outlook on current developments.

On behalf of all authors, I would like to express our sincere thanks to Mr Christoph von Friedburg and Ms Olivia Wilkins from the IET publishers. Christoph von Friedeburg gave the initial impetus for writing this book and Olivia Wilkins has provided excellent support throughout the publication process. Additionally, I would like to thank all colleagues and readers who have contributed to the realisation of this book with their valuable comments as well as their continuous encouragement and support. As this is the first edition, there is a possibility that there are some printing errors that have been overlooked despite several thorough reviews. I welcome feedback on such errors as well as suggestions for improvements in the event that a second edition should be published.

Michael Finkel, March 2024

About the editor

 **Michael Finkel** is a professor for high voltage engineering and electric power supply systems at Augsburg Technical University of Applied Sciences, Germany. After his Ph.D. at the Technical University of Munich, he worked for the Maschinenfabrik Reinhausen in Regensburg and for the SWM Infrastruktur GmbH, the grid operator in the city of Munich. His research focus is on local islanded power supply with distributed generation units, innovative planning methods and innovative equipment, especially for distribution grids. Michael is a member of IEEE, Cigre, and the VDE.

About the contributors

Nayeemuddin Ahmed is a Research Associate at the Institute of Electrical Power Engineering, University of Rostock. He completed his Master's in Electrical Engineering in 2019. Between 2019 and 2021, Nayeemuddin was a part of the project Netz-Stabil, where the goal was to develop a stable futuristic grid for Mecklenburg-Western Pomerania involving wind, solar and bioenergy with storage and flexible loads considering optimal sector coupling. Currently, he is working on his Ph.D., which introduces a novel strategy based on the control of voltage angles for an electrical grid with 100% renewable energy infeed. His research focus involves dynamic modelling of conventional and renewable power plants and storages, large-scale storage of electrical energy via hydrogen and grid-forming control for inverter-dominated networks.

Petros Aristidou is an Assistant Professor in Sustainable Power Systems at the Cyprus University of Technology. He got his Diploma from the Department of Electrical & Computer Engineering at the National Technical University of Athens in 2010 and his Ph.D. at the University of Liege in 2015. After his Ph.D., he worked at ETH Zurich as a postdoctoral researcher and at the University of Leeds as Lecturer. His expertise is in power system dynamics, planning, and control, and he has participated in several working groups looking into the challenges of low-inertia systems.

Holger Becker works in a management role at the Fraunhofer Institute for Energy Economics and Energy System Technology (IEE) in Kassel, Germany. After graduating from Christian-Albrechts-Universität in Kiel in 2005, he worked at SMA Technology AG and Nordex Energy GmbH in the field of grid integration of distributed generation systems. At M.O.E GmbH, he was responsible for the certification of generation units and photovoltaic systems as deputy head of the certification body. His research at Fraunhofer currently focuses on the topics of power grid resilience and grid restoration. Besides this, he works as a lecturer at the Technische Hochschule (University of Applied Sciences) in Lübeck.

Christoph Brosinsky is Head of the Distribution Management System Department at TEN Thüringer Energienetze GmbH & Co. KG, a distribution grid operator in East Germany. He previously worked at P&M Power Consulting GmbH, a specialized consulting company for electrical power systems engineering with a focus on the grid integration of renewable power generation, dynamic stability studies, grid code compliance, power quality, and ancillary services by distributed generation units. Christoph received his Ph.D. from Ilmenau University of Technology for his work on a framework for the next generation of energy management systems based on an adaptive power system digital twin. His current work focuses on distribution system operation, Redispatch 2.0, and ISMS compliance. He is a member of IEEE, Cigré and the VDE.

Enrique Romero Cadaval received the M.Sc. degree in Industrial Electronic Engineering from the Universidad Pontificia de Comillas, Madrid, Spain, in 1992, and the Ph.D. degree from the Universidad de Extremadura, Badajoz, Spain. He works at the University of Extremadura and belongs to Power Electrical and Electronic Systems (PE&ES) R&D Group in the School of Industrial Engineering. His areas of interest are power electronics applied to power systems covering power quality, active power filters, electric vehicles, smart grids, energy storage, and renewable energy resources. He has coordinated several projects related with smart inverters, smart transformers, energy storage systems, and virtual power plants.

Leonel de Magalhães Carvalho was born in Espinho, Portugal, in 1985. He received his B.Sc., M.Sc., and Ph.D. degrees in Electrical Engineering from the Faculty of Engineering of the University of Porto (FEUP), Portugal, in 2006, 2008, and 2013, respectively. In 2011 he was a Visiting Researcher at the Institute of Electric Systems and Energy of the Federal University of Itajubá (UNIFEI), Brazil. He is a senior researcher at INESC TEC, responsible for the System Planning and Reliability area of the Centre for Power and Energy Systems. His research interests include Power System Reliability Assessment and the application of Artificial Intelligence in Power Systems.

Christian Hachmann leads the Power System Dynamics Group in the Department of Energy Management and Power System Operation at the University of Kassel, Germany. He is also with the Transmission Grid Planning Group at the Fraunhofer Institute for Energy Economics and Energy System Technology (IEE) in Kassel, Germany. After graduating from Paderborn University in 2010, he worked at SMA Technology AG in the field of automated type testing and grid integration of residential photovoltaic inverters. His current research topics are power system restoration, emergency islanding and power system resilience with a focus on the role of distributed renewable generation and battery storage.

About the contributors   xxvii

Georg Kerber got his Diploma in Electrical Engineering and information technology from the Technical University of Munich. In 2011, he completed his doctorate on the 'capacity of low-voltage distribution grids for the infeed from photovoltaics' at TUM. He worked for many years as a planner for low-, medium-, and high-voltage grids and as head of the asset management at EnBW and LEW-Verteilnetz. Since 2010, he was involved in technical committees at VDE and Cenelec. Among other things, he helped to formulate the technical connection rules for medium voltage in 2018. In the course of these activities, he dealt with unintended island grids due to decentralized feed-in at an early stage and later carried out the LINDA project. Since 2021, he has been Professor of Electrical Power Grids at the Munich University of Applied Sciences.

Andreas Knobloch received his Dipl.-Ing. degree in electrical engineering and information technology with focus on automation and control at the Technical University of Darmstadt in 2010. In his current position as system architect for industry solutions, he is responsible for the technical design of the SMA Energy System Large Scale portfolio. In his previous positions as senior R&D engineer and project manager at SMA, since 2011, he was involved in the development of the SMA Grid Forming Solution and many other products for large-scale, commercial and residential applications. His fields of interest include the design, control and grid integration of power electronic systems for photovoltaic and storage applications as well as the stability of electricity grids with a high renewable share.

Holger Kühn received his Degree in Electrical Engineering from the Technical University of Karlsruhe in 1980. He worked for several years as a planning and commissioning engineer for control and protective systems at Brown, Boverie & Cie in Mannheim before moving to the electric supply company PreussenElektra in Hannover, which later became e.on Netz GmbH in Bayreuth and then TenneT TSO GmbH. There he was responsible for the protective systems in EHV in HV-grids. He started working on the grid-conforming behaviour of renewable generators in the year 2000. At VDE, he was involved in defining the 'Technical Requirements' from the very beginning. After his retirement, he is Consultant at Kühn – Netz und Systemschutz.

Tobias Lechner has been a research associate at Augsburg Technical University of Applied Sciences since October 2019. His research focus is on local island power supply with distributed energy resources. He graduated with a master's degree in electrical engineering and mechatronic systems from Augsburg University of Applied Sciences and Ulster University in Northern Ireland. He has been conducting a cooperative doctorate at the Chair of Electric Power Transmission and Distribution at the Technical University of Munich. Tobias is a member of CIGRE.

João Peças Lopes is Full Professor at Porto University (FEUP) where he teaches in the graduation and post-graduation areas and is Director of the Sustainable Energy Systems Ph.D. program. He is presently Associate Director and Coordinator of the TEC4Energy initiative at INESC TEC, one of the largest R&D interface institutions of the Unversity of Porto. He is also a member of the accompanying national committee of the Portuguese recovery and resilience plan. His main domains of research are related with large-scale integration of renewable power sources, power system dynamics, storage systems and security of supply, microgeneration and microgrids, smart metering, electric vehicle grid integration and exploitation of P2P using hydrogen. He is co-editor and co-author of the book 'Electric Vehicle Integration into Modern Power Networks' edited by Springer. He is a Fellow from IEEE. He was Adjoint Professor of the Iowa State University and Visiting Professor of the University Pontificia Comillas in Madrid.

Hugo Morais received B.Sc. and M.Sc. degrees from the Polytechnic of Porto in 2005 and 2010, respectively, and Ph.D. degrees in 2012 from the University of Tras-os-Montes e Alto Douro, all in the field of the electric power system. He is a senior researcher at INESC-ID and an Associate Professor at the University of Lisbon. His research interests are in the development and deployment of smart grid methods and tools. He participates in the definition of operational planning tools used by the French DSO and the development of hardware-agnostic solutions, based on IEC standards, to be used to control distributed energy resources. He also works actively in the development of strategies for electric mobility integration in power systems.

Carlos Moreira received a Ph.D. in Electric Engineering (Power Systems) from the University of Porto in November 2008. He is currently a Senior Researcher and Scientific Domain Leader on 'Static and Dynamic Analysis of Power Systems' at the Centre for Power and Energy Systems of INESC TEC. In February 2009, he joined the Department of Electrical and Computer Engineering at FEUP, holding presently the position of Associate Professor. His main research interests include microgrid operation and control, dynamics and stability analysis of electric power systems with increasing shares of converter interfaced generation systems, and grid code development.

Mohamed Mostafa works as Industry Lead Storage at SMA Solar Technology in the Utility Scale segment. After studying B.Sc. in Electrical Engineering and Master in Mechatronics, he started his career in R&D of industrial control system and power electronics inverters for wind, solar, and energy storage systems. During his 20 years of professional experience, he managed many R&D and power plant projects, focusing on hybrid and energy storage plants, from conceptual design to global commissioning.

About the contributors xxix

Agnes Marjorie Nakiganda is currently a Postdoctoral researcher with the Center for Electric Power and Energy at the Technical University of Denmark (DTU). She received a B.Sc. degree in Electrical Engineering from the Makerere University, Uganda, in 2012. She obtained an M.Sc. degree in Electrical Engineering and Renewable Energy Systems and a Ph.D. in Electronic And Electrical Engineering in 2022 from the University of Leeds, Leeds, UK. Her research interests include dynamics, control and optimization of networks with a high penetration of converter-interfaced units. Her current research focus is related to the data-driven operation and planning of power systems to ensure security and resilience through the application of mathematical optimization and machine learning methods.

Sebastian Palm is an Application Engineer at DIgSILENT GmbH and a lecturer at BA Bautzen (University of Cooperative Education). He previously worked at the Chair of Electrical Power Supply at TU Dresden and for SachsenNetze, a distribution grid operator in East Germany. His current work focuses on the modelling and integration of electric vehicles, heat pumps and renewable generation in distribution grids. Sebastian holds a Ph.D. from the Technical University Dresden, Germany, for his work on the detection and prognosis of unintentional electrical islands.

Franciso José Pazos (Industrial Electronics Engineer 1989) joined the distribution area of Iberdrola in 1990, where he worked in the Department of Substations, Protection before being in charge of Technical Assistance. Currently, he is in charge of Technological Transfer and Projects in the Innovation Department of i-DE. He has participated or led projects concerning power quality, earthing systems, power electronics for distribution networks, and DER connection of protection systems. He has been also active in standardization participating in several national, Cenelec and IEC working groups of publications on the same areas. His patents are related to the application of power electronics for fault extinction and predictive maintenance in high-voltage networks.

Lucas Pereira earned his Ph.D. in Computer Science from the University of Madeira, Portugal, in 2016. Since then, he has been at ITI/LARSyS, where he leads the Future Energy research line. Since 2019, Lucas has held the position of Assistant Researcher at Instituto Superior Técnico, University of Lisbon. His research endeavours encompass the realms of data science, machine learning, and human-computer interaction, strategically aimed at bridging the divide between laboratory-based innovations and their real-world applications for sustainable development goals (SDGs). With a particular focus on future energy systems and sustainable built environments, his work often involves the practical deployment and assessment of monitoring technologies and software systems.

Eva González-Romera received an M.Sc. degree in industrial engineering and a Ph.D. degree in electrical engineering from the University of Extremadura, Badajoz, Spain, in 1998 and 2005, respectively. She is currently an Associate Professor with the Department of Electrical Engineering, Electronics, and Control, and belongs to the Research Group in Power Electrical and Electronic Systems (PE&ES), University of Extremadura. Her primary areas of interest are power quality, distribution grids, and smart grids. She has coordinated research projects about load forecasting, integration of photovoltaic power plants in distribution networks, and control of nanogrids.

Herwig Renner holds a position as associate professor at the Institute for Electrical Power Systems at Graz University of Technology. His focus in research and teaching is on electrical power system control, stability and power quality. He is involved in numerous research projects with transmission and distribution system operators and acts as a consultant for the Austrian regulation authority. Herwig was active in several CIGRE working groups and is a member of the CIRED technical committee.

Peter Schegner is a Professor at the Technische Universität Dresden since 1995. Currently, he is serving as the Director of the Institute for Electrical Power and High Voltage Engineering. After completing his studies, he worked as a System Engineer in the field of network control technology at AEG. He assumed the role of leading the development department for network protection systems at AEG in Frankfurt. He continues to lead numerous research projects in the field of protection and is a senior member of the IEEE.

Sebastian Seifried is a research associate at Augsburg Technical University of Applied Sciences since October 2019. His research primarily concentrates on analysing the aggregated behaviour of low-voltage grids during voltage and frequency fluctuations and in addition on planning future grids. He graduated with a master's degree in electrical engineering and mechatronic systems from Augsburg University of Applied Sciences and Ulster University in Northern Ireland. He is conducting a cooperative doctorate at the Chair of Electric Power Transmission and Distribution at the Technical University of Munich.

Christoph Steinhart holds the position of a senior expert in a conceptional department at the SWM Infrastruktur GmbH & Co. KG, the distribution grid operator of the city of Munich. His current work focuses on the integration of large generation units, heat pumps and electrical vehicles in the distribution grid as well as the grid restoration concept. He previously worked at Augsburg University of Applied Sciences as a research associate in the field of local island power supply with distributed generation units in case of major power failures. He received his Ph.D. degree from the Technical University of Munich in 2020. He

studied electrical engineering at Augsburg University of Applied Sciences and subsequently obtained a Master of Mechatronic Systems and a Master of Electrical Engineering at Augsburg University of Applied Sciences and the University of Ulster in Northern Ireland as part of a double-degree programme.

Dominik Storch is a research associate at Augsburg Technical University of Applied Sciences. His research focus is on innovative planning methods for distribution grids and local island power supply with distributed generation systems. He received his M.Sc. degree in Applied Research on Electronic Engineering from Augsburg University of Applied Sciences in 2021 and is conducting a cooperative doctorate at the Chair of Electric Power Transmission and Distribution at the Technical University of Munich since then. Dominik is a member of CIGRE and VDE.

Harald Weber has been a Senior Professor at the University of Rostock since April 2020. Before this period, he was a Professor and the Director of the Institute of Electrical Power Engineering at the same university for 23 years. His research is focused on multiple areas, including load flow and stability calculations in large networks, dynamic power plant modelling, integration of decentralized renewable energy resources, black start and network restoration. Currently, his specialization is concentrated on the hydrogen storage power plant and sector coupling with hydrogen in grids with high renewable penetration. As a Professor, he was involved in multiple German and international projects involving the transition and advancement of smaller networks, Mecklenburg-Western Pomerania (Netzstudie) and larger grids, ENTSO-E (Netz-Stabil). In addition, he is a member of the working groups IFAC, CIGRE, VDI, VDE, ETG-GMA, and ASIM.

Ziqian Zhang (Member, IEEE 2018–) received M.S. degree in electrical engineering in 2014 and a Ph.D. degree from Graz University of Technology, Austria in 2017. Currently, he is a Postdoctoral Researcher with the Institute of Electrical Power Systems at Graz University of Technology, Austria. His main research interests are modelling and control of grid-connected converters, as well as test-oriented stability analysis and hardware-in-the-loop test systems.

Thanks to our supporters!
We would like to express our sincere thanks to all the experts who supported us with numerous interesting discussions, as well as many comments and suggestions during the preparation of the book. Additionally, we like to thank Franziska Hämmerle for the development of the design concept, the preparation of a design guide and the creation/revision of numerous figures.

xxxii *Intended and unintended islanding of distribution grids*

Acknowledgement

Grateful acknowledgement is given to the publishers and copyright owners for permission to reprint their materials. In particular, we would like to thank CIGRE, CIRED, DIN VDE, and VDE FNN. The corresponding publications are labelled with an attribution footnote. Full information on the other publications used can be found in the reference list at the end of each chapter.

Part I

Principles of island grid formation and their stability requirements

Chapter 1

Introduction

Michael Finkel[1] and Sebastian Palm[2]

Our current interconnected electric power systems have emerged from the interconnection of small local isolated/islanded grids. Nevertheless, these continental and supra-regional electric power grids are also islanded grids in the broader sense. In addition, there still exist smaller local isolated/islanded grids, as they can only be connected to the supra-regional network with considerable effort, e.g., due to their geographical location.

However, the basic physical relationships are identical in our interconnected power system (IPS) and in small isolated/islanded grids. Depending on the network size, the load, and the generation situation, effects can be relevant in small grids that are not the focus of the normal power system operation.

Because of the energy transition, more and more distributed generation units (DGs) have been installed in the distribution network and provide ancillary services. This means that intended and unintended islanded grids are now also possible at the distribution grid level, which results in new challenges and questions.

The term 'electrical island' has not yet been clearly defined in the literature. Therefore, different degrees of grid decoupling are presented, and a definition of the term 'electrical islands' has been derived for this book. Additionally, a subdivision into different types of electrical islands has been created.

Furthermore, the operational steps of forming sub-grids and islanded grids in the transmission and distribution system and the transitions in-between the different states are explained using a fictitious example.

Sub-grids and islanded grids can only be operated stable if an adequate number of generating units with grid-forming and system-supporting properties are operated in the sub-grid. In addition, the grid stiffness which indicates the capability of the grid to preserve its voltage and frequency stability during and after the presence of a disturbance and the SCR are important parameters used to estimate to which extent non-grid-forming generation units remain connected to the grid in the stationary state as well as in the event of a fault.

The operation of 'electrical islands' is currently being avoided at the distribution level. In addition to all the risks and challenges that the operation of island

---

[1] Department of High Voltage Engineering and Electric Power Supply Systems, Augsburg Technical University of Applied Sciences, Germany
[2] DIgSILENT GmbH, Gomaringen, Germany

grids entails, it also offers chances, e.g. for power supply of remote rural areas or in the case of natural disasters. Therefore, this chapter ends with a discussion of the opportunities and threats to islanded grids.

1.1 Historic development

In the early days of electricity, electric energy systems were small and localised. The first power stations, such as the DC Pearl Street Station in New York City (opened in 1882) and the first practical AC system at Great Barrington, Massachusetts (opened in 1886), supplied only few customers in their immediate vicinity.

Very soon, these first electric power systems were optimised with regard to the 'energy policy target triangle' (Figure 1.1): security of supply, economic efficiency, and environmental compatibility. Consequently, the supplied areas became larger with multiple generation units and many consumers. Due to the use of transformers in the AC system, the voltage levels in different parts of the system could be adjusted efficiently, and thus, the inherent power losses associated with distribution could be minimised. About 4000 companies with a total installed capacity of 2100 MW (500 kW on average) operated isolated distribution grids in Germany in the year 1910. [1].

Even if these networks only had a small spatial extent, the following technical and organisational questions had to be clarified:

- resource planning, reserve management;
- power sharing between the generating units;
- power system stability.
- protection of equipment, staff, consumer;
- grid restoration process in the case of a blackout.

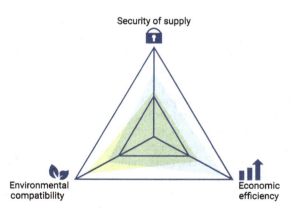

Figure 1.1 Balancing the 'Energy Trilemma'. The green and blue triangles illustrate different possible characteristics of the goals in the energy policy target triangle.

The first two questions were answered by establishing load forecasts, load dispatching centres, and implementing droop control curves. Power system stability was first recognised as an important problem in the 1920s [2]. With slow exciters and noncontinuously acting voltage regulators, power transfer capability was often limited by steady-state as well as transient rotor angle instability due to insufficient synchronizing torque. To analyse system stability, graphical techniques such as the equal area criterion and power circle diagrams were developed. These methods were successfully applied to early systems which could be effectively represented as two machine systems [3].

To ensure a reliable and safe operation of the power grid, the majority of the protection principles currently employed in protection relays were developed within the first two decades of the last century – Overcurrent (1901), Differential (1908), Directional (1910), and Distance (1923) [4].

At the beginning of electrification, all power stations had to be able to start by themselves and could be operated in islanded grid mode. Later on, this task was assigned to fewer and special power plants, e.g. hydro or combustion turbine units. In Germany, 4.7 GW of power plant capacity with black-start capability is currently under contract for grid reconstruction; this corresponds to approximately 5 per cent of the installed power plant capacity [5].

Power plant failures in these local grids led to an immediate interruption of the power supply to the connected consumers. Due to an increasing demand for electricity caused by advancing industrialisation and a higher required security of supply, gradually, individual power stations were connected at the low-voltage level, resulting in more extensive network structures in rural areas. Later on, the electrical industry was moving towards higher voltage, larger generation units, and larger grids. The characteristics of the synchronous machine in the power plant and the power demand of the customers have defined the requirements for the grid and its geographical expansion. Maps illustrating the development of the US and British power networks can be found in [6,7]. The boom created vast benefits from economies of scale and greater security of supply and has made it possible to supply power to large areas, regardless of the location of the production and consumption.

From the 1950s, the expansion of cross-border electricity exchange resulted in the interconnection of the networks of 11 countries to the Union for the Co-ordination of Production and Transmission of Electricity (UCPTE) in 1951 [1]. Today, the synchronous grid of Continental Europe covers the territory of the ENTSO-E Continental Europe regional group (UCTE*) and some neighbouring countries such as Albania and the Maghreb countries (Morocco, Algeria and Tunisia). In April 2015 the grid of Türkiye and in March 2022 the grid of Ukraine and Moldova were synchronised with the European grid (Figure 1.2).

Parallel to the formation of the interconnected continental European system, the Scandinavian Nordel system was created and also the IPS of the former Soviet Union and the countries in its sphere of influence. The latter is still the world's

---

*UCTE, which until July 1999 was known as UCPTE (Union pour la Coordination de la Production et du transport de l'Électricité), was incorporated into ENTSO-E on 1 July 2009 and continues to exist as the 'Regional Group Continental Europe'.

6 *Intended and unintended islanding of distribution grids*

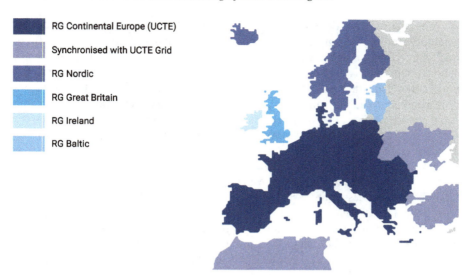

Figure 1.2   The main synchronous grids of Europe

geographically most extensive synchronised system in the world [8]. The UPS/IPS is still not interconnected with the ENTSO-E grids (except for a weak high voltage DC (HVDC) link to Nordel). A slightly different approach was adopted in North America and India: Although synchronous systems were created covering several states, synchronous operation was not extended across the continent/country. Today, there are four synchronously operated areas interconnected by HVDC couplings in North America and five in India [9]. Presently, China has the largest synchronised power system in the world (in terms of power) [8]. Finally, let us look at Australia: The networks of Western Australia and the Northern Territory are isolated from the electricity grid on the east coast.

The key driver for forming such large IPSs was a further optimisation with regard to the 'energy policy target triangle': security of supply, economic efficiency, and environmental compatibility (adapted from [10,11]):

- increased effective capacity of the power system due to the diversity of load demands;
- reduction in standby capacity due to joint provision of reserve power;
- increased reliability of supply;
- economical operation due to the combination of different types of power plants (e.g. thermal power plants and hydropower plants);
- increase in the size of generating sets, thereby reducing the capital and operating costs;
- load balancing between different load centres, therefore optimum utilisation of the available plant capacity and transmission facilities
- mutual support in the event of malfunctions (inertia, control reserve market)

Figure 1.3 Levels of grid connection (adapted from [12])

Even though the global electric power system has been propagated for many decades [9], the size of the present IPSs is limited due to the:

- geographic location
  - size determined by oceans, mountain ranges, deserts, etc.
  - distance to the neighbouring network – connection to the main grid is only possible by weak interconnections (e.g. Alaska) or at high costs (e.g. mountain villages/huts)
- potential political liabilities
- technical boundary conditions

Nevertheless, viewed from space the existing electric power systems – from microgrids to continental/supra-regional transmission grids – are just many isolated/islanded networks of different sizes in which the same technical boundary conditions and physical basics (see Chapters 2 and 3) apply as in the first local isolated networks.

1.2 Energy transition changes the power grid

Since the beginning of the new millennium, many countries have moved towards intensive promotion and support of new renewable energy sources, especially solar and wind. The goal of 'well below two degree Celsius', signed by all states in the 2015 Paris Climate Agreement has helped to accelerate this development. The massive expansion of renewable generation units in the last 20 years has led to the following:

- An increasing number of weather-dependent, renewable inverter-based resources (IBR) – primarily wind and photovoltaics – replacing synchronous and always available fossil generators (Figure 1.4).
- A significant share of generation is covered by a very large number of small IBR with the consequence that more and more generation capacity is connected to the distribution grid.
- A greater geographic separation between generation and consumption due to location-dependent primary energy sources such as wind and water. In the past,

8 *Intended and unintended islanding of distribution grids*

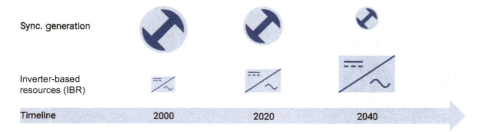

Figure 1.4 *Development of synchronous fossil generation and IBR*

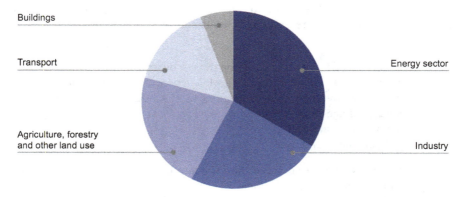

Figure 1.5 *Global net greenhouse gases emissions, 2019 by sectors (Data: [20])*

fossil or nuclear power plants were built close to load centres and thus balanced consumption and generation on a more regional level. While the average distance in Germany was about 80 km in the past, it now varies from almost 0 km for distributed generation (PV on the roof) to over 800 km for large-scale transport of wind power from northern to southern Germany [13].

This development can be summarised under the keywords 'decarbonisation' and 'decentralisation' of the electric power system.

To achieve the goal of a carbon-free energy supply, the energy transition must also be implemented in the sectors buildings (heating and cooling), transport, and industry (Figure 1.5). The idea of replacing fossil fuels by renewable power (e.g. replacing cars with combustion engine by electric vehicles; replacing oil/gas heating by heat pumps) inevitably leads to a higher 'electrification' degree in these sectors and an increasing electricity demand. This can be illustrated with the findings in [14]. To become greenhouse gas neutral by 2045, it is expected that the German electricity consumption will be almost doubled [15,16]. A very similar development is also expected in other countries [17–19].

Introduction 9

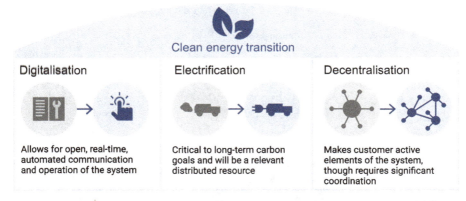

Figure 1.6   The three pillars of the clean energy transition (adapted from [22])

However, the number of solar and wind power plants that can be built is limited in many countries. Experts (cf. [21]) expect a maximum realisable renewable electricity generation in Germany of around 800 TWh. These experts also stated that not more than 50 per cent of the world energy consumption can be electrified. Consequently, there will also be energy trading (electricity, hydrogen, green fuels, etc.) on an international level in the future.

The last keyword is 'digitalisation'. The fluctuating generation of wind and solar energy leads to faster and larger fluctuations on the supply side, which are only predictable to a limited extent. In an almost carbon-free energy system, greater flexibility is required to balance generation and consumption. If a lot of renewable energy is fed into the power system, the excess energy must be curtailed, stored, exported or flexible loads must be switched on. If there is a lack of wind and sun, then the load can currently be supplied by conventional power plants. However, this is no longer possible in a completely regenerative energy system. In this case, the load must be reduced or the energy must be provided by storage systems. Digitalisation enables more control, including automatic, real-time optimisation of consumption and production, and interaction with customers.

The statements above can be summarised in Figure 1.6.

Additionally, this transition towards renewable energies has triggered fundamental changes in the transmission and distribution grids (adapted from [15,23–25]; cf. also Figure 1.7):

- Strong growth in distributed generation, especially renewable energy and combined heat and power (CHP) systems, and the resulting network expansion not only in the distribution but also in the transmission grids.
- Significant increase in self-generation or self-consumption by private customers: 'consumers' become 'prosumers'.
- Unidirectional energy transport is replaced by bidirectional energy transport (if the feed-in into the grids exceeds the regional consumption).

10 Intended and unintended islanding of distribution grids

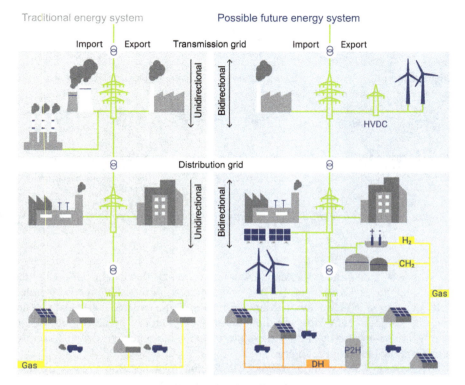

Figure 1.7 Comparison of a traditional (left) and possible future energy system (right); legend: electric grid in green; gas grid in yellow; district heating (DH) grid in orange (adapted from [26])

- As a result of weather-induced fluctuations, electricity generation in national and regional networks will become more inconsistent, which will have an impact on grid stability and congestion management in the distribution grid.
- Increasing cross-border trade of electric energy.
- Consumers adapt their electricity consumption to the current production or the current electricity price – demand-side management made possible by load-variable tariffs and smart meters.
- Sector coupling: coupling the electricity, mobility, gas and heat sectors through electrolysers, fuel cells or power to heat (P2H).
- Transparency and observability: The energy grids must be digitalised and equipped with sensors and actuators to make the state of the grid observable everywhere and to control the electricity demand/supply of the end customers in a targeted manner in critical situations.

The central challenge that the future electricity system has to overcome is that electricity supply and demand have to be matched much better than before, which requires flexibility at all levels. As a result, the conventional roles of a number of actors or participants in the electricity market are changing and the operation of the electricity grids is also changing as a result of these new players. In addition, a large number of new players are appearing who perform certain functions that are only possible, or even necessary, in the new electricity system: electricity traders for the direct marketing of renewable electricity, aggregators who bundle distributed systems into pools and use their abilities for the provision of system services, energy cooperatives that promote self-generation, smart meter administrators, and others [23].

Figure 1.8 now focuses on ancillary services that have so far been provided primarily in the transmission grid. In the past, the distribution grid was in a supporting role. Since the larger portion of the DGs are connected to the distribution grids, ancillary services that have so far been provided by large power plants have to be provided increasingly by DGs.

A prominent example is the 50.2 Hz problem: While in the 1990s, the approach was primarily still being pursued – in the event of larger frequency deviations, the 'disturbing' DGs are switched off and the large power plants control the problem – the 50.2 Hz problem has made it clear that the specified behaviour for individual generation units is system relevant. Until 2011, the technical connection rules for photovoltaic systems (PV) meant that they were simultaneously disconnected from the grid at an over-frequency of 50.2 Hz. In an extreme case, this would have corresponded to a sudden power reduction of 9 GW in the German power grid and thus significantly exceeded the available reserves [27].

Therefore, in many countries, the grid connection rules (compare Chapter 4) have been adapted continuously by the requirements listed below, so that the DGs

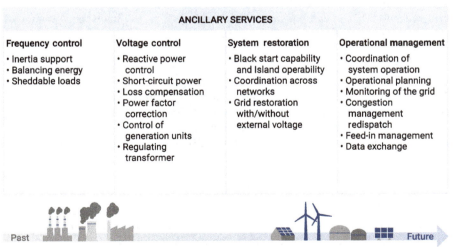

Figure 1.8 Relocation of ancillary services from a few large power plants in the transmission grid to many DGs in the distribution grid

12 *Intended and unintended islanding of distribution grids*

can increasingly replace the large power plants in their leading role in the electric power system [24,28–30]:

- Over- and under-frequency response and control requirements;
- Voltage support through the provision of reactive power;
- Dynamic grid support;
- Start-up and reconnection requirements;
- Ride through requirements.

This adjustment process is still ongoing. Additional requirements such as synthetic inertia, primary control and black-start capability are implemented, if at all, in the higher voltage levels. In addition to these technical questions, organisational questions must also be clarified. Ancillary services were previously assigned to the transport system operator. However, since the DGs are connected to the distribution grid, the responsibilities of the transmission system operators (TSOs), distribution system operators (DSOs) and power plant operators must be redefined.

As modern societies are highly dependent on electricity, power outages have a major social impact and can lead to huge economic losses. More details can be found, for instance in [31,32]. The methodology and application possibilities of a blackout simulator are presented in [33]. For this reason, many private households and industrial companies (e.g. research project INZELL [34]) have upgraded their systems so that they can be operated as an islanded grid, which increases the security of supply and facilitates system coupling. On a larger scale, these approaches are considered in cellular energy systems and respectively microgrids.

Here the above statements are briefly summarised: due to the required dynamic grid support and the associated non-immediate separation of the generating plants from the grid, the probability of unintended and undetected islanded grids increases, especially at the medium-voltage level. The continuous further development of the grid codes for DGs in combination with storage systems facilitates the operation of intended islanded grids in the distribution grid.

1.3 Definitions

The term 'electrical island' has not yet been clearly defined. Different publications, standards and guidelines introduce their own definition. In the following sections, different degrees of grid decoupling are shown and definitions from various sources regarding electrical islands are presented. Finally, a definition of the term electrical island has been written for this book.

1.3.1 Disconnection of sub-grids

The electrical power system can be separated into several sub-grids due to planned or unplanned switching operations. Previous power import and export into and from these sub-grids are spontaneously omitted. With the formation of a sub-grid, important system parameters change, including:

- power equilibrium with a corresponding change in frequency;
- short-circuit power;

Introduction   13

- inertia of the system;
- the system voltage according to magnitude and phase angle.

These islanded sub-grids can only be operated with the help of suitable protection and control capabilities. To avoid a breakdown of the inter-connected power system (large-scale blackouts), all players must jointly support the system's stability as long as possible. For this reason, the European network codes provide automatic 'system rescue' measures [35]. In Europe, sub-grid operation is therefore currently being avoided at the distribution level and identified sub-grids (see Chapter 7) are switched off at short notice by the network operator – if that is possible at all. This will also be permissible in the future and makes sense in many cases, as these grids can be quickly supplied again from the transmission network.

1.3.2   Degrees of decoupling sub-grids

In addition to islanded electrical grids that are completely separated from the rest of the utility grid, there are other grid operating modes that result in some form of decoupling from the interconnected system. Figure 1.9 shows possible operating modes.

Connected grid operation is currently the normal state of electrical energy supply. In this case, all DG and consumers operate in a common interconnected system. In this mode of operation, power, energy and system services of sub-grid and upstream grid are coupled.

In an **energy self-sufficient grid**, only energy decoupling takes place. In the sub-grid, the actual power balance is not necessarily balanced at all times, instead, the energy flow is regulated to ensure a balanced energy budget (over minutes, quarter hours, etc.). Not only the system services such as the provision of reactive power and short-circuit power but also the general power management continue to be determined and provided by the interconnected system.

On the other hand, there is the possibility of building **modular grids** [36], which primarily achieve the decoupling of system services. In this case, voltage quality, short-circuit power and reactive power are guaranteed and controlled by a power electronic grid interconnector. At the same time, the interconnector has to ensure the grid operation of the sub-grid. Energy decoupling is not necessary and an active power exchange with the upstream grid is still possible. Decoupling of the energy balance is possible but requires the controllability of DG or loads or the use of electrical storage in the decoupled sub-grid. The frequency within the modular grid can be used as a highly available communication channel for this purpose [37]. Further information on the modular grid can be found in [38].

In contrast to the **electrical island**, where power, energy and also system services are completely decoupled, these new approaches aim to reduce or equalise and control energy exchange with the upstream grid. This should lead to more effective utilisation of existing grid resources and thus reduce or at least delay the need for grid reinforcements or expansions.

14 *Intended and unintended islanding of distribution grids*

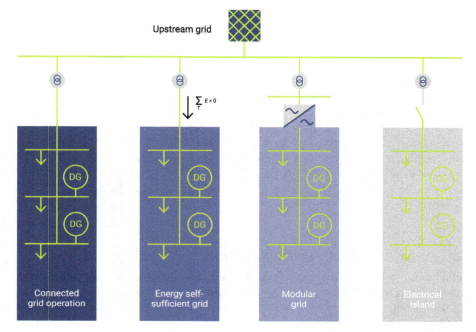

Figure 1.9 Classification of different operating modes according to the type of decoupling [39]

1.3.3 Definition of electrical islands in the literature

In the following, some definitions of the term 'electrical island' from the literature are presented to show that no two definitions are quite alike. However, all of them contain valuable information that will be used to develop a consistent definition for this book.

Electrical island in IEEE 1547 [40]
'island: A condition in which a portion of an Area EPS [electric power system] is energised solely by one or more Local EPSs through the associated point of common coupling (PCC) while that portion of the Area EPS is electrically separated from the rest of the Area EPS'.
'island, intentional: A planned island'.
'island, unintentional: An unplanned island'.

Electrical island in Mrugowsky [41]
Translated from German: 'In contrast to the interconnected system, an island is an electrical energy system that is spatially clearly limited, supplied by only one power plant or even only one generator, and is not permanently connected to other grids by transformers. Because of the usually small spatial extent, the extra-high and also the high-voltage levels, and in many cases even a separate medium-voltage grid, can be

dispensed with, so that then only a three-phase voltage level exists, i.e. the electrical island is a pure low-voltage grid'.

Electrical island in DIN VDE V 0126-1-1 [42]

Translated from German: 'The operation of an electrical island is the condition of a sub-grid separated from the larger rest of the network, in which the distributed generation units cover the consumption of the connected loads. Causes of disconnection include switching actions by the grid operator, tripping of protective devices, or equipment failures'.

[...] 'In the case of unintentional islanding, this process [the occurrence of an island] takes place outside the control of the grid operator. Voltage and frequency of the islanded sub-grid are not influenced by the grid operator'.

Intentional island in IEC 62116 [43]

Island that is intentionally created, usually to restore or maintain power to a section of the utility grid affected by a fault. The generation and loads may be any combination of customer-owned and utility-owned, but there is an implicit or explicit agreement between the controlling utility and the operators of customer-owned generation for this situation.

1.3.4 Definition of electrical islands in this book

Definition 1 (Electrical island). *The term electrical island describes the system state in which a sub-grid of the energy supply system is disconnected from all upstream grids, but shows an active and reactive power balance and is not de-energised as a result. In extra-high-voltage networks, complete disconnection from neighbouring networks of the same voltage level can also lead to the occurrence of an electrical island. The disconnection means both an energetic decoupling and a decoupling of the system services, since the short-circuit power and reactive power of the upstream grid or separated neighbouring grids are no longer available within the electrical island.*

A subdivision into different types of electrical islands is carried out:

- **Unintentional electrical island:** The electrical island is a sub-grid that does not turn into a de-energised state after a disconnection either operational or as a result of external influences (e.g. conductor break), although this was the aim of the disconnection.
- **Intentional electrical island:** The electrical island is a sub-grid in which the partial or complete continuation of supply is ensured after an operationally induced disconnection. This sub-grid contains at least one grid-forming and thus power managing system, i.e. a control that determines voltage and frequency.
- **Regular electrical island:** The electrical island is a sub-grid that is not connected to the upstream grid during normal operation (on-board networks, geographical islands, mountain grids, etc.).

1.4 Formation of sub-grids and islanded grids

This section describes the formation of sub-grids and islanded grids and illustrates the transitions in between the different states for a fictitious example. Sub-grids and islanded grids can basically form on all voltage levels. At the beginning, we focus on the transmission grid before we also consider islanded grids on lower voltage levels.

An interconnected electric power system can be split into two or more sub-grids, e.g., during planned switching operations or, more typically, during unplanned protection trips (due to faults, equipment overloads, power system imbalance). Due to these intentional or unintentional switching processes in the feeding power system, various sub-grids can form (Figure 1.10), which may either

(i) be de-energised if it is not equipped with a frequency and active power control or the generation units cannot balance the resulting power deficit, or
(ii) stabilise as an islanded grid, in which the connected consumers can be further supplied by the connected generation units.

In the transmission grid, a stable sub-grid is usually not referred to as an islanded grid but this state is described as a system split and the size of the separated system and the number of customers concerned is rather random.

In our example, the initial interconnected system can be re-established by (for further information see Chapter 5)

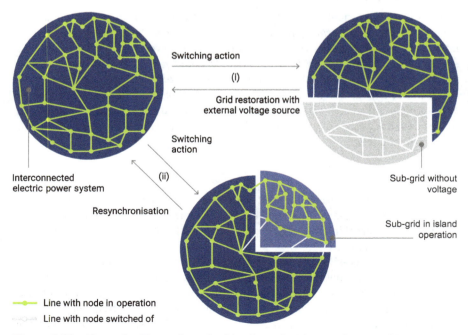

Figure 1.10 Example: Formation of a (i) sub-grid without voltage and (ii) a sub-grid in island operation

Introduction 17

(i) using external voltage sources (top-down re-energisation) and picking up load successively, or by

(ii) resynchronising the separated sub-grid with the remaining interconnected electric power system.

This process can also be summarised in a flow chart (Figure 1.11) which contains two further paths ('start' and 'shut down'). If in the case of a blackout, the grid cannot be restored with an external voltage source, the separated area has to be started from black and a start-up grid in island operation (Table 1.1) has to be established.

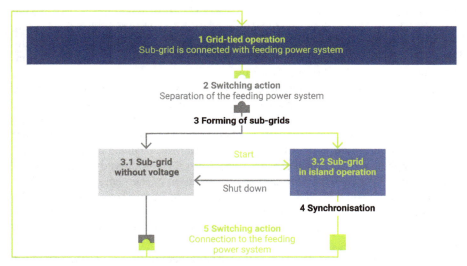

Figure 1.11   Formation of sub-grids and islanded grids; flowchart adapted from [11]

Table 1.1   Bottom-up grid restoration process; based on [44–46]

Preparation	Assessment of the post-blackout status of the system. Definition of a suitable sectionalising strategy. Energisation of the black-start units.
System restoration	Activate black-start power plants (e.g. hydropower plants) to energise the first segments of the electrical grid and supply cranking power to non-black-start thermal power plants (establish a start-up grid). Reconnection of load blocks to stabilise frequency and voltage. The aim is to rapidly energise the grid infrastructure and start thermal power plants, providing both power and inertia.
Load restoration	Restoration of load in each island as fully and quickly as reasonable. Load with lagging power factor should be picked up to control over-voltages.

18 *Intended and unintended islanding of distribution grids*

In general, the bottom-up strategy can be a problem with the presence of distributed generation since the value of load changes in the two following situations risking the failure of the restoration path [46]:

- After a blackout phase, there is a loss of dispersed generation. The value of the load increases, and this can be a challenge for the black-start units.
- During the restoration phase, the automatic reconnection of DG can present a challenge because it can change the load/generation equilibrium of the restoration path.

If the sub-grid in island operation cannot be stabilised or if there is a major failure resulting in a frequency $f < 47.5\,\text{Hz}$, the generating units must be disconnected from the grid; otherwise, natural oscillations would destroy their turbines (path: 'shut down'). The frequency limit of 47.5 Hz marks the point up to which the feeding power system must be supported as long as and as well as possible but also when the sub-grid must be disconnected in time to avoid improper system perturbations from the transmission grid and to avoid damages.

Next, we consider the distribution grid level. Even if the distribution grids are not designed for island grid operation, there is currently a trend that distribution grid operators or individual municipal utilities are striving to create a local island grid capability of small energy cells (e.g. companies, critical infrastructure, entire districts), so that in the event of a large-scale blackout, islanded distribution grids can be supplied with electrical energy [5,45,47,48]. The required dynamic ancillary services and the associated non-immediate separation of the generation units from the grid, as well as the various control mechanisms that are required to stabilise the network in inter-connected operation, such as the active power reduction for frequencies $f > 50.2\,\text{Hz}$ and the $Q(U)$ control, favour the creation and stable operation of islanded grids (see Chapter 7.1).

For a smaller sub-grid with decentralised generation (low inertia and short-circuit power), critical network conditions must be weighed up (islanding by planned switching operations): once again, the feeding power system must be supported for as long as and as well as possible within the defined limits; on the other hand, the sub-grid must be disconnected in time. The operation of a sub-grid is a challenge and can lead to further risks and other critical conditions [49].

Up to now, it was assumed in the operation of distribution grids that separate grid areas would become unsupplied sooner or later. Due to the lack of decentralised generators, in the past, disconnected generation units were automatically de-energised. Due to the increasing decentralisation of energy generation and the dynamic grid support of the DGs required by the standards, unintended islanded grids can remain stable. In a medium-voltage distribution grid operated by the network operator Iberdrola, e.g., unintended islanded grids repeatedly remained stable during maintenance work at a distribution network station [50]. For further information, see Chapter 13.1.

Such unintended islanded grids can be shut down by [11]:

- 'Doing nothing' and waiting for a sufficient power imbalance due to the volatility of loads and feed-ins.

Introduction 19

- Creating a power imbalance by controlling the DGs (cf. Chapter 7.2).
- Causing a power imbalance by switching an earthing switch (due to the load on the equipment this should be done only in an emergency).

Now, we can summarise the discussion above in a final Figure 1.12[†] of our fictitious example. Unplanned protection trips separated the original IPS into three parts: two parts could be stabilised, but in one sub-grid there was a blackout. In the de-energised sub-grid, two further sub-grids were formed:

- An intended electrical island to supply, e.g., critical infrastructures.
- A start-up grid for the bottom-up grid restoration process.

Intended and unintended islanded grids can also be formed in the other two grids even though the operation of an intended islanded grid in the separated stable sub-grid is not desirable from the point of view of solidarity. The formation of an unintended islanded grid within the black sub-grid is very unlikely since this requires switching actions on the distribution grid level at the time of the blackout.

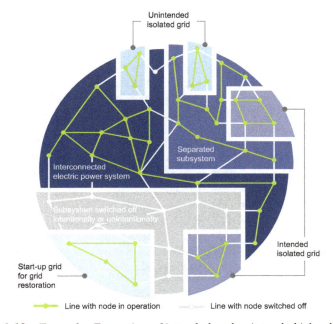

Figure 1.12 Example: Formation of intended and unintended islanded grids

---

[†]Note: Even though the different grids are shown as meshed grids, the basic considerations also apply to other voltage levels and grid structures.

20   *Intended and unintended islanding of distribution grids*

1.5 Stable operation of sub-grids and islanded grids

Sub-grids and islanded grids can only be operated stably if an adequate number of generating units is able to maintain a stable operating point with constant voltage and frequency. Transient and stationary grid stability is currently ensured primarily due to the inherent physical properties of large power plants and synchronous machines. For this reason, a few basic terms describing these physical properties of generating units will be introduced here. A detailed presentation with the corresponding equations can be found in Chapters 2 and 3.

1.5.1 Power system strength, rotating inertia and short-circuit ratio

A characteristic variable, when considering the stability limits of grid-connected systems, is the power system's strength or grid stiffness at the connection point.

Definition 2 (Power system strength). *Power system strength in general is defined as the capability of the grid to preserve its voltage and frequency stability during and after the presence of a disturbance.*

Power system strength can be defined from two aspects:

1.   Its mechanical rotating inertia, which is related to frequency stability.
2.   Its short-circuit current which is related to voltage stability.

Rotating inertia
System inertia through machine rotating masses defines the system's capability of maintaining a stable frequency in the short term and also the capability to stabilise the voltage angle variations [51]. This inherent provision of active power dampens the resulting frequency changes instantaneously with the occurrence of the disturbance, and also limits the frequency gradient and frequency deviations until other frequency-supporting mechanisms come into effect. Other mechanisms include the frequency-dependent power consumption of rotating loads connected directly to the grid – grid self-regulation effect (see Chapter 2) – which also occurs instantaneously, as well as primary and secondary control reserve (see Chapter 3) [52].

The interaction of the various mechanisms are illustrated in Figure 1.13 using a power plant failure in the continental European interconnected grid of 3 GW during a low-load situation (150 GW). The relative disturbance is thus 2 per cent. The three curves (a)–(c) show the effects of the acceleration time constant, self-regulation effect and primary control.

In the event of a power failure, the stored energy of the rotating masses cannot prevent the frequency drop and only reduces the frequency gradient (a). Without the intervention of other mechanisms, the frequency deviation would be 1 Hz after 10 s and would continue to increase (a). The self-regulation effect of loads leads to a new, stable operating point being reached at a reduced grid frequency (b) at 49.0 Hz, since the example assumed the self-regulating effect to be 2 per cent/Hz of the current grid load. The drop in load at $\Delta f = -1$ Hz thus exactly compensates for the original power deficit of 2 per cent [53].

Introduction 21

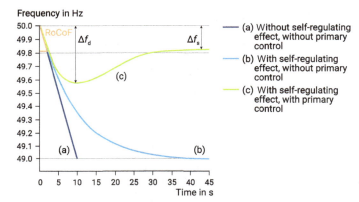

Figure 1.13 Frequency–response characteristic in a power system with mainly conventional power generation units ($T_N = 10\,s$) after a power plant failure (adapted from [53])

Further measures require active intervention in the operational management of the power plants. By providing more power, the so-called primary control reserve, the drop in frequency and the frequency deviation can be reduced to a stationary end value (c) that is significantly higher than in (b). The frequency is reduced over a longer period of time by the secondary control (not illustrated). The dynamic frequency deviation Δf_d (at the frequency nadir) thus depends – along with other influencing factors such as T_N or the magnitude of the imbalance – on the time behaviour of the primary control [53].

The stationary frequency deviation Δf_s results from the network power frequency characteristic λ, which is between 16,000 (low load) and 18,000 MW/Hz (high load) in the continental European interconnected grid [54] which leads to a stationary deviation of 0.19 Hz.

Short-circuit ratio
Traditionally, the short-circuit ratio (SCR) metric has been deployed to describe the power system strength, as it is directly related to the available short-circuit current. The SCR is given by the following equation:

$$SCR = \frac{S_{SCC}}{P_r} \tag{1.1}$$

where S_{SCC} is the short-circuit capacity at the point of connection (PoC) in the existing network before the connection of the new generation source, and P_r is the rated value of the new connected source. It can be concluded that a low SCR value indicates a lower amount of short-circuit current in a system, or a node with a high network impedance. Therefore, a system with a low SCR value ('weak' system) is very sensitive to active/reactive power injections (or absorptions), i.e. the system voltage changes rapidly as the amount of reactive power injected (or absorbed) changes. It is, therefore, difficult to stabilise the system voltage (magnitude and phase angle) in a weak system. Whereas, a high SCR ('strong') system is largely unresponsive

22   *Intended and unintended islanding of distribution grids*

Table 1.2   Classification of grid strength

Strong grid	Weak grid	Very weak grid
$SCR > 3$	$2 \leq SCR \leq 3$	$SCR < 2$

to active and reactive power injections (and absorptions) and the system voltage is not significantly influenced by changes in the network. A weak system generally requires a voltage control system with supplementary stabilisation control [55]. In [56], a classification of the power system strength depending on the SCR is given (Table 1.2).

1.5.2   *Grid-forming and system-supporting properties of generation units*

The grid-forming and system-supporting properties of a power-generating module can be defined as follows:

Definition 3 (Grid-forming characteristics). *Grid-forming characteristics of a power-generating module refer to the fundamental capability of maintaining a stable operating point with constant voltage and frequency during hypothetical stand-alone operation. The stability must also be maintained for defined disturbances with steady-state and dynamic deviations from the operating point. [57]* [‡]

Definition 4 (System-supporting characteristics). *System-supporting characteristics of a power-generating module mean designing control devices for active and reactive power balancing at the connection point such that the plant supports network stability beyond the connection point without having grid-forming characteristics. These characteristics must be provided by other power generation units without being usually impaired by the system-supporting power generation unit. Exclusively, system-supporting characteristics are only permissible to a very limited extent. [57]* [‡]

Up till now, the grid codes do not require grid-forming properties or sufficient system-supporting properties for DGs. This has been tolerable in the past because type 1 generation modules[§] with suitable control have inherent grid-forming properties due to their natural properties and the relevant reserves in the interconnected system had been still sufficiently high. With the change in the generation structure, however, the risk of a loss of stability increases when applying the existing grid connection rules. In addition, the situation can change spontaneously after a fault-related

---

[‡]Text reprinted with permission from FNN VDE, grid-forming and system-supporting behaviour of power-generating modules, © December 2021 .

[§]Here, we use the distinction from the VDN Transmission Code, 2007: A type 1 generating module is a synchronous generator directly coupled to the grid. Type 2 are all other generating modules. At European level (ENTSO), generator types are divided into four classes (types A, B, C, and D) depending on the connected load of the generator and the voltage level of the grid connection (cf. Chapter 4).

separation of the interconnected system, depending on the generation structure in the subsystems that have arisen. This depends on the fault history and cannot be predicted. Therefore, to control subsystems – but in the long term also with regard to the stability of the overall system – the grid-forming capabilities previously provided by type 1 high-power-generating modules must in future increasingly be provided by decentralised units. [57][‡]

A certain level of penetration of the interconnected system or subsystems separated from the interconnected system with generating units without grid-forming properties is still tolerable as long as the overall stability is guaranteed. However, these generating units shall then at least have system-supporting properties. In any case, controller configurations and parametrisations shall be specified on the basis of defined stability criteria. The focus is on frequency control (active power control) and voltage control (reactive power control) [57][‡]. The required behaviour of the generating modules can only be ensured if the relevant grid connection rules imply corresponding stability criteria.

1.6 Opportunities and threats of islanded grids

As already described in the sections above, the massive expansion of the DGs has meant that on the one hand local unintended islanded grids are becoming more and more likely and, on the other hand, it is being discussed whether and how these DGs can be used to operate local intended islanded grids in the event of a blackout. This trend was accelerated by the transfer of auxiliary services to the DGs.

Even if locally islanded grids offer opportunities, they also bear risks. Additionally, the previous optimum in the energy policy target triangle has been abandoned in favour of the parameter 'security of supply'. Therefore, the opportunities and threats of (un)intended islanded grids are summarised in this section.

1.6.1 Intended islanding

The planning and operation of islanded grids, in which private customers are also connected, is currently not common practice in Europe. However, the increasing

Figure 1.14 Opportunities and threats of islanded grids

24   *Intended and unintended islanding of distribution grids*

installation of generation units close to areas of consumption offer the following opportunities:

- Customer or sub-grids can be supplied if there is a problem in the IPS.
  - Increased security of supply and improved reliability [58] at the expense of economic efficiency.
  - Concept for emergency power supply in the case of natural disasters (e.g. flooding, storms), long-lasting blackouts, etc. and bottom-up grid restoration.
- Electrification of remote rural areas is possible.

For such an island operation, in addition to the technical questions also organisational rules must be established at an early stage that clarify

- whether and when islanded grids should be formed,
- which mutual responsibilities exist in which operating condition and
- under what conditions and how the resynchronisation with the inter-connected power system takes place.

The intended operation of islanded grids also has some disadvantages and contains some threats:

- Generally, a less robust stability than an inter-connected system;
- Very dependent on the control mechanisms;
- Increased voltage fluctuations;
- Greater frequency volatility;
- Possible risks, e.g., in the area of network protection (cf. Chapter 6);
- The islanded grid (including its power plants) does not help to solve the problems in the IPSs → desolidarisation;
- Lower overall energy efficiency;
- Additional expenses for the black start and island operation capability of the power plants;
- Higher costs for maintenance, regular training of employees and functional tests.

Finally, legal issues also need to be considered. A prioritised supply of certain loads is fundamentally in conflict with the legally regulated, non-discriminatory access to electrical energy. An exception can only be made on the basis of objectively justified reasons [5].

Summarising the discussion above, intended islanded grids can increase the security of supply and resilience. However, from the technical side, the operation of these grids is more challenging than in the IPS. Therefore, it must be decided on a case-by-case basis whether and to what extent intended islanded grids make sense.

If the regional electrical energy supply is destroyed, e.g. after a flood or a storm, then the option of local island grid operation has to be rated very positively. However, the question of the probability of occurrence of such events and the costs – investment and maintenance costs – to mitigate the consequences must be critically scrutinised and whether the consequences of such an event can also be prevented or mitigated by other measures.

Practical example: electric energy supply of an airport

The electrical power supply of an airport, as a critical infrastructure, is essential for safe operation. Therefore, there are various measures to ensure the power supply of the vital areas of the airport – e.g. air traffic control, runway lighting – in the event of a power failure or even blackout in the upstream network. Often, the airport is supplied with power from two or more substations to protect the facility against a single failure leading to an outage. On the other hand, there are decentralised emergency power generators that can provide a large share of the necessary power. In less critical areas such as the airport shops, the power demand can be significantly reduced by a partial shutdown.

Many examples of power failures around the world show that these measures are effective in protecting airports against power outages and that a single event might fully pay for the cost of power redundancy protection. Although the measures described above were also implemented at Atlanta Hartsfield–Jackson International Airport, a fire in a tunnel, that housed both the main power lines and a backup supply, shut down the Hartsfield–Jackson Atlanta International Airport for nearly 11 hours on Sunday, stranding hundreds of thousands of passengers on aeroplanes and in terminals. Redundancy was provided on the electrical side but not on the tunnel side. America's biggest airline, Delta Air Lines, has announced that it lost between USD 25 million and 50 million pre-tax income during this power outage [59].

1.6.2 Unintended islanding

The unintended operation of an islanded grid, although it should be de-energised by planned switching operations, is undesirable and bears high risks and no opportunities. Unintended islanding is in stark contrast to the safe operation of the grid and can lead to numerous problems [39]:

- The absence of voltage after disconnection is not guaranteed.
- Voltage and frequency cannot be controlled or influenced by the grid operator.
- Compliance with the step and contact protection voltage is not guaranteed.
- The probability of success of the automatic re-closure as a result of a switching operation by the selective protection is reduced.
- Liability for damage incurred in the context of island operation must be clarified.
- Asynchronous reconnection of an islanded grid can damage equipment.
- Reduction in the quality of supply, e.g. due to the occurrence of resonances or a reduction in the short-circuit power.

For these reasons, the rapid detection and safe shutdown of unintended islanded grids is desirable. The investigations in [39] showed that islanded grids can occur with all types of decentralised generation systems, even if the control system of the DGs is only aimed at the delivery of constant active and reactive power. The active power reduction required by the DGs in the event of over-frequency increases the probability that stable islanded grid can be formed, as this can lead to balancing out generation and demand in some cases. While this effect is desired when operating

26   *Intended and unintended islanding of distribution grids*

DGs with the IPS, it must be classified as critical in the case of unintended islanded grids. There are different approaches to detect unintended islanded grids, which are presented and discussed in detail in Chapter 7.

References

[1]   Beginn der Elektrifizierung in Deutschland [Homepage on the Internet]. Heidestraße 2, 10557 Berlin: 50Hertz Transmission GmbH [cited 2022 May 23]. Available from: http://www.50hertz.com/de/50Hertz/Ueber-uns/Historie.

[2]   Kundur P. Power System Stability and Control. *Power System Engineering Series*. Palo Alto, CA: McGraw Hill, Inc.; 1994.

[3]   Csanyi E. Historical Review of Power System Stability Problems [Homepage on the Internet]. EEP – Electrical Engineering Portal [cited 2023 June 14]. Energy and Power | Transmission and Distribution. Available from: https://electrical-engineering-portal.com/historical-review-of-power-system-stability-problems.

[4]   Bo ZQ, Lin XN, Wang QP, *et al.* Protection and Control of Modern Power Systems. EEP – Electrical Engineering Portal. 2016:1–6.

[5]   Steinhart C. Lokale Inselnetz-Notversorgung auf Basis dezentraler Erzeugungsanlagen mit Fokus auf die Frequenzstabilität [Dissertation]. Technical University of Munich. Arcisstraße 21, 80333 München, Germany; 2020.

[6]   Sporn P. Interconnected Electric Power Systems. *Electr Eng*. 1938;57(1): 16–25.

[7]   Cohn J. When the Grid Was the Grid: The History of North America's Brief Coast-to-Coast Interconnected Machine. *Proc IEEE*;107(1):232–243.

[8]   Kreusel J. Entering A New Epoch – A Brief History of the Electric Power Supply. *ABB Review*; 2014;4(14):46–53.

[9]   Imdadullah, Alamri B, Hossain MA, *et al.* Power Network Interconnection: A Review on Current Status, Future Prospects and Research Direction. *Electr Eng*. 2021;10(2179):1–29.

[10]   Mariani E and Murthy SS. In: *Introduction*. London: Springer London; 1997. pp. 1–9.

[11]   Englert H and Hoppe-Oehl H, editors. VDE-Studie, Schutz-und Automatisierungstechnik in aktiven Verteilnetzen, Modul C – Netzleittechnik. Energietechnische Gesellschaft (ETG), Stresemannallee 15, 60596 Frankfurt am Main: ETG; 2016.

[12]   Shaver L. Implementation of a DC Microgrid [Thesis]. University of Wisconsin-Madison; 2017.

[13]   VDE. Mit dem Ausbau des Übertragungsnetzes kann die Energiewende gelingen [Interview with R. Joswig]. Berlin und Frankfurt: VDE Verband der Elektrotechnik Elektronik Informationstechik e.V.; 2012.

[14]   acatech, Leopoldina, and Akademienunion. Coupling the different energy sectors – options for the next phase of the energy transition. Series on science-based policy advice position paper. München and Halle (Saale) and Mainz: acatech – National Academy of Science and Engineering

and German National Academy of Sciences Leopoldina and Union of the German Academies of Sciences and Humanities; 2018. Available from: https://www.acatech.de/publikation/sektorkopplung-optionen-fuer-die-naechste-phase-der-energiewende/.

[15] NEP. Szenariorahmen zum Netzentwicklungsplan Strom 2037 mit Ausblick 2045, Version 2023 – Entwurf der Übertragungsnetzbetreiber.

[16] Prognos, Öko-Institut, and Wuppertal-Institut. Towards a Climate-Neutral Germany. Executive Summary conducted for Agora Energiewende, Agora Verkehrswende and Stiftung Klimaneutralität [Study]. Agora Energiewende, Agora Verkehrswende and Stiftung Klimaneutralität; 2021.

[17] Deloitte M, DSO E, and Euroelectric. Connecting the Dots: Distribution Grid Investment to Power the Energy Transition [Presentation]. Deloitte Consulting, S.L.U.; 2021.

[18] IEA. World Energy Outlook 2021. France: International Energy Agency; 2014. Available from: https://iea.blob.core.windows.net/assets/88dec0c7-3a11-4d3b-99dc-8323ebfb388b/WorldEnergyOutlook2021.pdf.

[19] IRENA. *Global Energy Transformation: A Roadmap to 2050*. Abu Dhabi: International Renewable Energy Agency; 2014. Available from: https://www.irena.org/-/media/Files/IRENA/Agency/Publication/2018/Apr/IRENA_Report_GET_2018.pdf.

[20] IPCC. Synthesis Report of the IPCC Sixth Assessment Report (AR6). Geneva 2, Switzerland: Intergovernmental Panel on Climate Change (IPCC); 2023. Available from: https://www.ipcc.ch/report/ar6/syr/downloads/report/IPCC_AR6_SYR_LongerReport.pdf.

[21] VDE. Zielvorgabe Klimaneutral – Ein Expertengespräch. VDE dialog, Technologie-Magazin des VDE, 04/2020. 2020;12–17.

[22] Astarloa B, Kaakeh A, Lombardi M, *et al.* The Future of Electricity – New Technologies Transforming the Grid Edge [Paper]. World Economic Forum, Cologny/Geneva, Switzerland: World Economic Forum, REF 030317; 2017.

[23] Grünwald R. TAB-Arbeitsbericht Nr. 162: Moderne Stromnetze als Schlüsselelement einer nachhaltigen Energieversorgung. Neue Schönhauser Straße 10, 10178 Berlin: Büro für Technikfolgen-Abschätzung beim Deutschen Bundestag (TAB); 2014.

[24] FNN. VDE FNN Roadmap: Zum Klimaschutznetz bis 2030 – Aktionsschwerpunkte und Handlungsbedarf [Paper]. Berlin: Forum Netztechnik/Netzbetrieb im VDE (FNN); 2022.

[25] Merk P, Schulz S, Becker H, *et al.* Wer startet das Netz der Zukunft nach einem Blackout? *ETG Journal* 2022;2022(2):18–21.

[26] Bayer J, Benz T, Erdmann N, *et al.* Zellulares Energiesystem – Ein Beitrag zur Konkretisierung des zellularen Ansatzes mit Handlungsempfehlungen. VDE, Energietechnische Gesellschaft (ETG), Frankfurt am Main: VDE; 2019.

[27] BSW, EnBW, VDE, *et al.* Press release 20 September 2011 – Retrofitting of PV Systems – To solve 50.2 Hz problem planned [Press release]. Berlin, Stuttgart, Frankfurt, Berlin, Dortmund, Bayreuth, Cologne, Stuttgart: BSW – Bundesverband Solarwirtschaft e.V; EnBW Transportnetze AG; VDE e.V.;

28   *Intended and unintended islanding of distribution grids*

50 Hertz Transmission GmbH; Amprion GmbH; TenneT TSO GmbH; Ecofys Germany GmbH IFK; 2011.

[28]   Forsyth T. Draft World Grid Codes [Paper]. Wind Advisor Team; 2020.

[29]   Ingram M. Grid Code Essentials and Streamlining Process for Interconnections [Webinar]. USAID; NREL; 2020.

[30]   IRENA. Grid Codes as Enablers of the Energy Transition [Presentation]. IRENA; 2017.

[31]   Shuai M, Chengzhi W, Shiwen Y, *et al.* Review on Economic Loss Assessment of Power Outages. *Procedia Comput Sci.* 2018;130:1158–1163. *The 9th International Conference on Ambient Systems, Networks and Technologies (ANT 2018)/The 8th International Conference on Sustainable Energy Information Technology (SEIT-2018)*/Affiliated Workshops. Available from: https://www.sciencedirect.com/science/article/pii/S1877050918305131.

[32]   Andresen A, Kurtz L, Hondula D, *et al.* Understanding the Social Impacts of Power Outages in North America: A Systematic Review. *Environ Res Lett.* 2023, 05;18:1–15.

[33]   Schmidthaler M and Reichl J. Assessing the Socio-economic Effects of Power Outages Ad Hoc. *Comput Sci – Res Develop.* 2016, 08;31:157–161.

[34]   Forschungsprojekt INZELL [Homepage on the Internet]. Regensburg: OTH Regensburg; [cited 2022 May 23]. Forschungsstelle für Energienetze und Energiespeicher. Available from: https://forschungsprojekt-industriezell e.de/.

[35]   ENTSO-E. P5-Policy 5: Emergency Operations [Policy]. ENTSO-E, Belgium: ENTSO-E; 2017.

[36]   Schnelle T, Schmidt M, and Schegner P. Power Converters in Distribution Grids – New Alternatives for Grid Planning and Operation. *IEEE Eindhoven PowerTech.* 2015;1–6.

[37]   Schnelle T, Schweer A, and Schegner P. Islanded Operation of Modular Grids. *CIRED.* 2017;1264–1268.

[38]   Schnelle T, Schweer A, and Schegner P. Control of Modular Microgrids by Varying Grid Frequency. *IEEE Manchester PowerTech.* 2017;1–6.

[39]   Palm S. Untersuchung und Bewertung von Verfahren zur Inselnetzerkennung, -prognose und -stabilisierung in Verteilnetzen [Dissertation]. TU Dresden. Dresden, Germany; 2019.

[40]   IEEE. IEEE Standard for Interconnecting Distributed Resources with Electric Power Systems. IEEE Std 1547-2003. 2003;1–28.

[41]   Mrugowsky H. Drehstrommmaschinen im Inselbetrieb. Wiesbaden: Springer Fachmedien Wiesbaden; 2014.

[42]   VDE. DIN VDE 0126-1-1 Selbsttätige Schaltstelle zwischen einer netzparallelen Eigenerzeugungsanlage und dem öffentlichen Niederspannungsnetz [Standard]. Berlin und Frankfurt: DIN Deutsches Institut für Normung e.V. und VDE Verband der Elektrotechnik Elektronik Informationstechik e.V.; 2013.

[43]   IEC. IEC 62116:2014-02, Utility-interconnected photovoltaic inverters – Test procedure of islanding prevention measures [Standard]. Switzerland: IEC Central Office; 2014.

[44] Quirós-Tortós J, Panteli M, Wall P, *et al.* A Sectionalising Methodology for Parallel System Restoration Based on Graph Theory. *IET Generation, Transmission and Distribution.* 2015;1–23.

[45] Braun M, Brombach J, Hachmann C, *et al.* The Future of Power System Restoration: Using Distributed Energy Resources as a Force to Get Back Online. *IEEE Power Energy Mag*, 2018;16(6):30–41.

[46] CIGRE. Technical Brochure No. 712: System restoration procedure and practices [Technical Brochure]. CIGRE, Paris, France: CIGRE WG C2.23; 2017.

[47] Brunken E, Mischinger S, Willke, J. Systemsicherheit 2050 [Study]. Deutsche Energie-Agentur GmbH (dena), Chausseestraße 128 a, 10115 Berlin: dena; 2020.

[48] Katiraei F, Abbey C, Tang S, *et al.* Planned Islanding on Rural Feeders – Utility Perspective. *IEEE PES 2008 General Meeting (Panel on Microgrids),* July 20–24, Pittsburgh, Pennsylvania. 2008; pp. 1–6.

[49] FNN. Solidarität im Verbundsystem [Info]. Bismarckstr. 33, 10625 Berlin: Forum Netztechnik/Netzbetrieb im VDE (FNN); 2018.

[50] Pazos FJ. Operational Experience and Field Tests on Islanding Events Caused by Large Photovoltaik Plants. Cired, Frankfurt, Paper 0184. 2011; pp. 1–4.

[51] Urdal H, Ierna R, Zhu J, *et al.* System Strength Considerations in a Converter Dominated Power System. *IET Renew Power Gener,* 2015 9;1: 10–17.

[52] Knoll F, Antoni J, Gierschner S, *et al.* Momentanreserve in einem überwiegend EE-basierten Stromsystem – Eine interdisziplinäre Einführung unter Berücksichtigung technischer, ökonomischer und juristischer Aspekte [Report]. Greifswald, Rostock, Stralsund: Universität Greifswald, Universität Rostock, Hochschule Stralsund; 2021.

[53] Henninger S. Netzdienliche Integration regenerativer Energiequellen über stromrichtergekoppelte Einspeisenetze mit integrierten Energiespeichern [Dissertation]. Friedrich-Alexander-Universität Erlangen-Nürnberg. FAU University Press Erlangen, Germany; 2019.

[54] Schwab A. Elektroenergiesysteme: Erzeugung, Übertragung und Verteilung elektrischer Energie. Berlin: Springer; 2009.

[55] CIGRE. Technical Brochure no. 671: Connection of wind farms to weak AC networks [Technical Brochure]. CIGRE, Paris, France: CIGRE WG B4.62; 2016.

[56] Gleadow J, Love G, Saad H, *et al.* TF-77 – AC Fault Response Options for VSC HVDC Converters. In: Papailiou KO, editor. *Cigre Sicence & Engineering*, vol. 15, October 2019. Paris, France: Cigre; 2019. pp. 105–110.

[57] FNN. VDE FNN Guideline: Grid-forming and system-supporting behaviour of power-generating modules [Paper]. Berlin: Forum Netztechnik/Netzbetrieb im VDE (FNN); 2021.

[58] Gauthier M, Abbey C, Katirai F, *et al.* Planned Islanding as a Distribution System Operation Tool for Reliability Enhancement. In: CIRED, editor.

30 *Intended and unintended islanding of distribution grids*

19th International Conference on Electricity Distribution Vienna, 21–24 May 2007, Paper 0270. Liège (Belgium): CIRED; 2007: pp. 1–4.

[59] Jansen B. Delta: Atlanta airport power outage cost $25M to $50M in income [Homepage on the Internet]. USA Today; [cited 2023 June 14]. Today in the Sky. Available from: https://eu.usatoday.com/story/travel/flights/todayinthes ky/2018/01/03/delta-atlanta-airport-power-outage-costs/999761001/.

Chapter 2

Basics of power system dynamics and stability

Herwig Renner[1], Petros Aristidou[2] and Ziqian Zhang[1]

To operate an electricity grid in a stable manner, the same physical principles must be considered for interconnected and islanded operations. This chapter contains a brief overview of some of the basic concepts in power system dynamics and stability with a special focus on islanded grids.

In the first section, general information about dynamics in power systems, including the main characteristics of devices concerning system dynamics, is given. In the following section, the formal definition of power system stability is presented, and the various types of power system stability are classified. Concerning the original classification, the new stability class 'converter-driven stability' is introduced to cover the effects of the increasing penetration of fast-acting, converter-interfaced generation (CIG). In the following sub-sections, the different categories of system stability are presented.

The dynamic behaviour of the power system is directly influenced by inertia and system strength. The level of inertia influences the frequency gradient (rate of change of frequency = RoCoF) and transient frequency values during a system incident. The impact of reduced system inertia on system operation is discussed in the following section. This decreased overall system inertia is caused by a shift of generation from classical synchronous generation to power electronic-based non-synchronous generation. Islanded systems usually have significantly reduced inertia.

System strength is related to the inverse of the grid impedance. In classical power systems, dominated by synchronous machines, system strength corresponds to short-circuit capacity. In power systems with a high share of converter-based generation, short-circuit capacity as a measure of grid impedance during normal operation (close to nominal voltage) is different to short-circuit capacity during a fault. It strongly depends on control algorithms and the current limitation of connected inverters. Islanded systems usually have a significantly reduced system strength and inertia.

---

[1] Institute of Electrical Power Systems, Graz University of Technology, Austria
[2] Department of Electrical Engineering, Computer Engineering & Informatics, Cyprus University of Technology, Cyprus

32 *Intended and unintended islanding of distribution grids*

2.1 Dynamics in power systems

This chapter provides some fundamentals regarding dynamics in power systems. The basic idea of system state modelling is presented, introducing the more detailed content of Chapter 9 'Time-domain modelling'. The main dynamic properties of components in electrical power systems are marked out, assuming that basic properties necessary for steady-state analysis are well known.

However, electrical power systems are facing significant changes, also affecting power system dynamics and stability. Previous systems were dominated by synchronous machines with inherent, homogeneous properties like inertia and system strength. Modern systems with an increasing share of inverter-based generation reveal a huge variety of characteristics, determined by their individual controls. For instance, the internal phase-locked loop (PLL) is a crucial element.

Especially in islanded systems with limited extension, the penetration of renewable, converter-based generation must be considered. A relevant case is the Irish system, where a penetration of renewable generation of 60% was effectively hit in 2017 [1].

2.1.1 Dynamic model basics

As each system in engineering, a power system can be described mathematically by its system states and the time-varying state variables $x_1(t)$, $x_2(t), \ldots, x_n(t)$. The state variables are the minimum set of variables uniquely defining the system state [2]. Depending on the degree of modelling and the time/frequency range of interest, the number of state variables n may vary in a huge range. A basic model might describe electromechanical phenomena only, while an enhanced mode would include voltage and frequency control as well. System dynamics can be described and analysed in general by a set of first-order differential equations of the form

$$\dot{x} = f(x) \tag{2.1}$$

with x being the vector of state variables, $\dot{x}$ being the vector of their derivatives and f being a vector of – in general – nonlinear functions. For simplification, the system is often linearised in an operating point, leading to the well-known linear state–space description

$$\dot{x} = A \cdot x \tag{2.2}$$

with A being the 'system matrix' or 'state matrix', a square matrix with dimension $n \times n$. Equation (2.2) represents a set of linear first-order differential equations with the system matrix containing all necessary information for dynamic analysis in a linearised model.

The solution of the differential equations can be executed in time domain numerical with an appropriate solver. There is a huge variety of programs available, already tailored for power system dynamics. Besides solutions in the time domain, an analysis in the frequency domain can be performed. The eigenvalues of A correspond to the modes of the system with their real part representing the damping of the mode and the imaginary part representing the mode's oscillation frequency. More details can be found, for instance, in [2] or [3].

The number n of states of a model will vary from less than 10 for a single-machine infinite-bus model up to some ten-thousands for an interconnected system with synchronous generation and converter-based generation and all voltage and frequency controllers considered.

2.1.2 Time/frequency range of concern

The definition of state variables strongly depends on the scope of the analysis. Regarding the time and frequency range, according to [2], a broad classification into four categories can made according to [2]: wave, electromagnetic, electromechanical, and thermodynamic. This classification considers classical power systems with the generation relying mainly on synchronous generators (SGs). However, in modern power systems with higher penetration of inverter-based generation, the domain is enhanced (especially concerning control issues and fast dynamics). In Figure 2.1, an overview of the time frame of different phenomena and control mechanisms in power systems is shown.

2.1.3 Main components involved in power systems dynamics

In this section, a short description of the devices playing an important role in power system control and stability is given. Moreover, the basic characteristics concerning

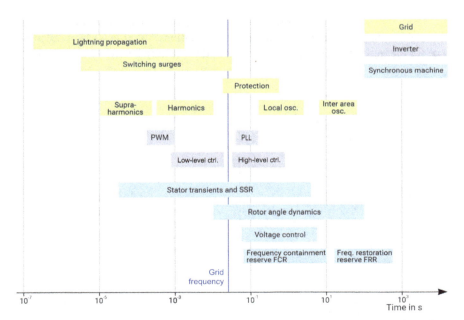

Figure 2.1   *Time frame of power system dynamics, characteristic timescales of different physical and control dynamics (based on [4,5])*

2.1.4 Synchronous generators

Synchronous generators (SG) have always played a significant role in the transformation of mechanical power into electrical power.

SGs consist of the stator and the rotor. The rotor excitation winding produces a rotating flux, inducing AC voltages in the three-phase armature windings mounted in the stator. The voltage at the terminals of the armature winding is proportional to the rotor speed and the exciting flux. Since the rotor speed is kept almost constant in normal operation, the excitation current can be used to control the terminal voltage of the SG. The capability diagram of a SG, indicating the active and reactive power range, is shown in Figure 2.2. Active power is limited by the prime mover. Reactive power in overexcited mode is limited by the maximum field current. Reactive power in the under-excited mode is limited by the practical stability limit [6]. The reactive power capability has a significant impact on voltage stability.

The rotor also has damper windings (amortisseur), which are short-circuited. These windings have a significant effect on the dynamic behaviour of the SG as they activate to dump oscillations after a disturbance. The steady-state and transient behaviour of the machine is described by the flux linkage of the windings and the voltage–flux relationships. The electrical parameters have a large impact on voltage stability and short-circuit currents. Details regarding the mathematical representation can be found for instance in [3].

Besides the electrical characteristics of SGs, their mechanical properties are essential for analysing frequency stability. The equation of motion is of central importance for the description of the accelerating or decelerating behaviour of SGs. Using

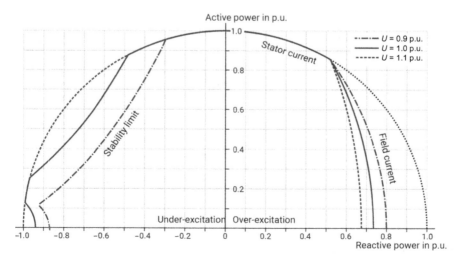

Figure 2.2 *Typical capability diagram of a synchronous generator*

Basics of power system dynamics and stability 35

the rotor angle as a reference variable, the linearised equation form has the form of an ordinary second-order differential equation, usually called the swing equation.

The moment of inertia J of the rotating elements is a machine constant and is assumed to be known. Thus, the stored kinetic energy can be calculated as

$$W_{kin} = \frac{J \cdot \omega_m^2}{2} \tag{2.3}$$

where
W_{kin} is kinetic energy stored in the SG
J is moment of inertia of generator and turbine
ω_m is mechanical angular speed

It is important to note that the stored kinetic energy is proportional to the mechanical speed squared. The mechanical angular velocity ω_m corresponds to the angular velocity of the rotor. It is determined from the electrical angular velocity ω and the number of poles p in the generator. The electrical angular velocity ω can be calculated from the grid frequency f.

$$\omega_m = \frac{\omega}{p/2} = \frac{2\pi f}{p/2} \tag{2.4}$$

Where
ω is electrical angular speed
f is grid frequency
p is number of pole pairs

Instead of the moment of inertia J, usually the inertia constant H is used.

$$H = \frac{J \cdot \omega_{m,n}^2}{2 \cdot S_r} \tag{2.5}$$

where
H is inertia constant
S_r is rated power of the generator
J is moment of inertia of generator and turbine
$\omega_{m,n}$ is a rated mechanical angular speed

The inertia constant H represents the kinetic energy at rated speed in the pu-domain.

$$W_{kin} = H \cdot S_r \cdot \left(\frac{\omega}{\omega_n}\right)^2 = H \cdot S_r \cdot \left(\frac{f}{f_n}\right)^2 \tag{2.6}$$

Sometimes the mechanical starting time T_m is used instead of the inertia constant.

$$T_m = 2 \cdot H \tag{2.7}$$

In case of a sudden unbalance between the mechanical power of the turbine and the electrical power of the load, the resulting power difference ΔP is compensated by

36   *Intended and unintended islanding of distribution grids*

a change in the kinetic energy, since the sum of the powers in an electrical network is always equal to zero.

$$P_{\mathrm{m}} - P_{\mathrm{e}} = \Delta P = \dot{W}_{\mathrm{kin}} = 2 \cdot H \cdot S_{\mathrm{r}} \cdot \frac{f}{f_{\mathrm{n}}^2} \cdot \dot{f} \tag{2.8}$$

Reformulating (2.8) results in an expression for the RoCoF in a system following a disturbance of the power equilibrium:

$$\mathrm{RoCoF} = \dot{f} = \frac{\Delta P}{2 \cdot H \cdot S_{\mathrm{r}}} = \frac{f_{\mathrm{n}}^2}{f} \tag{2.9}$$

If the RoCoF is only considered after a change on the load or generation side around the rated frequency f_n, (2.9) simplifies to:

$$\mathrm{RoCoF} = \dot{f} \approx \frac{\Delta P \cdot f_{\mathrm{n}}}{2 \cdot H \cdot S_{\mathrm{r}}} = \frac{\Delta P \cdot f_{\mathrm{n}}}{2 \cdot W_{\mathrm{kin}}} \tag{2.10}$$

A commonly used definition of system inertia constant based on [7] puts the rotating energy in relation to the system load. In this definition, only synchronous machines contribute to system inertia, synthetic inertia from inverters is not considered.

$$T_{\mathrm{N}} = \frac{\sum_1^n T_{\mathrm{m,n}} \cdot S_{\mathrm{r,n}}}{P_{\mathrm{load}}} \tag{2.11}$$

where
T_{N} is network acceleration time constant
$T_{\mathrm{m,n}}$ is mechanical starting time of synchronous generator n
$S_{\mathrm{r,n}}$ is rated power of synchronous machine n
P_{load} is system load

The simplest representation of a synchronous machine is the classical model, considering the swing equation and representing the electrical part by a constant voltage E' behind a reactance X'_{d}. All effects of synchronous machine field and damper dynamics are neglected. The state matrix is a 2×2 matrix with rotor angle and speed as state variables (Figure 2.3).

$$\begin{bmatrix} \Delta \dot{\omega} \\ \Delta \dot{\delta} \end{bmatrix} = \underbrace{\begin{bmatrix} -\dfrac{K_{\mathrm{D}}}{2H} & -\dfrac{K_{\mathrm{S}}}{2H} \\ \omega_0 & 0 \end{bmatrix}} \cdot \begin{bmatrix} \Delta \omega \\ \Delta \omega \end{bmatrix} \tag{2.12}$$

Considering the field and damper winding, the model is expanded. A commonly used model is the so-called 2.2 model, taking into account the field circuit and one equivalent damper circuit in the d-axis and two equivalent field circuits in the q-axis, resulting in a sixth-order machine model. A detailed description can be found in [3, 8]. Besides the electromechanical characteristics of the machine, the control of the terminal voltage has a significant influence on the dynamic characteristics. Details for voltage and frequency control can be found in Chapter 3.

A common model, including mechanical and electrical dynamics and a simple automatic voltage regulator (AVR), is the so-called Heffron–Phillips model. It is a linearised model of fourth order, suitable for small-signal analysis. $K_1 - K_6$ are the

Basics of power system dynamics and stability   37

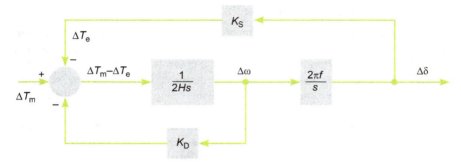

Figure 2.3   Classical model (second order) of the synchronous machine [3]

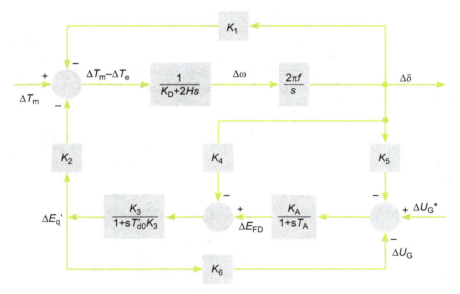

Figure 2.4   Heffron–Phillips model (4th order) of the synchronous machine and simplified AVR, based on [9]

Heffron–Phillips coefficient, depending on the synchronous machine parameters and the operating point (Figure 2.4) [9].

$$\begin{bmatrix} \Delta\dot{\omega} \\ \Delta\dot{\delta} \\ \Delta\dot{E}'_q \\ \Delta\dot{E}_{FD} \end{bmatrix} = \begin{bmatrix} -\dfrac{K_D}{2H} & -\dfrac{K_1}{2H} & -\dfrac{K_2}{2H} & 0 \\ \omega_0 & 0 & 0 & 0 \\ 0 & -\dfrac{K_4}{T'_{d0}} & -\dfrac{1}{K_3 T'_{d0}} & \dfrac{1}{T'_{d0}} \\ 0 & -\dfrac{K_A K_5}{T_A} & -\dfrac{K_A K_6}{T_A} & -\dfrac{1}{T_A} \end{bmatrix} \cdot \begin{bmatrix} \Delta\omega \\ \Delta\delta \\ \Delta E'_q \\ \Delta E_{FD} \end{bmatrix} \qquad (2.13)$$

2.1.5 Inverter-based generation

Inverters act as an interface between direct current (DC) based energy sources and the AC grid. Contrary to synchronous machines, the steady-state and dynamic characteristics are mainly determined by the inverter controls. Usually, the outer control loops are used to provide reference values for active and reactive power, while the inner control loops are responsible for realising the voltage and current at the inverter terminals.

Inverters do not have a rotating mass component; i.e. there is no inherent inertia. The prime mover behind the inverter might have inertia, but the inertia response must be implemented via the inverter controls (virtual inertia). There are strict limits in terms of maximum current through the power electronics devices, as well as maximum voltage. During faults, the fault current is usually limited to the rated current (or slightly above), thus allowing only limited contribution to fault currents and voltage support.

Regarding the control, a classification into 'grid-following' and 'grid-forming' units can be made. While for the former, the aim is to control in a fast way active and reactive power injection, independently of voltage magnitude and angle variations, the latter aims to keep the voltage magnitude and voltage angle stable. Grid-forming inverters exhibit a characteristic like SGs and are often controlled to behave as such (virtual synchronous machine).

Both types might have additional grid-supporting functionalities, adjusting the reference values for active and reactive power according to changes in voltage magnitude and frequency. Especially in islanded grids, these functionalities help to maintain stability.

Figure 2.5 shows an example of the principle control structure for the grid-following and the grid-forming inverters. Reference values are indicated with an asterisk.

There are two general approaches for the simulation of inverters: root mean square (RMS) models and electromagnetic transient (EMT) models. These simulation methods are explained in more detail in Section 2.4. Simplified inverter models (RMS models, also known as averaged models) provide the behaviour of the inverter ignoring fast switching transients and any control with very small-time constants compared to the simulation time steps and the phenomena considered.

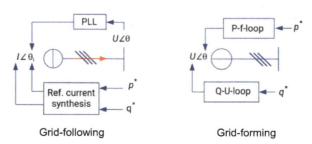

Figure 2.5 Basic structure of grid-following and grid-forming inverters

In stability analysis, there is usually no need to model the inverter in detail when its transients are likely to decay within the time step of the time-domain simulation. Only the fundamental frequency outputs of the inverter are modelled, mainly reflected in the electrical control model. Due to this averaged representation, the modelling of different technologies of inverter-based generators can be unified using generic sub-models for the basic components.

However, RMS-type models are not recommended for use in power system dynamic studies in weak grids, which is often the case in smaller islanded systems. Here, the use of EMT models is recommended. According to the level of detail required, one of the following models can be chosen:

- Discrete switch model
- Averaged switch model
- Simple source model

A comprehensive overview of the models can be found in [10].

2.1.6 Load

While in load flow calculations consumer loads are modelled with constant power, independent of voltage magnitude and frequency, this is not sufficient for dynamic studies. This is especially true for islanded grids with larger deviations in frequency and voltage.

The quasi-stationary power–frequency dependency of the load is often referred to in the literature as the self-regulation effect. This effect is mainly based on the frequency-dependent power consumption of motor-driven machines according to their speed-torque characteristic. The higher the rotational speed of the drive, the more active power is needed and vice versa. This effect is an inherent feature of the load itself and thus automatically contributes to frequency stabilisation.

With linearisation around the operating point, the load characteristic can be described by the parameter k_{pf}, which is also referred to as load self-regulating effect. The same procedure can be applied to reactive power:

$$k_{\mathrm{pf}} = \frac{\Delta P / P_0}{\Delta f / f_{\mathrm{n}}} \qquad k_{\mathrm{qf}} = \frac{\Delta Q / Q_0}{\Delta f / f_{\mathrm{n}}} \tag{2.14}$$

where
ΔP, ΔQ is Quasi-stationary power variation of the load due to a deviation from the nominal frequency f_0
P_0, Q_0 is active/reactive power at steady-state nominal frequency
Δf is deviation from the nominal frequency
f_{n} is nominal frequency

As can be seen in Tables 2.1 and 2.2, pumps hold a significant P–f-dependency, as long as their motor is directly connected to the grid. Therefore, directly connected storage pumps are considered in islanded operation during grid restoration. On the one hand, they provide a base load and on the other hand, they support frequency stability by their self-regulating effect since a drop in frequency leads to a significant reduction of the absorbed active power, thus helping to balance generation and load.

40 *Intended and unintended islanding of distribution grids*

Table 2.1 Frequency dependence of selected loads; adapted from [11,12]

Power	Typical loads
$P \approx$ constant	Light bulb, LED, ohmic loads, PC
	Drive with frequency converter, electric controlled circulating pump
$P \sim f$	Machine tool, hoist, coiler, extruder, piston pump, hoisting gear
$P \sim f^2$	Calander
$P \sim f^3$	Blower, centrifuge, radial pump, turbo-machine

Table 2.2 Load self-regulation effect of selected loads; adapted from [3,13–15]

Load type	k_{pf}	k_{qf}
Pumps in agriculture	$5.0 \ldots 5.6$	$4.0 \ldots 4.2$
Pumps in drinking water supply	4.5	2.5
Fan motor	2.9	1.7
Industrial motors	2.5	1.2
Mixed domestic load	$0.8 \ldots 1.0$	$-2.2 \ldots -1.5$
Mixed commercial load	$1.2 \ldots 1.5$	$-1.6 \ldots -1.1$
Mixed industrial load	2.6	1.6

The electrification of loads in the industrial, residential, commercial and transport sectors will increase the number of electric motors and drives in the grid and thus the electrical energy demand. However, a large number of these drives are nowadays operated via a frequency inverter resulting in a decreasing self-regulation effect. It must be mentioned that voltage and frequency sensitivity factors in literature are often based on the load characteristics 20 years or more ago. Comprehensive studies regarding the behaviour of actual loads, especially regarding the frequency characteristics, are not available. In total, the load self-regulation effect is expected to decrease in the future.

Besides the frequency, the effect of voltage dependency must be considered in dynamic studies. This is especially true for weak grids, as it is often the case in islanded operations. Here, a common way to model loads is the application of an exponential model or a ZIP model [16].

The exponential model relates the power and the voltage at a load bus by exponential equations.

$$\frac{P}{P_0} = \left(\frac{U}{U_0}\right)^{k_{pv}} \qquad \frac{Q}{Q_0} = \left(\frac{U}{U_0}\right)^{k_{qv}} \tag{2.15}$$

An exponent of 0 corresponds to constant power, an exponent of 1 corresponds to constant current and an exponent of 2 corresponds to constant impedance. Power

Basics of power system dynamics and stability   41

electronic devices usually will have a constant active power independent of the voltage ($k_{pv} = 0$). However, it is only valid in a limited voltage range around nominal voltage. Self-disconnection of load following a voltage dip will happen for voltages below 70–80% [17]. From the high voltage (HV) grid perspective, medium voltage (MV) loads connected via a transformer with an on-load tap changer (OLTC) and automatic voltage regulation behave as constant power loads for steady state. For composite loads, the exponent for active power ranges from 0 to 2; for reactive power, the range is typically between 1.5 and 6 [18].

An alternative representation is the polynomial load model, also known as the ZIP model. It represents the relationship between the voltage magnitude and power in a polynomial equation that combines constant impedance (Z, exponent 2), constant current (I, exponent 1), and constant power (P, exponent 0) components. The same approach can be used for reactive power.

$$\frac{P}{P_0} = \left(a_0 + a_1 \cdot \frac{U}{U_0} + a_2 \cdot \left(\frac{U}{U_0} \right)^2 \right) \tag{2.16}$$

$$a_0 + a_1 + a_2 = 1$$

where
a_2 is share of constant impedance load
a_1 is share of constant current load
a_0 is share of constant power load

Besides the static load model, a variety of dynamic load models are available. A well-known approach is Hill's model, proposed in [19]. A comprehensive overview of static and dynamic models is given in [17,20].

$$T_p \cdot \dot{P} + P = P_0 \cdot \left(\frac{U}{U_0} \right)^{\alpha_s} - P_0 \cdot \left(\frac{U}{U_0} \right)^{\alpha_t} \tag{2.17}$$

where
P is actual active power
P_0 is initial value before voltage change
U is actual voltage
U_0 is initial voltage before voltage change
α_s is steady-state active power exponent
α_t is transient active power exponent

A special effect of the load to be considered here is the cold load pick up (CLPU). When the power grid is re-electrified after a supply failure of longer than a few minutes, the residual load (the load that must be restored) can significantly differ from the values in regular operation before the outage.

Responsible for this effect is on the one hand start-up currents (time constant less than a few seconds), on the other hand, the effect of loss of static load diversity, which may last several minutes. The CLPU phenomenon was first described in the context of electric heating [13]. In general, the term is used to describe the different behaviour of loads when resupplied after an outage compared to normal operation. This is the case for almost any load that interacts with some kind of storage, i.e. thermal storage as in heating and cooling, mechanical as in pumping applications or

42   *Intended and unintended islanding of distribution grids*

electrochemical storage such as batteries. As a significant CLPU current magnitude may bring about many adverse effects on a distribution system, multiple models have been proposed to account for the effects of CLPU. This includes the delayed exponential model and various physical models [14]. The physical models typically rely on information on the amount of electrical heating and cooling, temperature, etc. Furthermore, a differentiation between non-controlled and manually controlled loads is relevant for adequate modelling [15].

2.1.7   High-voltage direct-current transmission

Although alternating current was the predominant method for transmitting electrical energy in the twentieth century, due to the rapid developments in power electronics technology, high-voltage direct-current (HVDC) links have turned out to be a better alternative in specific situations. Especially in the case of the connection of islands via submarine cables, HVDC is the only feasible option due to the reactive power flow of AC cables.

HVDC links can be classified into two basic groups, based on the power electronics technology used and the control behaviour.

Current source converters, usually using thyristors, are used in most existing point-to-point installations worldwide. The operation, especially the commutation process, relies on a strong AC grid. This type of converter is often referred to as a line-commutated converter (LCC) and has several operating restrictions. The converter always absorbs reactive power with the amount depending on the active power. Black-start capability, voltage support and frequency support are not provided by the device.

Modern HVDC links are realised as **voltage source converters** (VSC), which are self-commutated converters using devices suitable for high-power electronic applications, such as isolated gate bipolar transistors (IGBTs). They are able to provide a stable voltage and thus can act as a power source for grid restoration after a blackout (black-start capability) [21]. Reactive power can be controlled arbitrarily within the converter's current limits. Connected to a grid with generating units, the converter can operate in a grid-following or a grid-forming mode (see Chapters 2.1.5 and 2.7). In the **grid-following** mode, active and reactive power is injected into the main grid according to reference values. A PLL structure is used to guarantee that the converter is synchronised with the grid. The frequency and voltage output of the converter are governed by the main grid. A **grid-forming** converter can regulate its frequency and voltage. It does not rely on the PLL structure for its synchronisation with the electric grid. This results in a more robust controller for the converter when the power converter is connected to a weak grid with low inertia levels [22].

Both types of HVDC links are able to provide dynamic active power support. Depending on the control system, they can participate in fast frequency control [23] or emulate synthetic inertia, thus contributing to frequency stability (see Chapter 2.4).

However, in a scenario with low real inertia in the island grid, electromechanical interactions might cause frequency instability and must be studied in detail.

Basics of power system dynamics and stability 43

Table 2.3 Examples of HVDC links, connecting islands with the mainland

Island	Name	Voltage (kV)	Rated power (MW)	Technology	Remark
Ireland	Moyle HVDC	250	250	Thyristor	
Tasmania	Basslink	400	500	Thyristor	
Corse-Sardinia	SACOI	200	300	Thyristor	Multi-terminal
Sardinia	SAPEI	500	1,000	Thyristor	
Gotland	Gotland 2	150	130	Thyristor	
New Zealand	Inter-island 2	±350	1200	Thyristor	Inter-island link
Japan	Hokkaido-Honshu	250	300	Thyristor	
		250	300	Transistor	

Operation of the HVDC in a grid-forming mode may avoid interactions with the SGs [24].

A comprehensive overview of HVDC technologies can be found in [25].

Prominent examples of islands, supplied via HVDC from the mainland, are given in Table 2.3.

2.1.8 Power system protection

Stability analysis is often related to disturbances. Especially in islanded grids, voltage, frequency and RoCoF are likely to be close to or even beyond normal operational limits. In that case, it is important to consider the actions of under-/over-voltage and under-/over-frequency protection relays. Internal device protection as well as system protection will have a significant impact on the system behaviour. More details can be found in Chapter 6.

2.1.9 Wide area monitoring and stability assessment

Continuously monitoring the grid is essential for ensuring a stable system. However, traditional supervisory control and data acquisition (SCADA) systems provide only a limited picture of the system dynamics due to limited data transmission speed, rather large data integration intervals and missing information about voltage angles. To tackle these problems, an increasing number of phasor measurement units (PMUs) are deployed in control centres, building a wide area measurement system (WAMS). PMUs are high-accuracy measurement units, providing voltages and currents as phasors with magnitude and phase angle. The units are GPS-synchronised through GPS, enabling a common reference for the angle information. Typical applications are steady-state and dynamic state estimation, control and protection [26] – especially for monitoring inter-area. In islanded grids, the wide-area-aspect is of less interest. However, PMUs can still provide accurate information about frequency and RoCoF

44 *Intended and unintended islanding of distribution grids*

with rather high resolution in time. In [27], the estimation of actual inertia in Great Britain, using PMUs, is presented.

2.2 Definition and classification of power system stability

2.2.1 General definition of power system stability

Stability
Power system stability is the ability of an electric power system, for a given initial operating condition, to regain a state of operating equilibrium after being subjected to a physical disturbance, with most system variables bounded so that practically the entire system remains intact [28].

Classical definitions of power system stability were mainly based on systems with synchronous generation. Recently, the definitions were extended in [28] to cover inverter-based generation as well. A graphical representation of the classification tree is shown in Figure 2.6. In the following, the main stability aspects are explained.

The mathematics of active and reactive power flow can be described by a two-port network as shown in Figure 2.7.

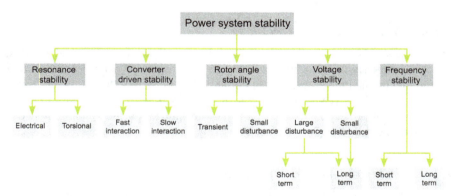

Figure 2.6   Classification of power system stability [28]

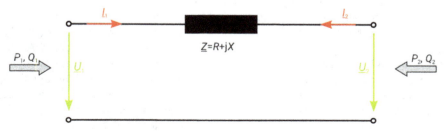

Figure 2.7   Principal illustration of active and reactive power transport

Basics of power system dynamics and stability 45

For the complex voltage, current and apparent power of the two ports in Figure 2.7, the following equation can be formulated:

$$\begin{pmatrix} \underline{I}_1 \\ \underline{I}_2 \end{pmatrix} = \begin{pmatrix} \dfrac{1}{\underline{Z}} & -\dfrac{1}{\underline{Z}} \\ -\dfrac{1}{\underline{Z}} & \dfrac{1}{\underline{Z}} \end{pmatrix} \cdot \begin{pmatrix} \underline{U}_1 \\ \underline{U}_2 \end{pmatrix} \tag{2.18}$$

$$\begin{pmatrix} \underline{S}_1 \\ \underline{S}_2 \end{pmatrix} = \begin{pmatrix} \underline{U}_1 & 0 \\ 0 & \underline{U}_2 \end{pmatrix} \cdot \begin{pmatrix} \underline{I}_1 \\ \underline{I}_2 \end{pmatrix}^* = \begin{pmatrix} \underline{U}_1 & 0 \\ 0 & \underline{U}_2 \end{pmatrix} \cdot \begin{pmatrix} 1/\underline{Z} & -1/\underline{Z} \\ -1/\underline{Z} & 1/\underline{Z} \end{pmatrix}^* \cdot \begin{pmatrix} \underline{U}_1 \\ \underline{U}_2 \end{pmatrix} \tag{2.19}$$

Splitting the complex power into real and imaginary parts and expressing voltage and impedance in polar coordinates (with ϑ_1 and ϑ_2 representing the voltage angles and ψ representing the impedance angle) results in (2.20) and (2.21).

$$P_1 = \frac{U_1^2}{Z} \cos(\psi) - \frac{U_1 U_2}{Z} \cos(\vartheta_1 - \vartheta_2 + \psi) \tag{2.20}$$

$$Q_1 = \frac{U_1^2}{Z} \sin(\psi) - \frac{U_1 U_2}{Z} \sin(\vartheta_1 - \vartheta_2 + \psi) \tag{2.21}$$

These two equations form the basis for angle and voltage stability analysis. The well-established models are

- a remote SG connected via a transmission line with a strong system (infinite bus), focus is put on rotor angle stability,
- a remote (passive) load connected via a transmission line with a strong system (infinite bus), the focus is put on voltage stability.

There is a strong correlation between voltage phase angle, respectively, its derivative, angular velocity or frequency and active power. On the other hand, a close relationship exists between voltage magnitude and reactive power.[*]

2.2.2 *Relevance of stability phenomena*

Table 2.4 indicates the relevance of stability phenomena for different grid categories. Most stability phenomena are related to large interconnected transmission systems. However, especially the question of frequency stability is inevitable for smaller islanded grids.

2.3 Rotor angle stability

Rotor angle stability describes the property of maintaining synchronism in a power system with several synchronous machines connected. It is split into small-disturbance or steady-state stability and transient angle stability.

---

[*]To be precise, the statement is true for grids with a high X/R ratio, respectively, ψ being close to 90°, which is in fact true for HV grids mainly.

Table 2.4 Relevance of categories of power systems stability in electric power systems

Categories	Transmission grid interconnected/islanded	Distribution grid interconnected/islanded
Rotor angle stability	✓/✓	–/–
Inter-area oscillations	✓/–	–/–
Frequency stability	✓/✓	–/✓
Voltage stability	✓/✓	✓/✓
Inverter-driven stability	–/✓	✓/✓

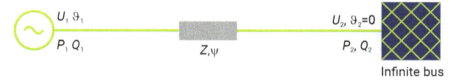

Figure 2.8 Basic model for rotor angle stability analysis

2.3.1 Steady-state rotor angle stability

Steady-state rotor angle stability deals with the ability to keep synchronism in case of a small imbalance in load and generation. It is strongly connected with the maximum power that can be transferred.

The typical single-machine infinite-bus model to describe angle stability is shown in Figure 2.8, which corresponds topologically to Figure 2.7.

With (2.22), the dependency of ϑ_1 on active power P_1 and the magnitude of the inner voltage U_1 of the synchronous machine can be expressed. With (almost) lossless lines, as is often the case in high-voltage systems, the phase angle ψ of the impedance is close to 90° and the equation can be simplified. The voltage of the infinite bus U_2 is assumed to be constant with an angle ϑ_2 equal to 0 (reference bus)

$$P_1 = \frac{U_1^2}{Z}\cos(\psi) - \frac{U_1 U_2}{Z}\cos(\vartheta_1 - \vartheta_2 + \psi) \approx \frac{U_1 U_2}{X}\sin(\vartheta_1) \qquad (2.22)$$

In Figure 2.9, the power-angle characteristic is shown. The operating point results from the intersection of mechanical power P_{mech} provided by the turbine (independent of rotor angle) and the electrical power according to the characteristic.

In general, describes the ability to transfer active power between two nodes with fixed (controlled) voltage magnitude.

There exists a maximum transferable power that depends on the voltage magnitude and the line impedance. Neglecting the ohmic losses and line capacitance, the maximum power transfer occurs at a rotor angle of 90°, which is the stability limit for steady-state rotor angle stability. Operating points at lower angles are stable

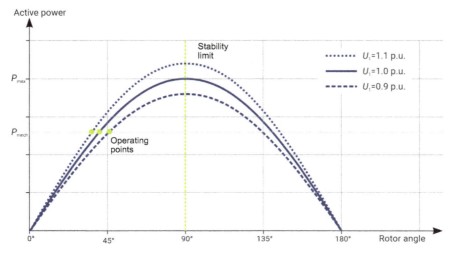

Figure 2.9 P–ϑ curves with U_1 as parameter with the operating points indicated; active power scaled to P_{max} at $U_1 = 1.0$ p.u.

whereas operating points above 90° are unstable [3]. Thus, the stability criterion can be formulated as

$$\frac{dP}{d\vartheta} > 0 \qquad (2.23)$$

Since only small disturbances are considered, the system's state description can be linearised around the operating point (equilibrium point). The state matrix's eigenvalues describe the oscillation modes of the synchronous machines' rotor angles.

Exceeding maximum transferable power may lead to system splits and unintentional islanding. A prominent example is the Italian blackout in 2003 [29]. After several contingencies, the Italian import power exceeded the remaining maximum transferable power. Consequently, the voltage phase angle of the Italian system drifted apart from the rest of Europe, leading finally to a complete separation of the Italian grid and leaving it in islanded operation. However, due to a power imbalance, the system collapsed a few minutes after separation.

In general, rotor angle stability is typically a problem in interconnected systems with long distance between generation units. In islanded grids with limited extent, small-disturbance stability is normally not an issue.

2.3.2 Transient rotor angle stability

While steady-state rotor stability deals with the evaluation of whether a certain operating point is stable (small-signal stability), transient rotor angle stability describes the ability to return to a stable equilibrium after a major disturbance, typically a short circuit. During the short circuit, the mechanically supplied power cannot be transmitted electrically, the rotor accelerates and the rotor angle increases. After clearing the

48 Intended and unintended islanding of distribution grids

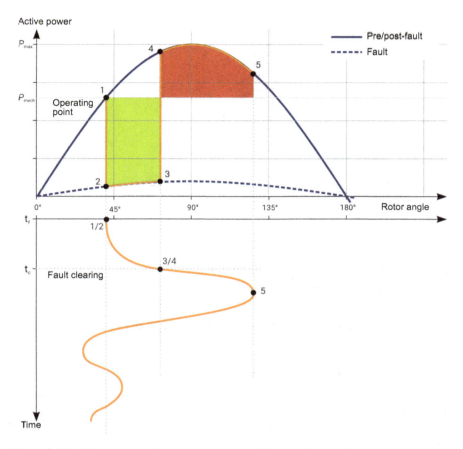

Figure 2.10 Illustration of transient rotor stability analysis with equal-area-criterion

fault, the rotor will be decelerated and the increase of the rotor angle will slow down and – in a stable case – reverse to the pre-fault operating point (equilibrium point).

Using the classical synchronous machine model and neglecting the AVR and the turbine governor, transient rotor angle stability can be analysed analytically by applying the equal-area criterion. The principle is illustrated in Figure 2.10.

The pre-fault operating point (1) is determined by the intersection between mechanical turbine power and electrical characteristic I. With the fault occurring at t_f, the electric characteristic changes due to the voltage sag to characteristic II (2). The power surplus now accelerates the rotor, increasing the rotor angle. At fault clearing time t_c (3), the original characteristic is restored and the rotor is decelerated, slowing down the speed of the rotor angle increase. The maximum rotor angle is reached at t_m (5) and decreases from now on. Finally, the rotor angle will settle at the pre-fault operating point (1). The damping of the post-fault oscillation is determined by the damper winding. The green is proportional to the accelerating energy while

Basics of power system dynamics and stability 49

the red area is proportional to the decelerating energy. Thus, the equal-area criterion as defined in can be used as a necessary (but not sufficient) criterion.

$$\int_{\vartheta=\vartheta_0}^{\vartheta_m} (P_{el} - P_m) \cdot d\vartheta = 0 \tag{2.24}$$

Since this method neglects electrical machine dynamics and the effect of control systems, a more accurate result can be achieved by time-domain simulations using the Heffron Philips model of the generation unit. The main parameters determining transient stability are the fault clearing time and the inertia constant of the synchronous machine and grid impedance as well as the amount and type (grid forming vs. grid following) of inverter-based generation (see (2.6)). A typical task in these studies is the determination of the critical clearing time (CCT), which is the maximum clearing time ensuring transient stability. The topic is closely related to the fault ride-through (FRT) capability of generation units. Challenges of stability assessment of parallel operation of grid-forming inverters and SGs are presented in [30].

The controlled islanding of a power system can be a measure for avoiding blackouts due to cascading loss of synchronism of generators. In case of appropriate timing of controlled islanding, the island's generators will keep their synchronism; thus, the island may reach a new stable equilibrium point [31–33].

2.4 Frequency stability

In electric power systems, huge emphasis is placed on the active power balance in the grid and a considerable effort is made to keep up the synchronism of each SG as well as the frequency stability of the whole grid.

Therefore, a characteristic of a high-quality power supply is an almost constant frequency. Imbalances (e.g. due to load fluctuation, power plants being ramped up or down, major forecast inaccuracies, power plant failures) between power generation and consumption cannot be avoided and result in a change of the network frequency. Therefore, the frequency is an excellent indicator of a power system's active power balance. The frequency is constant if the generation and consumption side, including system losses, is balanced (Figure 2.11).

It is often said that active power is a global quantity, in contrast to reactive power, which is a local quantity since reactive power cannot be transmitted over longer distances [34].

The system's frequency is reduced when the system load exceeds the (mechanical) turbine power. This can be due to an unscheduled increase in load or due to loss of generation (power plant tripping). Too large frequency reductions could lead to system collapse since a lot of equipment in the power stations, e.g. power supply systems, do not tolerate too low frequencies and are tripped by under-frequency protection.

Since the stored energy in the rotating machines is relatively small, the power balance must be restored within a short time. Frequency stability strongly depends

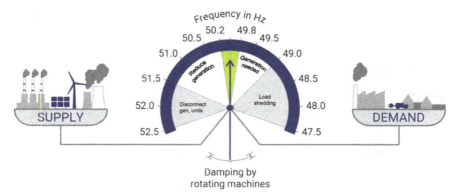

Figure 2.11 Balance between electricity generation and electricity consumption

on the system inertia, respectively, the stored rotating kinetic energy in relation to possible disturbances (see (2.5) and (2.6)).

In a similar approach, the rotating energy can be referred to a potential disturbance power. The resulting time constant is a rough indicator for the required frequency control dynamics. Especially in small islanded grids, this time constant is significantly reduced and frequency control is quite challenging. Usually, the smaller the islanded grid the smaller the kinetic energy in relation to potential disturbances, thus demanding faster frequency control to restore power balance.

$$\frac{W_{kin}}{\Delta P} = \frac{\sum_1^n H_n \cdot S_{r,n}}{\Delta P} \tag{2.25}$$

In the following, the reaction of a synchronous generating unit with activated frequency control on the frequency, following a stepwise increase of load, is analysed. In Figure 2.13, the time course of frequency f, turbine power, and generator power at the synchronous machine's terminal are shown.

Immediately after the incident, frequency will decrease with a RoCoF basically depending on-load change and rotating kinetic energy of the system. The higher the kinetic energy, the lower is the resulting RoCoF.

$$\text{RoCoF} = \frac{\Delta P \cdot f_n}{2 \cdot W_{kin}} \tag{2.26}$$

Applied to the continental European interconnected system data (Table 2.5), the RoCoF following the reference incident with ΔP being 3,000 MW will be in the range of 0.05 Hz/s. Actual measurements [35] show that for power outages of 1,000 MW, the resulting RoCoF is in the range of 0.005–0.01 Hz/s. However, load steps in an islanded grid during a grid restoration test caused RoCoF in the range of 0.4 Hz/s (Figure 2.12) [36].

The response speed of the turbine power is limited by the turbine governor, respectively, the primary energy source of the generating unit (see Chapter 3). Therefore, the initial power at the generator's terminal, covering the change in load ΔP,

Basics of power system dynamics and stability 51

Table 2.5   System inertia and potential disturbance for some selected grids

Grid	System load	Kinetic energy W_{kin}	Disturbance ΔP	$W_{kin}/\Delta P$	RoCoF
Continental Europe	150 GW	1,500 GWs	3,000 MW[1]	500 s	0.05 Hz/s
Ireland	6 GW	23 GWs	500 MW[2]	46 s	0.54 Hz/s
Tasmania	1.6 GW	3.8 GWs	450 MW	8 s	3 Hz/s
Cyprus	1.3 GW	3.0 GWs	120 MW[3]	25 s	1 Hz/s
Islanded grid (110 kV)	50 MW	0.6 GWs	20 MW[4]	30 s	0.83 Hz/s
Islanded grid (MV)	0.5 MW	6 MWs	300 kW[5]	20 s	1.25 Hz/s

[1] Continental Europe reference disturbance.
[2] Trip of Moyle HVDC.
[3] Loss of largest generation unit.
[4] Loss of generating unit.
[5] Connection of LV load.

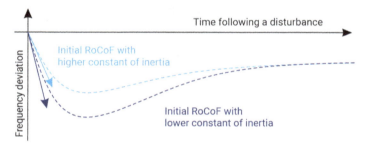

Figure 2.12   Influence of inertia on RoCoF

originates from the kinetic energy, which will be released due to the frequency gradient (Figure 2.13).

$$\Delta P = \text{RoCoF} \cdot \frac{2 \cdot W_{kin}}{f_n} \tag{2.27}$$

Inverter-based generation will not contribute to system inertia inherently. Thus, replacing conventional synchronous generation with inverter-based generation will decrease system inertia and kinetic energy and complicate maintaining frequency stability.

However, inverters can provide fast frequency control with an appropriate prime mover. Battery storage systems already participate in the market for frequency containment reserve. They can also provide new functionalities like 'virtual inertia' or 'extra fast frequency reserve'. It is not limited to an emulation of the behaviour of synchronous machines but can offer 'smart variable inertia' with defined time delay and consideration of frequency gradient and deviation from nominal frequency. In certain cases, it could be advantageous to deactivate virtual inertia, during a positive frequency gradient but negative frequency deviation.

52 Intended and unintended islanding of distribution grids

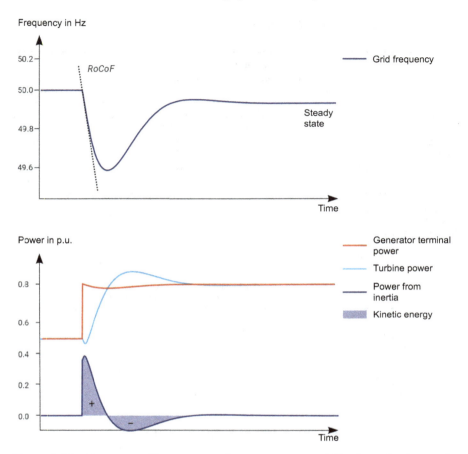

Figure 2.13   Response of frequency and generator power following a load step of 0.3 p.u.

In summary, the frequency stability of a system will be determined by the real and virtual inertia of a system, the dynamics of turbine governors of units participating in frequency control (see Chapter 3) and the inherent self-regulation effect of the load. As shown in Table 2.5, small islanded grids are significantly more prone to frequency instability since potential power imbalance is higher in relation to rotating kinetic energy.

2.5   Voltage stability

Voltage stability refers to the ability of a power system to maintain steady voltages at all buses in the system after being subjected to a disturbance [4]. This is strongly connected to the provision of reactive power.

Figure 2.14 Generic model for voltage stability analysis

One branch of voltage stability analysis is dedicated to small-disturbance stability or static stability, focusing on small disturbances. Regarding the time frame usually long-term stability problems are addressed. The analysis is mainly based on the evaluation of $U(P)$ and $U(Q)$ curves. The ability to transfer reactive power from reactive power sources like synchronous machines or compensation devices to loads is a major issue of voltage stability.

The other branch is dedicated to large disturbance analysis or dynamic stability, in most cases addressing short-term stability problems. Here, dynamics of fast-acting load components such as induction motors, electronically controlled loads, HVDC links and inverter-based generators, following severe disturbances like short circuits are considered.[†]

Multiple aspects like system strength (system impedance), voltage dependency of loads, automatic var control in power systems or operation of tap changers of distribution transformers will affect voltage stability.

Static voltage stability analysis deals with the voltage drop along equipment impedances. In this section, the dependencies of the voltage at the load on the active power P and reactive power Q will now be presented in more detail.

In Figure 2.14, a generic model to investigate voltage stability is shown. From the mathematical point of view, this corresponds to Figure 2.7 if line capacitance is neglected. With an elimination of ϑ_1, the dependency of voltage magnitude U_1 on the load P_1 and Q_1 can be expressed.

$$U_1 = \sqrt{\frac{A \pm \sqrt{A^2 - 4 \cdot Z^2(P_1^2 + Q_1^2)}}{2}} \qquad (2.28)$$

with

$$A = U_2^2 - 2 \cdot Z \cdot (\cos(\psi) \cdot P_1 + \sin(\psi) \cdot Q_1) \qquad (2.29)$$

The visualisation of the result is usually presented as P–U curves for constant values of Q (or constant power factor $\cos(\varphi)$) or as Q–U curves for constant values of P. The maximum power is determined by the power factor, respectively, the reactive power demand Q. The critical points ('nose points') are defined by the maximum power and the corresponding voltage. The locus of critical points limits the area of satisfactory operating points [3].

[†] In case of large disturbance stability analysis, a clear distinction between rotor angle stability and voltage stability is not always possible.

54 *Intended and unintended islanding of distribution grids*

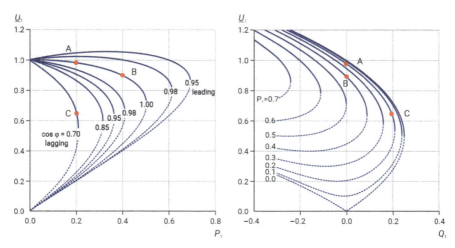

Figure 2.15 P–U curves with constant power factor (left) and Q–U curves with constant active power (right), normalised to U^2/Z with $R/X = 0$

In Figure 2.15, those curves are shown exemplarily. The solid curves indicate the area of stable operating points, while the dashed curves stand for the unstable region. Active and reactive power are normalised to U^2/Z for generalisation.

To highlight the effect of active power and reactive power on the voltage, three operating points are marked in the diagrams. The initial point is (A), representing a load of $P_1 = 0.2$ p.u. and $Q_1 = 0$ p.u. (power factor $\cos(\varphi) = 1.0$) at a voltage $U_1 = 0.98$ p.u. An increase of P_1 by 0.2 p.u. leads to a voltage decrease down to 0.90 p.u. (B). On the other hand, an increase of $Q_1 = 0.2$ p.u. leads to a voltage decrease down to 0.64 p.u. (C), which is close to the stability limit.

In **small-disturbance or long-term voltage stability** studies, slow changes (e.g. increase in load) are investigated. The main outcome is the distance of an operating point from the critical point. Load characteristics, as described in Section 2.1.6, have a significant impact on voltage stability.

Voltage stability problems are often associated with highly loaded lines, leading to high reactive power demand in addition to the load. Limited (local) reactive power reserve in generators (see synchronous machine's capability diagram in Section 2.1.5) may lead to critical voltage conditions in the grid and finally to a voltage collapse.

Short-term voltage stability includes the fast control behaviour of automatic voltage control at generators, inverter control, transformer tap changers, flexible AC transmission systems (FACTS), and HVDC. The dynamic behaviour of loads must be considered (controlled loads, induction motors) as well as disturbing events (tripping of compensation devices, faults). Besides voltage decrease, voltage increase might be an issue as well. The operation of a SG in a weak grid (for instance, an islanded system during grid restoration), with low-loaded lines, providing a surplus of reactive power, the AVR will lower the excitation of the generator to

Basics of power system dynamics and stability   55

maintain voltage. Reaching the minimum excitation, the system can become unstable with an uncontrolled rise of voltage during self-excitation of the generator [37]. Simplified, this can be seen as a kind of resonance between the generator reactance and the capacitive load. This effect, only limited by saturation of the generator's reactance, can cause severe damage to equipment due to overvoltage stress.

2.6 System strength

Especially in the context of voltage stability, system strength, usually defined by the short-circuit ratio at a specific bus, plays an important role. The short-circuit ratio (SCR) is calculated as the ratio of the short-circuit level at the point of connection (POC) and the rated power of a device connected thereto. This definition originates from the connection of DC systems and is used to estimate the impact of the connected device to the grid's voltage level.

The short-circuit level accounts for the short-circuit current flowing to the ground during a fault at this bus. In classical power systems, supplied by synchronous machines, this is also indirectly a measure of the grid's impedance. A low SCR corresponds to a weak system with high grid impedance, being sensitive to active and reactive power generation and load. A high SCR corresponds to a strong system with low grid impedance, being immune to active and reactive power generation and load. System strength is therefore strongly connected to voltage stability.

Classification of system strength is not completely consistent in the literature. Based on [38], a system with a wind park connected

- is considered to be weak if SCR at the POC is less than 5 or if SCR at the medium-voltage collection grid is less than 4 (whichever of these criteria is met).
- is considered to be very weak if SCR at the POC is less than 3 or if SCR at the medium-voltage collection grid is less than 2 (whichever of these criteria is met).

Similar values, but slightly lower, are given in [39].

According to the above definition, grids dominated by inverter-based generation will be classified as weak grids, since the current contribution in the case of a fault is rather low, respectively, the grid impedance is rather high.

However, in normal operation with the system's voltage close to the nominal voltage, the equivalent grid impedance will be determined by the inverter's controller behaviour. As long as the inverter's current limits are not violated, the devices will be capable of maintaining the voltage, thus behaving like a strong grid with low grid impedance. In case of faults, the inverter will run into current limitation, which is equivalent to a weak grid with rather high impedance.

This means that the representation of the grid at the POC will differ according to the stability phenomena that are under investigation. The same grid can behave like a strong grid for small-disturbance stability and at the same time as a weak grid for large disturbance stability. Therefore, care must be taken when the SCR is being used as a stability indicator based on the approximated impedance of the system to

56 *Intended and unintended islanding of distribution grids*

avoid misleading assessments. This is demonstrated for instance in [40]. Alternative concepts like the weighted short-circuit ratio (WSCR), composite short-circuit ratio (CSCR) and equivalent circuit-based short-circuit ratio (ESCR) are compared in [38]. Those definitions aim to consider the interactions of connected inverters.

In islanded grids, system strength might become an issue in connection with transformer inrush currents or motor starting currents.

2.7 Impact of high penetration of inverter-based generation on power system stability

2.7.1 Differences between grid-connected inverters and synchronous generators

Grid-connected inverters are gradually replacing SGs in the power system. Nevertheless, even if they mimic the dynamic characteristics of SGs through their control strategies [41], grid-connected inverters cannot present the same stability characteristics as SGs due to physical hardware limitations [42].

This comes from the difference in their hardware components. An inverter converts electrical energy through power semiconductors. Its overcurrent capability is then much smaller than that of a generator made of copper and iron. A transient overcurrent can permanently damage the power semiconductors. Therefore, the inverter must strictly limit its output current's magnitude [42], which changes the inverter's dynamic characteristic and prevents it from mimicking a SG.

Second, the stored energy in an inverter is limited. Without an additional energy storage unit, the DC side's capacitance of a voltage-source inverter is generally in the hundreds to thousands of microfarads. The stored energy in a DC capacitor is far from enough to meet the needs of primary frequency regulation [43].

In addition, the inverter's dynamic characteristics are dominated by its control strategy and control parameters. These control strategies and parameters can be adjusted in real-time as required. For example [44], during FRT, the inverter's dynamic characteristics can be improved by temporarily changing the damping control parameter, etc. The dynamic performance of a generator is dominated by its mechanical hardware structure and cannot be changed. This brings more flexibility to the inverter, but on the other hand, it poses more challenges for the prediction of the inverter's behaviour.

There are two dominant concepts of grid-connected inverter control: Grid-following [45] and Grid-forming [42]. In the **grid-following** mode, active and reactive power is injected into the main grid according to reference values. A PLL structure is used to guarantee that the converter is synchronised with the grid. The frequency and voltage output of the converter are governed by the main grid.

The grid-following inverter's reference phase is determined by calculating the terminal voltage, usually utilising a PLL technique, as illustrated in Figure 2.16. Next, the reference current is obtained from the reference power and terminal voltage. The current reference signal is then restricted by the saturation unit. As a result, the inner loop control of the grid-following inverter only manages the current. In

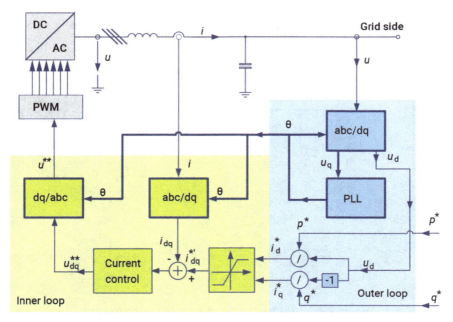

Figure 2.16 Block diagram of a grid-following inverter

Figure 2.16, the bold part represents the reference phase-dependent control unit. Reference values are indicated with an asterisk (*), and the final reference values for the PWM are indicated with a double asterisk (**).

A **grid-forming** converter is able to regulate its own frequency and voltage. It does not rely on the PLL structure for its synchronisation with the electric grid. This results in a more robust controller for the converter when the power converter is connected to a weak grid with low inertia levels.

The reference phase for a grid-forming inverter is not determined through a direct calculation of the terminal voltage. Instead, it relies on an active power–frequency (P–f) control loop, as shown in Figure 2.17. This P–f loop can take the form of a simple first-order power droop control or simulate the rotor equation of a SG.

On the other hand, the reference voltage amplitude of a grid-forming inverter is obtained through a reactive power–voltage amplitude (Q–U) control loop. To ensure proper control of the output waveform in conjunction with voltage and current control, the reference signals for both voltage and current are restricted by the saturation unit. This precaution is taken to prevent any potential damage to the inverter caused by over-voltage or overcurrent scenarios. In Figure 2.17, the bold part represents the reference phase-dependent control unit. Reference values are indicated with an asterisk (*), and the final reference values for the PWM are indicated with a double asterisk (**).

58 Intended and unintended islanding of distribution grids

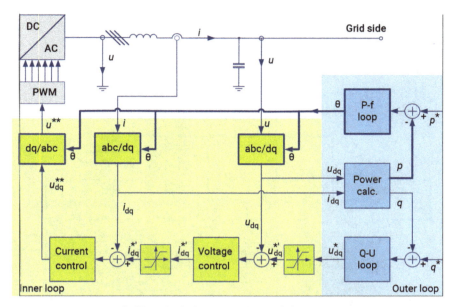

Figure 2.17 Block diagram of a grid-forming inverter

In addition, there are grid-supporting [46] and grid-tracking [47] concepts, etc., which can be considered as derivatives of the grid-following and grid-forming concepts.

2.7.2 Small-signal stability of inverters

For the grid-following inverter that behaves as a controlled current source, the current loop is mandatory. Since the current loop is the main responsible unit for reducing its robustness in the small-signal sense [48], the grid-following inverter is prone to unstable in weak grids with SCR < 1.5. This unstable phenomenon manifests as power oscillations at specific frequencies, current amplitude oscillations, etc. The oscillation frequencies range covers from sub-synchronous [49] to harmonics [50], where sub-synchronous oscillations are found in high-voltage and medium-voltage grids, and harmonic oscillations are found in low-voltage grids. This unstable phenomenon can be explained by the impedance-based stability theory [51].

In the case of the grid-forming inverter, it is possible to operate without a current loop due to their controlled voltage-source characteristics, thus greatly improving their small-signal stability and enabling them to operate stably in weak grids [52].

2.7.3 Large-signal stability of inverters

If a large disturbance causes a destabilisation of the inverter's synchronous control, which in turn causes a reference phase oscillation, it will directly trigger an output

Basics of power system dynamics and stability 59

power oscillation. This not only affects the power balance within an inverter but may also change the power flow and affect the stable operation of the power system.

The grid-following's synchronisation control calculates the terminal voltage's phase and guides its output current's phase. In a strong grid case, i.e. small grid impedance, the output current's phase has a limited influence on its terminal voltage's phase. However, in a weak grid case, fluctuations in the output current's phase can cause oscillation in the terminal voltage's phase through a larger grid impedance, which in turn affects the stability of the grid-following's synchronous control [53]. This instability occurs during large disturbances in the terminal voltage, such as voltage dips caused by a short-circuit fault, or voltage recovery after a fault clearing.

The grid-forming synchronous units, on the other hand, regulate the output voltage's phase by feedback from the output active power. Therefore, its dynamic characteristics are more like a SG. However, due to hardware limitations, the grid-forming inverter cannot behave like a SG in all cases. Limited by current amplitude, the power-angle curve of the grid-forming inverter changes relative to the SG and becomes nonlinear. This is also accompanied by a change in the stable equilibrium point.

Regardless of the active power–frequency loop control strategies for a grid-forming inverter, the power–angle relationship is following:

$$P = \frac{E \cdot U}{X_g} \cdot \sin(\theta) \tag{2.30}$$

As presented in, when the terminal voltage E, as well as the grid voltage U, are kept constant, the smaller the grid impedance X_g, the larger the change in active power caused by the change in voltage phase θ. Therefore, for the grid-forming inverter, in a strong grid environment, i.e. small grid impedance, the output phase angle oscillation leads to greater active power oscillation, and such positive feedback leads to destabilisation of the inverter. In addition, due to the limited current, the output phase angle oscillation will make the inverter keep switching between saturated and unsaturated current states, thus making it impossible to control the inverter effectively [54].

When a fault occurs with the grid topology's changing, the operating point of an inverter diverges during the fault, due to the non-existence of a stable equilibrium point. When a commonly used synchronous reference frame PLL is applied in a grid-following inverter, the change of the operating point in the $d\theta/dt$ direction in the phase plane is unbounded. Since the damping of its nonlinear differential equation is a dependent variable [55]. This leads to an inability to converge quickly, or even to divergence, even after the fault clearing.

For a grid-forming inverter, the non-existence of the stable equilibrium point also leads to the failure of the operating point to converge during the fault. However, suppose a standard virtual synchronous strategy, or equivalent droop control with a first-order low-pass filter, is applied in a grid-forming inverter. In that case, its non-linear differential equation has a positive constant damping [55]. So that the change of the operating point in the $d\theta/dt$ direction in the phase plane is bounded during the fault. After the fault clearing, fast convergence and stabilisation are facilitated

60 *Intended and unintended islanding of distribution grids*

because the operating point is closer to the stable equilibrium point in the phase plane.

2.8 Simulation and analysis models and method

This chapter provides an overview of models and methods for the analysis of power system dynamics. It shall serve as a guide for the selection of appropriate methods and tools for specific stability phenomena. More details regarding the methods can be found in Chapter 9.

Power system studies rely mostly on the use of computational tools that can analyse the security and stability of the systems and provide insights regarding their response to abnormal conditions and disturbances. Depending on the time frame that the power engineer requires to analyse and the type of phenomena that need to be captured, three types of models are usually used. EMT models, RMS models (also referred to as positive sequence modelling) and power-flow models (also referred to as static models).

The first two types of models are also called time-domain models since they capture the time-domain evolution of the system when subjected to disturbances. Thus, they are both modelled with sets of differential algebraic equations (DAEs) and solved using integration methods over a discretised time horizon. Power-flow equations depict the steady-state response of power systems and thus are modelled only using algebraic equations.

Regarding the time frame analysed by each method, EMT models can capture the response of power systems in the range of micro or even nano-seconds, RMS models capture the behaviour of power systems in the range of milliseconds to seconds, while power-flow models only depict the steady-state response (minutes-hours).

A separate category of analysis methods, such as modal analysis, allows to study the power system dynamics in the frequency domain. While the methods rely on a snapshot of the power system dynamic model, they can provide insights into the time-domain dynamic response of the system.

In this section, we will briefly discuss the models used for each of the tools and their benefits and weaknesses.

2.8.1 EMT-type models

In EMT-type models, the power system components are modelled to capture high-frequency (short-time frame) dynamics. Thus, the differential equations of the electrical network are modelled in detail for the sinusoidal voltages and currents, for all frequencies. To cover the necessary bandwidth, the simulation methods use small integration time steps of the order of 50 μs or less. This allows to model power system components (such as generators, loads and controllers) with high accuracy and makes EMT simulations valuable when studying the effects of power electronic devices on system behaviour and analysing very fast dynamics (Table 2.6). Moreover, these models are also used in real-time simulation platforms where high accuracy of

Basics of power system dynamics and stability 61

Table 2.6 Type of phenomena and studies typically performed using EMT-type models [10]

Type of phenomena	Type of studies
Behaviour in response to large deviations	Short-term voltage stability
	Transient stability
	Provision of short-circuit current
	Low/high-voltage ride-through
Other phenomena and studies	Controller interactions, switching transients, harmonic studies

the simulation at a short time frame is necessary to be able to couple with external, physical and hardware.

However, the necessary small integration time steps and high-detail models make EMT simulations much slower than RMS. This is one of the main limitations of EMT-type simulation tools. It is thus common practice to divide large-scale systems into a study zone and an external system encompassing the rest of the system. Moreover, in many cases, the power system component models are not available or inaccessible with the high detail required by EMT-type simulation tools.

2.8.2 RMS-type dynamic models

Power system dynamic simulation tools using RMS-type models were mainly developed to study electromechanical oscillations of power systems consisting of large generators and motors. Phasor simulation methods are used when only the fundamental frequency behaviour is of interest and thus use RMS positive sequence phasor equations to represent the electrical network, i.e. it is not necessary to solve all the differential equations of the network resulting from its R, L and C elements. Thus, time-varying complex numbers are used to represent the sinusoidal voltages and currents at system frequency, expressing their value either in Cartesian (real and imaginary) or polar coordinates (amplitude and phase).

In general, RMS-type simulations consider phenomena in a bandwidth of at most 10 Hz (e.g. periods no higher than 0.1 s). Thus, the required numerical integration times steps are in the range of 1 ms or higher. This allows for much faster simulation times compared to EMT-type simulations with larger systems simulated and longer time horizons. Due to the bandwidth captured by RMS-type simulations, simplified models of the power system components are usually used compared to EMT-type studies. For instance, averaged models of inverters (instead of switching models) or the averaged response of the fast control loops. Moreover, only the fundamental frequency output of the components is depicted in the models. Overall, the modelling requirements (both in terms of accuracy and parameter availability) are much lower for RMS-type studies than EMT-type (Table 2.7).

62   *Intended and unintended islanding of distribution grids*

Table 2.7   Type of phenomena and studies typically performed using RMS-type models [10]

Type of phenomena	Type of studies
Behaviour in response to frequency deviations	Frequency regulation and transient stability
Behaviour in response to large voltage excursions	Short-term voltage stability
	Transient stability
	Short-circuit current provision
	Low/high-voltage ride-through
Behaviour in response to smaller but longer voltage deviations	Long-term voltage stability
Modelling simplifications for small-disturbance stability analysis	Modal analysis

Table 2.8   Type of phenomena and studies typically performed using power-flow models

Type of phenomena	Type of studies
Behaviour in response to loss of generation, load or network component	$N-1$ analysis, static security assessment
Behaviour of system in steady state	Congestion analysis, overloading analysis

2.8.3   Power-flow models

Power-flow analysis is the most frequently used analysis in electric power systems. It only considers the steady-state response of the system, i.e. the values after all the dynamics of the system have settled. Power-flow models use RMS positive sequence phasor equations to represent the electrical network and steady-state equations to model the response of other power system components. This leads to a set of algebraic equations, usually solved using an iterative nonlinear solver (e.g. Newton–Raphson method).

This kind of study analyses a snapshot of the system and provides information about the steady-state response (Table 2.8). They offer the highest speed of simulation compared to the other simulation methods but also introduce many assumptions and simplifications. In this type of study, the power system component response is significantly simplified and trajectory-based or time-critical protection and control cannot be modelled.

Basics of power system dynamics and stability 63

Table 2.9 Type of phenomena and studies typically performed using modal analysis

Type of phenomena	Type of studies
Behaviour in response to small disturbances	Small-signal analysis, angle stability, oscillatory stability, power system stabiliser design

2.8.4 Modal analysis

Modal analysis methods focus on studying the power system dynamics in the frequency domain [56]. The analysis is based on the dynamic equations of the RMS-type models around a pre-defined operating point. A linearisation of the differential equations is performed using the first-order Taylor expansion leading to a linear state-space representation of the system modelling around that operating point. Finally, an eigenvalue analysis is performed on the Jacobian matrix of the system providing the modes of the system and other information concerning the participation factors, observability and controllability.

The modelling requirements and simplifications of modal analysis are the same as the RMS-type dynamic simulations. The computational cost varies depending on the type of the system and the requirement for evaluating all or part of the system modes. In small-scale systems, the analysis is very fast while in large-scale systems, modal analysis can be extremely demanding and special techniques are required to analyse the critical modes (Table 2.9) [57].

2.8.5 Comparison of models

As seen above, the different models and analysis methods employed have different modelling requirements (and thus can capture different phenomena) but at the same time have different computation costs.

EMT models are capable of incorporating significant levels of detail. They are typically more complex than RMS models and generally require advanced knowledge of the equipment component and control system design. They are generally unsuitable for large-scale studies due largely to the computational burden that comes with running complex models at time steps typically in the order of 'tens of microseconds', as well as the difficulty of post-processing more complex output data from detailed 3-phase EMT models.

On the contrary, RMS models are computationally efficient and allow large-scale simulations to be performed in minutes rather than hours. Furthermore, the data inputs and post-processing of output data for RMS models are far less burdensome. Nevertheless, RMS models have their limitations in the following circumstances (Figure 2.18):

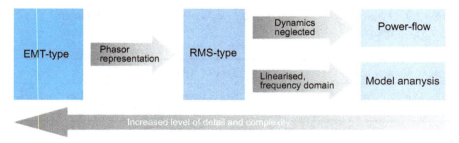

Figure 2.18 Models according to their simplification (more detailed on the left)

- Weak system conditions (typically characterised as having a very low SCR).
- For undertaking detailed inverter and collector system design.
- For performing certain system interaction studies, such as those involving sub-synchronous resonance (SSR) and sub-synchronous control interactions (SSCI).
- For analysing the response to unbalanced faults and resulting voltage phase angle shifts.

It remains the responsibility of the power system engineer to understand the benefits and limitations of each dynamic modelling method and tool. Judicious selection of model type is required if the power system dynamics of interest are to be properly identified and analysed. It is emphasised that the above remarks related to the selection of the model type are more critical, especially when the penetration of renewable energy systems becomes high.

References

[1] Milano F, Dörfler F, Hug G, Hill DJ, and Verbič G. Foundations and Challenges of Low-Inertia Systems (Invited Paper). *2018 Power Systems Computation Conference (PSCC)*, Dublin, Ireland, pp. 1–25, 2018, doi: 10.23919/PSCC.2018.8450880.

[2] Machowski J, Bialek JW, and Bumby JR. *Power System Dynamics: Stability and Control*. New York: Wiley, 2008.

[3] Kundur P. *Power System Stability and Control*. New York: McGraw-Hill Professional Publishing, 1994.

[4] Hatziargyriou N, Milanovic JV, Rahmann C, *et al*. Stability Definitions and Characterization of Dynamic Behaviour in Systems with High Penetration of Power Electronic Interfaced Technologies. *IEEE Power and Energy Society*, Tech. Rep. PESTR77, May 2020.

[5] Markovic U, Stanojev O, Aristidou P, Vrettos E, Duncan C, and Hug G. Understanding Small-Signal Stability of Low-Inertia Systems. *IEEE Transactions on Power Systems*, 2021;36(5):3997–4017.

[6] Walker JH. Operating Characteristics of Salient-Pole Machines. *Proceedings of the IEE – Part II: Power Engineering*, 1953;100(73):13–24.

Basics of power system dynamics and stability 65

[7] ENTSO-E. Frequency Stability Evaluation Criteria for the Synchronous Zone of Continental Europe. Belgium, 2016. Available from: https://eepubli cdownloads.entsoe.eu/clean-documents/SOC%20documents/RGCE_SPD_f requency_stability_criteria_v10.pdf.

[8] IEEE Guide for Synchronous Generator Modeling Practices and Parameter Verification with Applications in Power System Stability Analyses. *IEEE Std 1110-2019 (Revision of IEEE Std 1110-2002)*, pp. 1–92, 2020, doi: 10.1109/IEEESTD.2020.9020274.

[9] Heffron WG and Phillips RA. Effect of a Modern Amplidyne Voltage Regulator on Underexcited Operation of Large Turbine Generators. *Transactions of the American Institute of Electrical Engineers. Part III: Power Apparatus and Systems*, 1952;71(3):692–697, doi: 10.1109/AIEEPAS.1952.4498530.

[10] Modelling of Inverter-Based Generation for Power System Dynamic Studies, Technical Report, JWG C4/C6.35/CIRED, 2018.

[11] Brückl O. Zukünftige Bereitstellung von Blindleistung und anderen Maßnahmen für die Netzsicherheit, OTH Regensburg, 2016.

[12] IEC TS 62898-3-3 ED1. Microgrids – Part 3-3: Technical Requirements – Self-Regulation of Dispatchable Loads [Standard], IEC, 2021.

[13] McDonald J and Bruning A. Cold Load Pickup. *IEEE Transactions on Power Apparatus and Systems*, 1979;PAS-98(4):1384–1386.

[14] Gonzalez M, Wheeler KA, and Faried SO. An Overview of Cold Load Pickup Modeling for Distribution System Planning. *2021 IEEE Electrical Power and Energy Conference (EPEC)*, pp. 328–333, 2021.

[15] Hachmann C, Becker H, and Braun M. Cold Load Pickup Model Adequacy for Power System Restoration Studies. *Energies*, 2022;15:7675, doi:10.3390/en15207675.

[16] Arif A, Wang Z, Wang J, Mather B, Bashualdo H, and Zhao D. Load Modelling – A Review. *IEEE Transactions on Smart Grid*, 2018;9(6):5986–5999.

[17] Milanovic JV, Matevosiyan J, Gaikwad A, *et al.* Technical Brochure 566: Modelling and Aggregation of Loads in Flexible Power Networks [Technical Brochure], Cigre, WG C4.605, 2014.

[18] Concordia C and Ihara S. Load Representation in Power System Stability Studies. *IEEE Transactions on Power Apparatus and Systems*. 1982;PAS-101(4).

[19] Hill DJ. Nonlinear Dynamic Models with Recovery for Voltage Stability. *IEEE Transaction on Power Systems*. 1993;8:166–176.

[20] IEEE Task Force on Load Representation for Dynamic Performance, Bibliography on Load Models for Power Flow and Dynamic Performance Simulation. *IEEE Transaction on Power Systems*, 1995;10(1):523–538.

[21] Jiang-Hafner Y, Duchen H, Karlsson M, Ronstrom L, and Abrahamsson B. HVDC with Voltage Source Converters – A Powerful Standby Black Start Facility. *IEEE PES T&D Conference*, Chicago, 21.–24.4, 2008.

[22] Lourenço LFN, Perez F, Iovine A, Damm G, Monaro RM, and Salle MBC. Stability Analysis of Grid-Forming MMC-HVDC Transmission Connected to Legacy Power Systems. *Energies*, 2021;14:8017, doi: 10.3390/en14238017.

66 *Intended and unintended islanding of distribution grids*

[23] Castro LM and Acha E. On the Provision of Frequency Regulation in Low Inertia AC Grids Using HVDC Systems. *IEEE Transactions on Smart Grid*, 2016;7(6):2680–2690, doi: 10.1109/TSG.2015.2495243.

[24] Collados-Rodriguez, Prieto-Araujo E, Cheah-Mane M, *et al.* Interaction Analysis in Islanded Power Systems with HVDC Interconnections. *3rd International Hybrid Power Systems Workshop*, 2018.

[25] Jovcic D and Ahmed K. *High Voltage Direct Current Transmission*. New York: Wiley, 2015.

[26] Phadke AG and Thorp J S. *Synchronized Phasor Measurements and Their Applications, Power Electronics and Power Systems*, 2nd edition. Berlin: Springer, 2017.

[27] Ashton PM, Saunders CS, Taylor GA, Carter AM, and Bradley ME. Inertia Estimation of the GB Power System Using Synchrophasor Measurements. *IEEE Transactions on Power Systems*, 2015;30(2):701–709, doi: 10.1109/TPWRS.2014.2333776.

[28] Hatziargyriou N, Milanović J, Rahmann C, *et al.* Definition and Classification of Power System Stability Revisited and Extended. *IEEE Transactions on Power Systems*, 2021;36(4):3271–3281, doi: 10.1109/TPWRS.2020.3041774.

[29] UCTE. Final Report of the Investigation Committee on the 28 September 2003 Blackout in Italy, 2004.

[30] Cheng H, Shuai Z, Shen C, Liu X, Li Z, and Shen ZJ. Transient Angle Stability of Paralleled Synchronous and Virtual Synchronous Generators in Islanded Microgrids. *IEEE Transactions on Power Electronics*, 2020;35(8):8751–8765, doi: 10.1109/TPEL.2020.2965152.

[31] Ding L, Guo Y, Terzija V, and Sun K. Identifying the Timing of Controlled Islanding Using a Controlling UEP based Method. *2019 IEEE Power & Energy Society General Meeting (PESGM)*, Atlanta, GA, USA, 2019, doi: 10.1109/PESGM40551.2019.8974089.

[32] Wang Z and Wang Z. Criterion and Timing Strategy of Intentional Islanding Based on Transient Stability Prediction. *IEEE International Conference on Advances in Electrical Engineering and Computer Applications (AEECA)*, 2022.

[33] Demetriou P, Quirós-Tortós J, and Kyriakides E. When to Island for Blackout Prevention. *IEEE Systems Journal*, 2019;13(3):3326–3336, doi: 10.1109/JSYST.2018.2866937.

[34] Anderson: Power System Analysis, *Lecture Notes*, ETH Zürich, September 2011.

[35] Sattinger W, Notter R, and Zerva M. Das kontinentaleuropäische Verbundnetz – Herausforderungen des heutigen Netzbetriebs, VSE Bulletin, Band104, 2013, Heft 12.

[36] Polster S, Schürhuber R, Renner H, *et al.* Best Practice Grid Restoration with Hydropower Plants. *20th International Seminar on Hydropower Plants*, Vienna, Austria, November 14–16, 2018.

[37] Nikkilä A-J, Kuusela A, Laasonen M, Haarla L, and Pahkin A. Self-Excitation of a Synchronous Generator During Power System

Restoration. *IEEE Transactions on Power Systems*, 2019;34(5):3902–3911, doi: 10.1109/TPWRS.2019.2909050.

[38] CIGRE WG B4.62, Connection of Wind Farms to Weak AC Networks, Technical Brochure 671, Paris, 2016.

[39] Gleadow J, Love G, Saad H, *et al.* TF-77 – AC Fault Response Options for VSC HVDC Converters. In: Papailiou KO (ed.) *Cigre Science & Engineering*, vol. 15. France: Cigre, pp. 105–110, 2019.

[40] Damanik O, Sakinci ÖC, Grdenic G, and Beerten J. Evaluation of the Use of Short-Circuit Ratio as a System Strength Indicator in Converter-Dominated Power Systems. 2022 IEEE PES Innovative Smart Grid Technologies Conference Europe (ISGT-Europe).

[41] Rosso R, Wang X, Liserre M, Lu X, and Engelken S. Grid-Forming Converters: Control Approaches, Grid-Synchronization, and Future Trends—A Review. *IEEE Open Journal of Industry Applications*, 2021;2:93–109.

[42] Qoria T, Gruson F, Colas F, Kestelyn X, and Guillaud X. Current limiting algorithms and transient stability analysis of grid-forming VSCs. *Electric Power Systems Research*, 2020;189:106726.

[43] Lin Y, Eto JH, Johnson BB, *et al. Research Roadmap on Grid-Forming Inverters (No. NREL/TP-5D00-73476)*. Golden, CO: National Renewable Energy Lab. (NREL), 2020.

[44] Zhang Z, Schuerhuber R, Fickert L, Friedl K, Chen G, and Zhang Y. Domain of Attraction's Estimation for Grid Connected Converters with Phase-Locked Loop. *IEEE Transactions on Power Systems*, 2021;37(2):1351–1362.

[45] Pattabiraman D, Lasseter RH, and Jahns TM. Comparison of Grid Following and Grid Forming Control for a High Inverter Penetration Power System. *2018 IEEE Power & Energy Society General Meeting (PESGM)*. Piscataway, NJ: IEEE, pp. 1–5, 2018.

[46] Meng X, Liu J, and Liu Z. A Generalized Droop Control for Grid-Supporting Inverter Based on Comparison Between Traditional Droop Control and Virtual Synchronous Generator Control. *IEEE Transactions on Power Electronics*, 2018;34(6):5416–5438.

[47] Gao M, Chen M, Zhao B, Li B, and Qian Z. Design of Control System for Smooth Mode-Transfer of Grid-Tied Mode and Islanding Mode in Microgrid. *IEEE Transactions on Power Electronics*, 2019;35(6):6419–6435.

[48] Lehmal C. Investigation and Validation of Stability for the Photovoltaic Integration into a Medium Voltage Grid Based on PHIL Testing, 2021.

[49] Chi Y, Tang B, Hu J, et al. Overview of Mechanism and Mitigation Measures on Multi-Frequency Oscillation Caused by Large-Scale Integration of Wind Power. *CSEE Journal of Power and Energy Systems*, 2019;5(4):433–443.

[50] Zhang Z, Gercek C, Renner H, Reinders A, and Fickert L. Resonance Instability of Photovoltaic E-bike Charging Stations: Control Parameters Analysis, Modelling and Experiment. *Applied Sciences*, 2019;9(2):252.

[51] Wang X, Li Y W, Blaabjerg F, and Loh PC. Virtual-Impedance-Based Control for Voltage-Source and Current-Source Converters. *IEEE Transactions on Power Electronics*, 2014;30(12):7019–7037.

68 *Intended and unintended islanding of distribution grids*

[52] Rosso R, Wang X, Liserre M, Lu X, and Engelken S. Grid-Forming Converters: An Overview of Control Approaches and Future Trends. *2020 IEEE Energy Conversion Congress and Exposition (ECCE)*. Piscataway, NJ: IEEE, pp. 4292–4299, 2020.

[53] Zhang Z, Schuerhuber R, Fickert L, and Friedl K. Study of Stability After low Voltage Ride-Through Caused by Phase-Locked Loop of Grid-Side Converter. *International Journal of Electrical Power & Energy Systems*, 2021;129:106765.

[54] Zhang Z, Lehmal C, Hackl P, and Schuerhuber R. Transient Stability Analysis and Post-Fault Restart Strategy for Current-Limited Grid-Forming Converter. *Energies*, 2022;15(10):3552.

[55] Fu X, Sun J, Huang M, et al. Large-Signal Stability of Grid-Forming and Grid-Following Controls in Voltage Source Converter: A Comparative Study. *IEEE Transactions on Power Electronics*, 2020;36(7):7832–7840.

[56] Okubo S, Suzuki H, and Uemura K. Modal Analysis for Power System Dynamic Stability. *IEEE Transactions on Power Apparatus and Systems*, 1978;PAS-97(4):1313–1318.

[57] Aristidou P, and Hug-Glanzmann G. Accelerating the Computation of Critical Eigenvalues with Parallel Computing Techniques. *Proceedings of 2016 Power Systems Computation Conference*, 2016.

Chapter 3

Control of electric power systems

Michael Finkel[1], Herwig Renner[2] and Ziqian Zhang[2]

There is only a very limited possibility of storing electric energy in any electric system. Therefore, the balance between generated and consumed electric energy has to be maintained continuously by control actions. Synchronous generation units play an essential role in controlling the frequency and node voltage since a dominating share of energy consumed by various loads in an electric power system is still produced by synchronous machines. This applies to larger power systems but not always in smaller systems like isolated/islanded grids or other smaller systems. In these systems, the energy is often supplied by a large share of inverter-based resources (IBR). Thus, knowing the characteristics and control of synchronous machines and IBR is very important.

The active power balance is controlled by the generation units. Therefore, in the first part of this chapter, the fundamental mechanisms for frequency control are described. The frequency-dependent behaviour of load is discussed in the second part of this chapter. In an emergency situation with underfrequency, active power consumption is reduced by disconnecting parts of the load. Finally, after a longer supply failure, the cold load pick-up has to be considered.

The following section deals with the basics of voltage control in electric power systems. First, essential relationships for the voltage control of electrical networks are derived with the help of a short-line model and the corresponding phasor diagram. Then, the static voltage control at the synchronous generator, in the distribution grid and through the provision of reactive power from DGs is discussed. Considerations on short-circuit current and dynamic voltage support complete this section.

In the previous sections, frequency and voltage control, as part of the ancillary services, were described and discussed more generally. In the case of islanded grid operation, the task of ensuring a secure and reliable grid operation is transferred from the TSO to the operator of the islanded grid. Therefore, the last section discusses the transition from interconnected to islanded grid operation in more detail and also describes essential aspects of operational management.

---

[1]Department of High Voltage Engineering and Electric Power Supply Systems, Augsburg Technical University of Applied Sciences, Germany
[2]Graz University of Technology, Institute of Electrical Power Systems, Austria

3.1 Frequency control and adjustment of generation to consumption

3.1.1 System frequency

A characteristic of a high-quality power supply is an almost constant frequency. Imbalances (e.g. due to load increases or decreases, power plants being ramped up or down, major forecast inaccuracies, power plant failures) between power generation and consumption cannot be avoided and result in a change of the network frequency. Therefore, the frequency is an excellent indicator of a power system's active power balance. The frequency is constant if generation and consumption, including system losses, are in balance (Figure 3.1). It can also be noted that the frequency in the entire system is equal at steady state[*] since active power can easily be transported in the system (which is why it was possible to build the power systems of today's size). It is often said that active power is a global quantity, in contrast to reactive power, which is a local quantity, since reactive power cannot be transmitted over longer distances[†] [1].

The system's frequency is reduced when a corresponding increase in the turbine power of the connected generators does not compensate for a load increase. The power deficit decelerates the generator rotors and consequently, the frequency is reduced. Frequency reductions also arise when generation capacities are lost, e.g. due to system failures, which results in disconnecting the failed equipment by the protection devices. Too large frequency reductions could lead to system collapse since a lot of equipment in the power stations, e.g. power supply systems, do not tolerate too low frequencies. A load reduction in the system, which is not compensated for by a reduction of turbine power, leads to a frequency increase [1].

Since the stored energy in the rotating machines is relatively small, the electrical energy must be produced in the same time as it is consumed by the loads. Additionally, a certain power reserve must constantly be available to compensate for load variations.

Basically, different measures are available, staggered in time and intensity, to comply with the frequency limits. They are used depending on the actual frequency

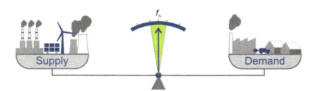

Figure 3.1 Balance between electricity generation and electricity consumption

---

[*]In a large network such as the European interconnected grid, this statement does not apply without restriction, as a certain, albeit small, location dependency of the frequency can occur.
[†]This is because X normally is much greater than R in a power system, at least for transmission and sub-transmission networks.

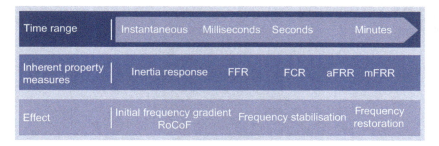

Figure 3.2 Schematic classification of the measures for frequency stability and their effect (adapted from [2,3])

deviation or represent an inherent system property. Figure 3.2 provides an overview of the different measures and the corresponding time ranges [2].

In the case of frequency deviations, the inherent properties of the energy system (inertial response and self-regulation effect of loads) are first provided instantaneously. The frequency stabilisation and frequency restoration are then carried out by further control power products staggered over time.

3.1.2 Inertia management

Inertia is an inherent physical behaviour of spinning machines (synchronous generators, synchronous condensers and some spinning demand loads) in a power system that resists changes in rotational speed. It is subject to daily [4] and seasonal fluctuations [5] due to the generation mix. Inertia plays an essential role in limiting RoCoF during disturbances and affects the eigenvalues and vectors that limit the stability and mode shape of transient stability response (see Chapter 2). It is a key determinant of the strength and stability of the power system [6].

Figure 3.3 compares the responses to a frequency excursion of two systems: a low and a high inertia system. Since the capacity of the primary control reserve, also known as frequency containment reserve (FCR)[‡], is the same in both inertia cases, the steady-state frequency after the event settles at the same value. However, the lower inertia in the system exhibits a lower frequency nadir and a faster RoCoF.

The frequency nadir can be improved by increasing the system inertia (green dashed curve) and/or by speeding up the control (yellow dashed curve). The lower the inertia of the system, the faster the frequency drops after an event. Consequently, a faster response of the primary reserve is required. However, the measurement of RoCoF has an inherent time delay, based on the time window of the measurement. Typically acceptable measurement times range from 100 ms to 500 ms. But, there is a trade-off between faster activation times and the accuracy of measurements. By definition, RoCoF requires time to calculate, so an overly fast response may result

---

[‡]The used abbreviations correspond to the technical terms of the Electricity Balancing Guideline of the ENTSOE [7].

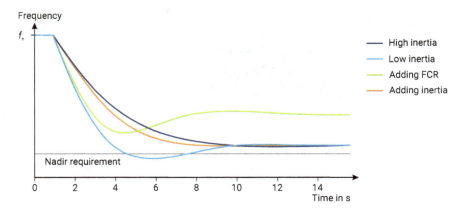

Figure 3.3   Impact of system inertia and FFR on frequency response after an event (adapted from [8])

in an inaccurate value of RoCoF [9]. An inaccurate RoCoF may result in over- or undershooting the needed level of change in inverter output [10].

The 'speeding up the control' approach exploits the inverter controls of power electronics to provide rapid power injections in response to events. This may be described as 'synthetic inertia' (also the terms 'virtual inertia' and 'emulated inertia' are used) or fast frequency response (FFR). Although these terms are often used interchangeably, a distinction may be drawn between synthetic inertia and FFR based on how the change in power injection is achieved. The synthetic inertia emulates the reaction of a direct-coupled system to an active power imbalance and thus supports the instantaneous reserve in the system. Wind turbines, where the mechanical inertia of the rotor does exist but is asynchronous to the system, can provide some synthetic inertia without curtailing the wind turbines.

Conversely, there is a need to curtail the generation of wind turbines or solar PV. While in theory, one could tap into the energy stored in the inverter's intermediate circuit, the capacity of such energy storage is relatively limited. Hence, FFR can be sourced from a variety of systems, including wind and PV installations, battery energy storage systems, HVDC interconnectors and other IBR. It is worth noting that in certain regions, synthetic inertia is exclusively designated for wind generation. Nevertheless, FFR and synthetic inertia serve similar functional roles, both offering solutions to challenges associated with low inertia operations. Realizing these capabilities necessitates the enabling of control functions and their subsequent provision of the desired response. The controls implemented in inverters provide significant additional flexibility that can be harnessed. Consequently, exploring methods to utilise these resources is not only essential for maintaining stability but also for enhancing it.

The available system inertia can be managed through TSO measures or grid codes. The first and most obvious one is a requirement for a certain minimum amount of inertia that always must be present to limit the RoCoF to an acceptable value. Such 'inertia floors' are imposed, e.g., in Great Britain and Ireland, but only present a short-term solution as they inevitably limit non-synchronous penetration and thus RES contribution. Additionally, the Irish grid operator EirGrid incentivises synchronous generators to decrease their minimum continuous operating level, to make way for more RES generation without having to disconnect synchronous generation and reduce inertia [11]. Finally, synchronous storage resources such as pumped hydroelectric storage or compressed air energy storage and synchronous condensers with enhanced inertia characteristics can be installed [10,12].

Moreover, the Irish grid operator, EirGrid, is actively encouraging synchronous generators to reduce their minimum continuous operating levels. This strategy allows for the increased incorporation of RES generation without the need to disconnect synchronous generation and diminish system inertia.

3.1.3 Types of control reserve and their provision

As described in the previous section, the system has a certain inertia due to the rotating masses of generators, steam turbines, motors, etc. In the short term, this instantaneous reserve, in combination with the self-regulating effect of the loads (Section 2.1.6), creates a certain stabilisation of the frequency in the case of power imbalances. The resulting deviation of the system frequency from its set-point value is then compensated for by the various types of control reserve (Figure 3.4).

A distinction is made between positive and negative control reserves. If the energy fed into the grid exceeds the energy withdrawn simultaneously, there is a

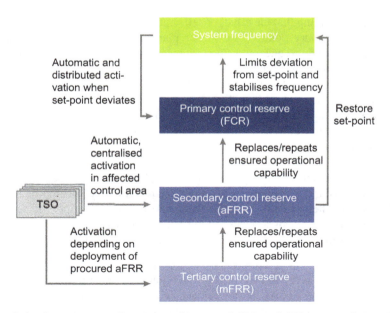

Figure 3.4 Overview: application and responsibilities of different qualities of reserve (based on [13])

74 Intended and unintended islanding of distribution grids

power surplus in the grid. In this case, the grid operator requires a negative control reserve from electricity consumers who withdraw electricity from the grid for a short time. In the case of unforecasted, increased electricity demand, a positive control reserve is required. In this case, the grid operator needs additional energy to be fed into the grid at short notice. Both positive and negative control reserves are provided in three different qualities:

- Primary balancing energy – necessary to quickly stabilise the grid within 30 seconds;
- Secondary control energy – must be available in full within 5 minutes;
- Minute reserve – is used to replace secondary control reserve and must be available within 15 minutes.

To guarantee a fast reaction on frequency deviations and to keep the share that has to be contributed by each unit as small as possible, the primary control reserve is activated in a non-selective manner from the total interconnected system. Any additional load flows that may occur are taken into account in the calculation of network capacities by means of safety margins. The primary control reserve is dimensioned in such a way that an outage of a power plant capacity of 3,000 MW in the continental European grid of ENTSO-E can be managed. Primary control is designed as proportional regulation and is triggered by decentralised controllers of the participating technical units. These are basically the turbine speed governors of power plants (Figure 3.6). The required control reserve is provided in thermal power plants by throttling the turbine inlet valves. A maximum quasi-stationary frequency deviation of ±200 mHz is permissible when the available primary control reserve is fully activated and assuming that the effect of self-regulation of the load is absent [14].

Following primary control, the power plants involved in secondary control reserve – or automatic frequency restoration reserves (aFRR) – take over the control power fed in by the primary control power plants completely and centrally controlled (Figure 3.4). As a result, the primary control power plants are again available for possible new disturbance events. In addition, the secondary control power plants reduce the control deviation remaining after primary control to zero and are usually provided by easily controllable and fully automatically switchable power plants, such as pumped storage power plants or gas turbines.

This process, known as secondary control, takes place in the minute range. After about 15 minutes, the secondary control power plants are replaced by the manually activated tertiary control power plants (Manual Frequency Restoration Reserves (mFRR)) so that they can also be available again for possible further incidents [15]. Finally, the various control services are put out to tender and awarded via auctions to power plants that meet specified prequalification requirements [16].

Primary control reserve is used to limit and quickly reduce frequency deviations after load changes, establish the power balance and thus stabilise frequency while secondary control reserve restores scheduled power exchanges and compensates for remaining frequency deviations.

Control of electric power systems 75

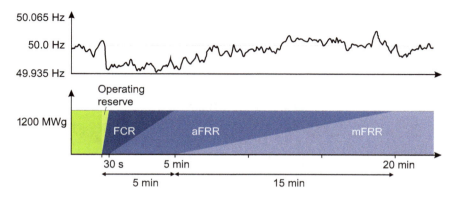

Figure 3.5 Example of a 1200 MW power plant outage (based on [17])

Figure 3.5 illustrates the interaction of the different control reserves. After a loss of generation capacity of 1200 MW, the primary control reserve had to stabilise network frequency as fast as possible. After 30 seconds, secondary control power is automatically called up and replaced after 15 minutes by tertiary control, provided by power stations contracted for this purpose.

3.1.4 Characteristics and control of synchronous generators

In this subsection, the properties and control of synchronous generators are examined in more detail, as they strongly influence the operation and stability behaviour of electrical grids.

3.1.4.1 Stand-alone generator

First, we consider a network with a stand-alone synchronous generator (islanded grid operation) that supplies the required active and reactive power. In Figure 3.6, the different components and control units of a synchronous generator are presented. The generator is connected to a turbine via a shaft, which is driven by a working fluid (e.g. steam or water). The supply of the working fluid and thus, the mechanical power can be controlled via a valve (speed governor). The second control variable of the generator's control is the excitation system, which is explained in more detail in Section 3.3.2.

In the case of an uncontrolled synchronous generator with constant excitation voltage U_E and constant turbine power, the speed n drops sharply when the active electrical power output is increased. These dependencies $n(P)$ are called natural characteristics of the synchronous generator (Figure 3.7).

If a controller with proportional behaviour is used, the characteristic curve can be improved (Figure 3.7). Since the proportional gain cannot be set to any desired level for stability reasons, the controlled variable always has a finite control deviation in the steady state [20].

If a control with proportional behaviour is supplemented by an integral component, its finite deviation can be made to disappear. This isochronous control is

76 *Intended and unintended islanding of distribution grids*

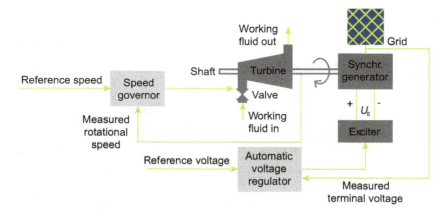

Figure 3.6 Overview of the components and control elements of a synchronous generator (adapted from [18,19])

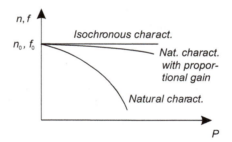

Figure 3.7 Speed power characteristics of the (un)controlled synchronous generator

suitable for islanded grid operation in which only one voltage source regulates frequency and voltage. Active and reactive power are then automatically adjusted according to the load demand. The generator only has to be designed for the maximum load.

Parallel operation of two machines, each with an isochronous control strategy, is problematic, even if the set-point frequency of both characteristic curves is chosen identically. Two set-point devices and two controllers can never be tuned completely identically so that two isochronous characteristic curves would always be parallel to each other – albeit with a minimal distance. In addition, there would be differences due to the different controller speeds. The result would be that the machine controlling the larger set-point would take over the entire output while the other machine would be completely unloaded. This is an unwanted situation, and this problem can be resolved by adding a droop control characteristic.

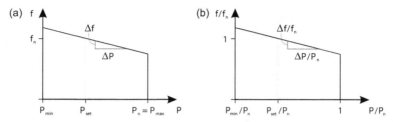

Figure 3.8 (a) Absolute and (b) relative speed/active power characteristic curve for defining the generator power frequency characteristic and droop [20]

3.1.4.2 Droop control characteristic and generator power frequency characteristic

The $f(P)$ control characteristic has a negative slope, and a lower active power output is established when the frequency rises. This sloping curve can be defined using either the absolute or relative speed/active power characteristic curve.

The droop s_{Gi} and generator power frequency characteristic K_{Gi} can be expressed as:

$$s_{Gi} = -\frac{\Delta f / f_n}{\Delta P_{Gi}/P_{Gi,n}} \quad \text{in \%} \quad K_{Gi} = -\frac{\Delta P_{Gi}}{\Delta f} \quad \text{in MW/Hz} \tag{3.1}$$

where s_{Gi} Droop or regulation of generator i
 K_{Gi} Power frequency characteristic of generator i
 Δf Frequency change in the system
 f_n Rated frequency
 ΔP_{Gi} Change in active power of generator i
 $P_{Gi,n}$ Rated power of generator i

In other words, the generator power frequency characteristic is the quotient of the power deviation ΔP responsible for the disturbance and the quasi-steady-state frequency deviation Δf caused by the disturbance. The droop is the frequency drop, in per cent of the rated frequency f, when the active power output of the generator rises from no load to full load (being the rated power).

Important note: When people talk about droop and generator power frequency characteristics, they often discuss them as positive values. However, both are actually negative values.

The principle of droop control is now explained in the left Figure 3.9. Operating point A describes the stationary operation with the set frequency f_n. If, e.g., the consumer load jumps by ΔP from P_{set} to P_2, the speed of the generator or the frequency drops according to the generator characteristic curve from point A to point B. To restore the equilibrium of the active power, the speed controller increases the mechanical power of the prime mover by ΔP. As a result, the new frequency f_2 at which the active power equilibrium is restored is lower than the original frequency.

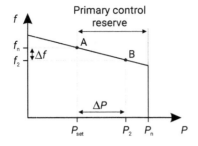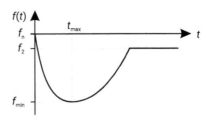

Figure 3.9 Left: Schematic illustration of the primary control; Right: Schematic characteristic of f(t) in the case of a sudden load increase

If we take a closer look at the transition from operating point A to the new steady-state operating point B in the case of a sudden load increase ΔP. In that case, the frequency drops relatively sharply immediately after the load step and then rises again to the new steady-state frequency f_2 (Figure 3.9 right). This relationship can be described with the following differential equation [21].

$$\Delta P + k_{pf} \cdot \Delta f(t) = \Delta P_R(t) - 4\pi^2 \cdot J \cdot f(t) \cdot \dot{f}(t) \qquad (3.2)$$

The frequency characteristic is influenced by the following four parameters:

- the magnitude of the initial load jump ΔP,
- the factor k_{pf} characterising the load self-regulating effect (Section 2.1.6),
- the primary control reserve $\Delta P_R(t)$, and
- the moment of inertia J.

To keep the maximum frequency deviation Δf_{max} (for a given factor k_{pf}) within the permissible limits, the magnitude of the load jump must be limited, the control speed of the primary control reserve $\Delta \dot{P}_R$ must be improved, or additional moment of inertia J (real or synthetic) must be introduced into the system.

In the case of grid-forming synchronous generators, the load jumps that can be controlled inherently and stably within the normally permissible frequency band also depend on the rated power of the synchronous machine and the drive machine technology – turbine or motor. A conservative value is $\Delta P/S_n = 0.05\ldots 0.1$, which can, however, be significantly higher in individual cases. In islanded grids with grid-forming battery storage systems, significantly higher power changes can be stably controlled.

Now a second generator with a flatter droop shall be operated in parallel (Figure 3.10). Since the steady-state maximum generator power is the nominal power, the droop control characteristics are limited upwards by the nominal power, which is expressed by the vertical lines at $P_{G1,n}$ and $P_{G2,n}$. Similarly, the steady-state power output is zero at a frequency greater than the no-load frequency. It is, therefore, essential to extend the control characteristics to frequencies beyond the linear falling section of the droop characteristic when operating several machines in parallel, as frequencies do not necessarily always have to lie in the linear falling section of the droop characteristics.

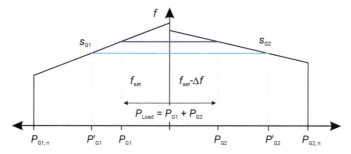

Figure 3.10 Load sharing in parallel operation of two generators with droop control characteristic [20]

Table 3.1 Droops of different power plants (based on [20,22–24])

Droop	Peak load	Medium load	Base load
s_G	2…4%	4…6%	6…8%

3.1.4.3 Parallel operation of two and more generators

If two generators supply the load, the frequency is set by both speed governors and the active power generation is shared by the two generators. When suddenly more active power ΔP is required by the load, the frequency drops with Δf (see Figure 3.10). With the droop control, a disturbance is thus compensated for by several power plants together.

The respective share of a power plant depends on its droop s_{Gi}. The smaller the value of the droop, the flatter the power droop control characteristic. This means that the respective generator takes over a correspondingly larger share of the active power difference in the grid. Peak load power plants have a flat characteristic curve, i.e. a small value for s_{Gi}, and therefore take over a high share of the disturbance (Table 3.1). With a droop of, e.g. 2.5%, they would change their power by 8% with a frequency deviation of $\Delta f = 100\,\text{mHz}$. By contrast, power plants with a higher value for s_{Gi} (baseload power plants) hardly change their output after a disturbance.

> **Example: two generators in parallel**
> Two generators are running in parallel, and we like to determine the power sharing between both generators. Generator 1 has a nominal rated power $P_{G1,\,n} = 400\,\text{MW}$ and a droop $s_{G1} = 6\%$. Generator 2 has also a nominal rated power $P_{G2,\,n} = 400\,\text{MW}$ and a droop $s_{G2} = 4\%$. When the load suddenly consumes more active power ΔP, the frequency drops by Δf. To restore the active power balance, the speed governors increase the prime mover power following the droop control characteristics:
>
> $$\frac{\Delta f}{f_n} = (\Delta P_{G1}/P_{G1,\,n}) \cdot s_{G1} = (\Delta P_{G2}/P_{G2,\,n}) \cdot s_{G2} \qquad (3.3)$$

$$\rightarrow \quad \frac{s_{G1}}{s_{G2}} = \frac{\Delta P_{G2}/P_{G2,n}}{\Delta P_{G1}/P_{G1,n}} \quad \rightarrow \quad \frac{\Delta P_{G1}}{\Delta P_{G2}} = \frac{s_{G2}/P_{G2,n}}{s_{G1}/P_{G1,n}} = \frac{2}{3} \qquad (3.4)$$

That means that the additional load increase $\Delta P = \Delta P_{G1} + \Delta P_{G2}$ will be distributed over the two generators with a ratio of 2:3; generator 1 takes 40% of the load increase while generator 2 (with the lower droop) takes 60%. When both generators have an equal droop, the load increase will be distributed over the two generators according to the ratio of their nominal rated powers.

3.1.4.4 Resulting droop of parallel generators

If more than two machines, each with a power frequency characteristic, operate in parallel, the interaction of these two machines can be examined with the help of the individual power frequency characteristics. However, if the number of machines working in parallel increases, their behaviour can no longer be determined with this simple approach. Therefore, it is inevitable to combine the power frequency control characteristics of the different machines into a single resulting power frequency characteristic that represents the totality of the generation units. The simple example of three power frequency characteristics in Figure 3.11 will explain how a resulting power frequency characteristic is formed.

It can be seen from the three individual power frequency characteristics that at frequencies greater than 1 Hz, the power set-point of all machines is zero. This fact is taken into account in the resulting generator power frequency characteristic by the straight line $P_{G,res} = 0$ for frequencies $f > 51$ Hz (Point A). At a frequency of

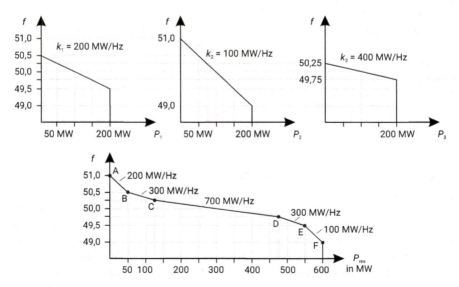

Figure 3.11   Determination of the resulting generator power frequency characteristic of three machines

Control of electric power systems   81

50.5 Hz, the power set-points of machines 2 and 3 are still zero, while the power set-point of machine 1 is 50 MW. So point B can be determined. At frequencies between 50.5 Hz and 50.25 Hz, the resulting generator power frequency characteristic results from the characteristics of machines 1 and 2. So we can determine point C at 125 MW and 50.25 Hz. At frequencies between 50.25 Hz and 50 Hz, all three machines feed into the grid. At 49.75 Hz, machine 3 reaches its nominal power output of 200 MW. If a generation unit is operating at its upper power limit, a decrease in the system frequency will not produce a corresponding increase in its power output. This consideration results in point D. Points E and F can be constructed in the same way as the previous points. At frequencies lower than 49 Hz all machines feed with their nominal power.

The resulting generator power frequency characteristic has six bending points, each at the frequencies at which one of the three individual power frequency characteristics has a bend. Exactly this fact is used in constructing the resulting generator power frequency characteristic: the bending points are determined and connected with each other. With this method, the resulting generator power frequency characteristic of any number of generators operating in a real system can be formed. Adding up all the individual characteristics would give a non-linear resulting characteristic consisting of short segments with different slopes.

A single droop or generator power frequency characteristic, as defined in (3.1), cannot be specified for a resulting generator power frequency characteristic since the gradient of the resulting characteristic is different in each section. However, for a given total system demand P_{Load} and due to the fact that the total system generation is equal to the system load, we get:

$$\Delta P_{\text{Load}} = \sum_{i=1}^{N_{\text{G}}} \Delta P_{\text{Gi}} \tag{3.5}$$

where, N_{G} is the number of generating units. With the definition of K_{Gi} in (3.1), we get:

$$\Delta P_{\text{Load}} = \sum_{i=1}^{N_{\text{G}}} K_{\text{Gi}} \cdot \Delta f \quad \rightarrow \quad K_{\text{G, res}} = \frac{\Delta P_{\text{Load}}}{\Delta f} = \sum_{i=1}^{N_{\text{G}}} K_{\text{Gi}} \tag{3.6}$$

Equation (3.6) describes the linear approximation of the generator power frequency characteristic calculated for a given total system demand (see also Figure 3.11). With the same considerations also the resulting droop $s_{\text{G, res}}$ can be determined:

$$K_{\text{G, res}} = \sum_{i=1}^{N_{\text{G}}} K_{\text{Gi}} = \sum_{i=1}^{N_{\text{G}}} \frac{1}{s_{\text{Gi}}} \cdot \frac{P_{\text{Gi, n}}}{f_{\text{n}}} \tag{3.7}$$

Real measured values of $K_{\text{G, res}}$ of an electric power system differ from the calculated generator power frequency characteristic since the load is also frequency-dependent (self-regulating effect of the load, see Section 2.1.6).

The sum of all generator and load characteristics results in the so-called network power frequency characteristic λ and depends on the number of generation units currently in use and the current load mixture. The network power frequency characteristic λ is, therefore, time-dependent.

For example, the overall power frequency characteristic set-point for the Continental Europe synchronous area is 26,680 MW/Hz [25]. The network power frequency characteristic allows the calculation of the frequency reduction in the interconnected grid in the event of the failure of a large generator unit. For the power plant outage in Figure 3.5, we get

$$\lambda = \frac{\Delta P}{\Delta f} = \frac{-1200\,\text{MW}}{-61.5\,\text{mHz}} \approx 24{,}600\,\text{MW/Hz} \qquad (3.8)$$

This lower value can be explained by summer/winter fluctuations [26,27].

In small power systems, such as in Cyprus, the power frequency characteristic λ is much smaller. The worst outage is round about 120 MW (largest infeed of generator; no interconnections) and the allowed frequency deviation in steady-state is 0.5 Hz. Thus, $\lambda = 240$ MW/Hz. That is the minimum value; it is much larger in summer due to the larger number of connected generators.

3.1.4.5 Secondary control

Due to the proportional characteristic of the FCR, a stationary frequency deviation remains after the power deficit or the power surplus of the generation has been compensated. This frequency deviation is brought back to the nominal frequency by the aFRR through increased feed-in of active power with a resulting parallel shift of the droop characteristic (Figure 3.12).

Secondary control is organised centrally. A centralised controller takes into account not only the grid frequency but also the power balance of the respective control area. The total control power required in the control area is determined and automatically called secondary controllable generators take over the power difference Δf. In this way, the remaining generators involved in primary control can return to their original operating points (OPs) at their droop characteristic and are ready for primary control again. In interconnected operation, the secondary control also ensures that the transfer powers between the different transmission systems in

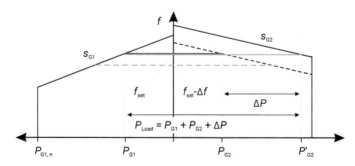

Figure 3.12 Restoration of the original frequency after a load step by adjusting the set-point on the speed controller [20]

the interconnected system are maintained in addition to the compensation of the frequency deviation.

3.1.5 Control behaviour of inverter-based resources

In this subsection, the control behaviour of IBR is examined in more detail, as they strongly influence the operation and stability behaviour of electrical islands.

Inverters can be classified according to three different characteristics. The hardware topology (Thyristor or IGBT), the terminal behaviour (grid-following or grid-forming) and the control behaviour determine the overall behaviour of the inverters on the electrical grid and, thus, also their possible contribution to the system services.

3.1.5.1 Control hierarchies

For stable operation on the grid, control of the inverter is required. The task of this control is to set the inverter output voltage with the required amplitude and angle difference synchronously to the grid voltage applied. Inverter control typically employs a hierarchical approach comprising both a superimposed and subordinate control loop (cf. Chapter 2). Figure 3.13 provides an overview of these control hierarchies and their respective control speeds.

The subordinate control loop, operating at a faster timescale, directly governs the instantaneous behaviour of the inverter. It controls in the time range up to a few period durations of the grid voltage. Typically, this control loop is designed to manage the inverter's output current, ensuring continuous monitoring and control over its behaviour. The fundamental capabilities to provide active and reactive power are commonly described by the inverter's P–Q diagram. Figure 3.14 shows the possible working areas of a device in a four-quadrant view and the limits in terms of maximum and minimum active and reactive power that can be delivered [28].

The setup illustrated in Figure 3.13 effectively safeguards against inverter overload and potential damage in the event of a short circuit. In certain scenarios, such as inverters operating in islanded mode, subordinate voltage regulators may be

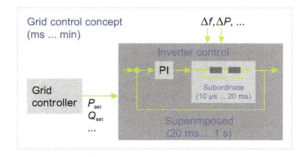

Figure 3.13 Control hierarchies in inverter systems [2]

84 Intended and unintended islanding of distribution grids

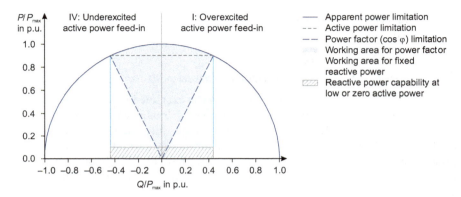

Figure 3.14 *P–Q diagram of a typical four-quadrant inverter (generator reference frame)*

employed. These regulators are tasked with maintaining a consistent output voltage while accommodating varying load currents. However, it is important to note that in this context, additional measures must be implemented to ensure overload and short-circuit protection. Conversely, the higher-level control loops operate at a slower timescale and are responsible for tracking various set-points, such as active and reactive power. They play a critical role in generating reference parameters for the lower-level control loop, operating within timescales of a few seconds.

Moreover, in alignment with the chosen grid control strategy, additional super-ordinate control loops can be integrated to achieve specific grid-supportive functionalities. These may encompass droop control mechanisms like $P(f)$ control or $Q(U)$ control, which dynamically adjust the set-points for active and reactive power based on variations in grid frequency and voltage [2].

3.1.5.2 Frequency-dependent active power reduction

An example of a $P(f)$ droop control mechanism is the frequency-dependent active power reduction at overfrequency (LFSM-O, cf. Chapter 4), which is currently required in many technical connection rules, e.g. [29,30].

If the frequency in the interconnected grid increases, this usually indicates a power surplus. The reason for this is the acceleration of the power plant generators. Depending on the inertia of the entire system, this leads to an increase in the grid frequency. To stabilise the frequency, the connected generation units have to reduce their fed-in active power from a programmable frequency threshold f_1, at least between and including 50.2 Hz and 52 Hz, with a programmable droop in a range of at least $s = 2\%$ to $s = 12\%$. The droop reference is P_{ref}.

- P_{ref} is the maximum continuous active power, in the case of synchronous generating technology and electrical energy storage systems.
- $P_{\text{ref}} = P_{\text{M}}$, the actual AC output power at the instant when the frequency reaches the threshold f_1, in the case of all other non-synchronous generating technology.

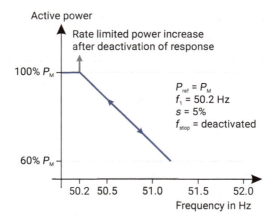

Figure 3.15 Example of active power frequency response to overfrequency (adapted from [30][I]).

This behaviour is described by (3.9). This results in the power characteristic curve in Figure 3.15.

$$\Delta P = \begin{cases} 0 & f \leq f_1 \\ \frac{1}{s} \cdot \frac{f_1 - f}{f_n} \cdot P_{\text{ref}} & f > f_1 \end{cases} \tag{3.9}$$

For a detailed description of the different methods for frequency-dependent active power reduction according to individual standards, please refer to Chapter 4.

3.1.6 Power sharing in islanded grids

Islanded grids exhibit some characteristics that considerably differ from large power systems (e.g. low inertia and large R/X-ratios) and complicate the direct implementation of droop control methods. The power-sharing issue has been addressed in the literature [31,32] and the methods are divided into two major classes: communication-based or droop-based methods (Table 3.2).

In communication-based methods, high reliability and redundant communication links between all converters and also high-bandwidth control loops are required. The communication-based methods offer the advantages of tight current sharing, high power quality, fast transient response, and reduction of circulating currents. But it has the disadvantage of being hard to expand the islanded grid, owing to the communications strategy, the need to have a full overview of the electric network, and the physical communication link costs.

[I]Extracts from DIN EN 50549-2 (VDE 0124-549-2), edition October 2020, are reproduced for the notified and limited circulation with permission 132.024 of DIN German Institute of Standardisation and VDE Association for Electrical, Electronic & Information Technologies. For further reproductions or editions, separate permission is necessary. The additional comments reflect the authors' opinions. Decisive for the usage of standards are the editions with the latest date of issue. These most recent edition can be ordered from VDE VERLAG GMBH, Bismarckstr. 33, 10625 Berlin, Germany, www.vde-verlag.de.

86   *Intended and unintended islanding of distribution grids*

Table 3.2   Advantages and drawbacks of communication and droop characteristic-based control methods (adapted from [31,32])

Method	Advantages	Disadvantages
Communication-based methods	• Accurate power sharing • High power quality • Circulating power elimination • Good transient response	• High-investment cost • Reduce the expandability and redundancy of the system • Current overshoot in the case of the master–slave method
Conventional $P(f)$, $Q(U)$	• Easy implementation without communication • High expandability, modularity and flexibility	• Poor voltage regulation • Inability to handle non-linear loads • Slow transient response • Circulating current among the DG units • Undesirable ripple in the output power

The distributed methods, normally based on droop control, allow a fully distributed and redundant strategy that uses local measurements of the electric network state variables, with several advantages in terms of expandability, modularity, flexibility and redundancy. Distributed control disadvantages include circulation currents between converters, frequency and amplitude deviations, and slow transient response. Additionally, some disadvantages of the conventional droop method are also the poor sharing of harmonics caused by non-linear loads, the line impedance mismatch between parallel converters (affects active and reactive power sharing) and the poor performance of renewable energy resources. Hence in the last years, there were several improvement suggestions to the conventional droop method to overcome some of the mentioned issues [33].

3.2   Frequency characteristics of loads

In addition to the inertia and the control speed of the primary control, the load characteristic influences the frequency stability in grid operation. Some loads react to frequency and voltage fluctuations with changing power demand. This effect is called the load self-regulation effect and is described in more detail in Section 2.1.6.

If the drop in frequency, e.g. after significant disturbances, cannot be compensated by the grid control, the grid load must be switched off stepwise by 'underfrequency load-shedding' (UFLS) schemes to stabilise the frequency. If a further drop in frequency cannot be prevented by this measure, the generation plants in the European interconnected power system disconnect from the grid at 47.5 Hz.

UFLS schemes are used as an emergency measure during operation in the event of unforeseeable and major faults. If an energy deficit is foreseeable, e.g. due to a lack of generation capacity, a specific form of load shedding, the so-called rolling blackout, is used.

Depending on the duration of the load shedding or (rolling) blackout, the so-called 'Cold Load Pick-Up' (CLPU) must be taken into account for the subsequent reconstruction of the grid. The CLPU is caused by the loss of diversity of loads and the simultaneous reconnection of equipment (e.g. electric heating). A second phenomenon is superimposed on the CLPU: when DGs do not feed their pre-blackout power values immediately upon system restoration but start to gradually increase their active power after monitoring the grid for a certain period.

3.2.1 Underfrequency load shedding

If the power imbalance between the total power fed into the system and the power that is consumed by the loads, including losses, is not too large, the generators participating in the frequency control will regulate the active power input from their prime movers and bring back the frequency deviation to acceptable values [1]. If the imbalance is too large, the frequency deviation will be significant with possible serious consequences. Therefore, shedding an appropriate amount of loads is used in underfrequency scenarios to restore the balance between energy generation and consumption. The distribution system operator can disconnect customers to contractual arrangements or do soft load shedding.

UFLS can be generally classified into three categories [34]: traditional stage-by-stage scheme, adaptive scheme and semi-adaptive scheme. The significant difference between the three schemes lies in the method to determine load-shedding amount.

For the *stage-by-stage scheme*, the loads to shed are predefined and divided into several stages. Each stage is tripped by frequency relays if the frequency drops beyond a certain threshold for a few seconds. It sheds loads based on the local frequency and is widely used in utilities because it is simple and does not require sophisticated relays, such as RoCoF relays whose accuracy is often questionable. The values of the thresholds and the relative amounts of load to be shed are decided offline, on the base of experience and simulations [35]. The typical frequency difference between stages is $0.2\ldots0.25$ Hz, and $5\ldots8$ stages are usually set [34]. As an example, Figure 3.16 shows the load-shedding scheme advised by ENTSO-E.

At the frequency of 49 Hz, it is recommended to shed at least 5% of the load. If the frequency drops further below 49 Hz, then a linear stepwise disconnection is recommended by this scheme. The suggested scheme takes steps of 10% such that at 48 Hz, 50% of the load has been shed. It is also allowed to do the first step of the shedding scheme already at 49.2 Hz [36].

The *semi-adaptive scheme* provides a step forward. In fact, it measures df/dt when a certain frequency threshold is reached. According to that value, a different amount of load is shed. In other words, this scheme also checks the speed at which the threshold is exceeded: the higher this speed is, the more load is shed. Usually, the measure of the RoCoF is evaluated only at the first frequency threshold, the following ones being traditional [35].

88 Intended and unintended islanding of distribution grids

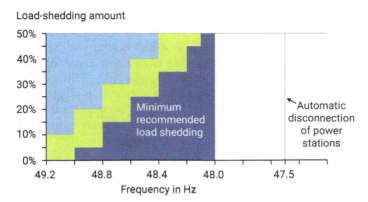

Figure 3.16 Recommendations for load-shedding scheme advised by ENTSO-E (adapted from [36])

The third type of load-shedding program is the *adaptive method*, which calculates the active power deficit on the basis of the measured initial-frequency derivative and sheds blocks of load at specific frequency rates of decline [37].

UFLS schemes can also be used in islanded grids to restore the balance between energy generation and consumption. However, it should be noted that it can be challenging to split the load steps sensibly, as often only discrete power steps are available (e.g. entire feeders or local network stations). To determine the maximum load step, it must be considered that the grid can still be stably operated. In principle, the load shedding can cause a rapid frequency reversal and an oscillation in the direction of overfrequency. In this case, it is advantageous for system stability if PV systems feed in. The instantaneous power reduction of the PV systems at overfrequency has a strong damping effect. With regard to the selection of suitable loads that can be used for load-shedding stages, the emergency supply case must be clearly distinguished from the interconnected operation. In general, the principle of non-discrimination must be assured in energy supply. This also applies in principle to UFLS. In the case of UFLS, however, the generation capacity is insufficient to supply all loads. Therefore, from a technical point of view, it is necessary to switch off individual loads to stabilise the entire system. From a societal perspective, critical infrastructures have the highest priority in this emergency situation to avert humanitarian and economic catastrophes. Therefore, it makes sense for emergency supply not to select loads with the highest priority for the society for underfrequency load shedding.

3.2.2 Rolling blackout

A specific form of load shedding is the so-called rolling blackout. A rolling blackout is an intentional power shutdown that stops electricity delivery for non-overlapping periods of time over different parts of the distribution region. It is a last-resort measure used by utility companies to avoid a total blackout of the power system when electricity demand exceeds the power supply capability. Rolling blackouts may be localised or more widespread and result from insufficient generation capacity or inadequate transmission infrastructure. They are also used to cope with reduced output beyond reserve capacity from power stations taken offline unexpectedly.

The causes of rolling blackouts vary. In some countries, generating capacity falls chronically short of demand due to inadequate investment. Temporary decreases in generating capacity may also occur due to power station outages or loss of renewable capacity caused by weather conditions. Natural disasters or a lack of fuel can abruptly reduce supply. Additionally, industrial accidents or poor maintenance can take generation capacity offline. Conflict can disrupt fuel supply and damage infrastructure. Demand spikes due to extreme weather can also cause blackouts. In well-managed systems, blackouts are planned in advance and schedules are published to allow people to work around them.

Businesses experience substantial economic consequences due to rolling blackouts, including loss of resources, reduced patronage, and curtailed production caused by the sudden cessation of electrical equipment like machinery, refrigeration, or lighting. In regions with regular blackouts, businesses may purchase backup power generation to mitigate these costs, but this entails additional expenses such as maintaining, fuelling, and acquiring generators.

When blackouts are planned beforehand, it becomes easier to manage them. The duration of the blackouts can be controlled by adjusting the speed at which they roll. For instance, in Italy, the national grid safety emergency plan (PESSE) restricts controlled blackouts to a maximum of 90 minutes. Similarly, in Canada, blackouts are planned so that no area remains without electricity for more than one hour.

3.3 Control of reactive power and voltage

3.3.1 Relationship between active and reactive powers and voltage

The operation of AC networks is characterised not only by frequency and active power but also by voltage and reactive power. The correlation between (re)active power and voltage can be explained using the one-line diagram and the phasor diagram of a short line (Figure 3.17). The line is modelled as ohmic-inductive, and line capacitances are neglected. Whether this simplification is acceptable depends on several factors: voltage level, overhead line or cable and the load on the line. In Figure 3.17, $\underline{U}_1$ and $\underline{U}_2$ are the phase voltages, while $\underline{I}_1$ and $\underline{I}_2$ are the currents at the line ends.

The absence of the shunt admittance leads to $\underline{I}_1 = \underline{I}_2 = \underline{I}$ in all points along the line. With the phase angle φ and the active (I_p) and reactive components (I_q) of the current $\underline{I}$, we can calculate the longitudinal ΔU and transversal δU components of the voltage drop:

$$\underline{U}_{12} = \underline{U}_1 - \underline{U}_2 = \Delta U + j\delta U \quad \text{with} \quad \begin{cases} \Delta U & = R \cdot I_p + X \cdot I_q \\ \delta U & = X \cdot I_p - R \cdot I_q \end{cases} \tag{3.10}$$

90 *Intended and unintended islanding of distribution grids*

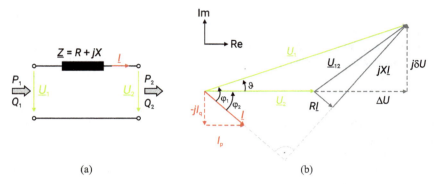

Figure 3.17 Short-line model: (a) one-line diagram; (b) phasor diagram

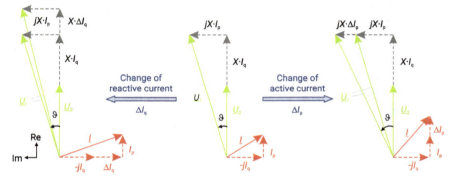

Figure 3.18 Influence of active and reactive current changes on $|U_1|$ and ϑ with fixed $\underline{U}_2$ and $R/X = 0$

With the single-phase complex power $\underline{S} = U_2(I_p + jI_q) = P_2 + jQ_2$, we get

$$\Delta U = \frac{R \cdot P_2 + X \cdot Q_2}{U_2}$$
$$\delta U = \frac{X \cdot P_2 - R \cdot Q_2}{U_2} \qquad (3.11)$$

Since $X \gg R$ for high-voltage transmission lines, it results:

$$\Delta U \approx \frac{X \cdot Q_2}{U_2}$$
$$\delta U \approx \frac{X \cdot P_2}{U_2} \qquad (3.12)$$

If the resistance R can be neglected, then a change in the reactive power influences almost exclusively the voltage magnitude $|U_2|$, a change in the active power almost exclusively increases the angle ϑ (Figure 3.18). Therefore, to reduce the voltage drop, the flow of reactive power should be avoided. In practice, this is possible by reactive power generation near the consumption area [38].

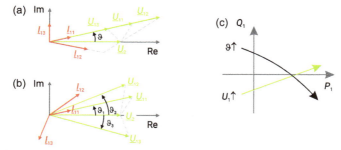

Figure 3.19 Effect of (a) voltage amplitude and (b) voltage angle changes on (c) active and reactive power for a MV line (R/X = 1) (based on [18])

Important relationships for the control of electrical networks can be derived from this. For example, an increase in the voltage angle (without changing the amplitude of U_1 and U_2) primarily increases the active power flow. By contrast, a change in the voltage amplitude U_1 (without changing the amplitude of U_2 and the voltage angle ϑ) primarily increases the reactive power flow.

For medium- and low-voltage (MV and LV) lines, the resistance can no longer be neglected. Equation (3.11) now shows a strong coupling between the different quantities. Figure 3.19 also shows the effects of varying the voltage amplitudes and voltage angles on the active and reactive power for a medium-voltage line with a typical ratio $R/X = 1$. The angle between current and voltage drop across the line (dashed coloured lines) is then 45°. In Figure 3.19(a), the voltage $\underline{U}_2$ and the angle ϑ were kept constant, resulting in varying voltages $\underline{U}_1$ and currents $\underline{I}_1$. This primarily influences the active power P_1 and only to a certain extent the reactive power Q_1 (Figure 3.19(c)). In Figure 3.19(b), the voltages $\underline{U}_1$ and $\underline{U}_2$ were kept constant. An increase of ϑ (indicated in Figure 3.19(c) by $\vartheta \uparrow$) no longer only increases the active power but also strongly influences the reactive power.

3.3.2 Generator control (AVR)

A synchronous generator in islanded operation is operated with the nominal terminal voltage at a given load. If the required inductive reactive power in the grid is higher than the injected reactive power, the voltage drop in the machine increases and the terminal voltage decreases. The automatic voltage regulator (AVR) increases the excitation (Figure 3.6) so that the original terminal voltage can be reached again. In doing so, the generator supplies exactly the additional reactive power that is required in the grid. To achieve a balance between the required and generated reactive power in the grid, the voltage at one point in the grid, e.g. at the terminals of the generator, must be kept constant [39]. Analogue to the frequency control (Section 3.1.4.1), a constant voltage characteristic can also be realised (Figure 3.20).

In this case, the AVR is tuned such that the generator terminal voltage remains constant and is independent of the generator loading. But with generators in interconnected power systems, the same difficulties exist as those already listed for power

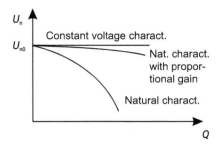

Figure 3.20 U–Q-characteristics of the (un)controlled synchronous generator

frequency control. With the constant voltage characteristic of all machines, the reactive load will also be distributed unevenly, with the consequence that some machines can be overloaded while other machines see no load. To create defined ratios here as well, the generators are given controllers with a droop characteristic. A mixed terminal voltage and load current variable is fed to the control amplifier as the actual value.

3.3.3 Voltage control in distribution grids

Voltage control in power systems is the process of controlling the voltage at different points in the electrical grid to maintain a stable and safe operating voltage range.

In this section, the theoretical principles of voltage distribution and the occurring voltage drops in the distribution grid are first presented, and the possible influencing variables are derived from their mathematical description. Afterwards, the allocation of the voltage bandwidth in the medium- and low-voltage grid is presented. Finally, conclusions for voltage control in islanded grid operation are drawn.

3.3.3.1 Fundamentals

Based on the principles presented in Section 3.3.1, the voltage drop along the distribution line shown in Figure 3.21 can be understood as the addition of individual voltage drops along the equipment.

To be able to display the voltage drops in the MV and LV grid uniformly in a diagram, they are normalised to the corresponding nominal voltage:

$$\Delta \underline{u} = \frac{\Delta \underline{U}}{U_n} \tag{3.13}$$

The transformers are viewed from the high-voltage side and hence, the impedance is assigned to the high-voltage side. The indices 3, 5 and 7 represent the network level (see Table 3.3). The letter m is an integer value and symbolises the automatic voltage control via the on-load tap changer (OLTC).

Starting from the voltage $\underline{u}_3$ at the substation, which can usually deviate from the set-point by a maximum of half the control bandwidth ($\pm m_4 \cdot \underline{u}_{st,4}$), a minimum voltage $\underline{u}_{\text{load, min}}$ results from the geometric addition of the voltage drops on the MV

Control of electric power systems

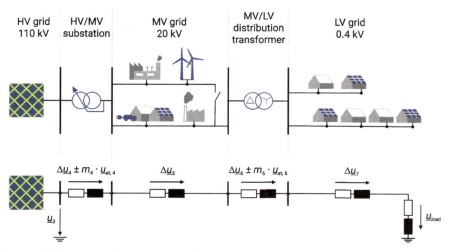

Figure 3.21 Typical structure of a distribution grid and the corresponding equivalent circuit (adapted from www.ront.info)

Table 3.3 Grid levels

Grid Level	Description
1	Extra-high-voltage level
3	High-voltage level
5	Medium-voltage level
7	Low-voltage level
2, 4, 6	Transformer levels

line ($\Delta\underline{u}_5$), across the local network transformer ($\Delta\underline{u}_6$) and along the low-voltage line ($\Delta\underline{u}_7$).

$$\underline{u}_{\text{load}} = \underline{u}_3 - [\Delta\underline{u}_4 \pm m_4 \cdot \underline{u}_{st,4}] - \Delta\underline{u}_5 - \Delta\underline{u}_6 - \Delta\underline{u}_7 \tag{3.14}$$

The resulting interlinked voltage vector diagram is presented in Figure 3.22. The black circles are the limits of the maximum permissible voltage bandwidth ($U_n \pm 10\%$). The control bandwidth of the substation transformer is illustrated by the dashed lines. For the heavy load case without feed-in (right part of Figure 3.22), a voltage at the lower threshold of the control bandwidth must be assumed on the high-voltage side of the substation transformer. In this case, it must be ensured that the minimum voltage bandwidth limit of 0.90 p.u. is not undercut. On the other hand (left part of Figure 3.22), the voltage must not exceed 1.10 p.u. at any point in the distribution network. This is based on the low-load case with maximum feed-in from DGs with an assumed voltage at the upper threshold of the control bandwidth.

94 *Intended and unintended islanding of distribution grids*

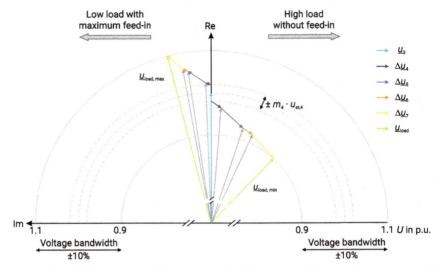

Figure 3.22 Voltage vector diagram for distribution grid for maximum permissible feed-in and load case (adapted from www.ront.info)

For reasons of clarity, only the OLTC in the substation was considered in (3.14) and Figure 3.22. In distribution grids, local network transformers are typically equipped with an off-load-tap changer, and in some specific cases, voltage-regulated distribution transformers (VRDTs) are used. It can be used to decouple the voltage of a complete low-voltage network from the medium-voltage network.

If the voltage drops are replaced by their product of current and impedance and the active and reactive components of the currents are expressed by the powers P and Q (cf. (3.11)), all principle measures to control the voltage can be derived:

- Direct voltage sources: this term refers to devices (usually based on the transformer principle) that influence the voltage level by means of an 'additional voltage in phase with the main voltage'. In the distribution grid, typically, transformers with variable transformation ratios are installed. These transformers can adjust the voltage amplitude and not the phase angle. In the transmission grid, phase-shifting transformers are also used to better control the power flows in the power grid. By default, there are two such direct voltage sources in every distribution network – the substation transformer with OLTC – usually in conjunction with automatic voltage controllers (AVCs) and a line drop compensator (LDC) – and the local network transformer with off-load-tap changer. In addition, the voltage level can be influenced in the MV or LV line by means of so-called single-line regulators. Another solution is the VRDT.
- Impedance reduction: impedance reduction aims at reducing the effective network or transformer impedance. Possible measures are the installation of

Control of electric power systems   95

Table 3.4   Typical R/X-ratios [40,41]

Grid level	R/X-ratios
1	$R \ll X$
3	$R < X$
5	$R/X = 0.25 \dots 1.25$
7	$R/X \approx 2.5$

parallel lines, the use of lines with a larger cross-section, the meshing of power supply units, the reduction of the effective impedances R and X, additional local network stations or substations (reduction of line lengths) or the use of a larger transformer (lower Z).

- Active power reduction: the active power reduction influences the voltage drop along the network components in the same way as the impedance reduction. The reduction of the active power to be transmitted can be achieved via storage systems, peak clipping for distributed generation units as well as load and feed-in management.
- Reactive power control: in addition to influencing the complex voltage drop, reactive power control also results in a rotation of the phase position. Inductive reactive power has a voltage-reducing effect and capacitive reactive power has a voltage-increasing effect. Influencing the reactive power flows in the distribution grid is possible via compensation systems (e.g. capacitors, reactors, power electronics) or generation systems (e.g. $Q(U)$ or $\cos\varphi(P)$ control by inverters, synchronous generators) but also loads (e.g. synchronous machines in phase-shifting operation) and converter-based storage systems.

If the R/X-ratios in the different voltage levels are also considered (Table 3.4), it becomes clear why different voltage control measures are used in the different grid levels.

3.3.3.2   Allocation of voltage bandwidth

According to DIN EN 50160 [42], the nominal voltage in the distribution grid must be maintained with a tolerance of $\pm 10\%$ at an averaging interval of 10 minutes. Even if it would be sufficient to maintain the voltage bandwidth in each voltage level itself, in practice the voltage bandwidth of $\pm 10\%$ from the low-voltage grid to the substation must be allocated as efficiently as possible. The reason for this is that conventional local grid transformers have a fixed transformation ratio, and in substations, both consumer-dominated and generation-dominated medium-voltage feeders must be connected. While there is a voltage increase in the direction of the low-voltage grid in feed-in-dominated grids (dark blue line), the voltage drops in consumer-dominated grids (cyan line). The grid voltage must thus be set to an appropriate set-point at the transformer in the substation with the help of the OLTC so that the entire voltage band of ±10% in the low- and medium-voltage grid can be utilised efficiently. In [43,44], a voltage set-point of 102% at the substation is proposed. The value was determined taking into account the following boundary conditions:

- Maximum deviation of the nominal voltage of ±10% according to DIN EN 50160.
- Maximum voltage increase in the low-voltage grid of +3% due to distributed generation units according to VDE-AR-N 4105 [45].
- Average voltage drop of ±1% at the local network transformer.
- Maximum voltage increase in the medium-voltage grid of +2% by distributed generation units according to VDE-AR-N 4110 [46].
- Control tolerance range of the transformer with automatic OLTC in the substation of ±2%.

The allocation of the permissible voltage bandwidth for the maximum generation scenario and the maximum demand scenario is schematically represented in Figure 3.23.

For the maximum generation scenario, which is relevant for expanding renewable energy systems, the guidelines [45,46] recommend a maximum voltage increase of up to 3% in the LV grid and of up to 2% in the MV grid. The grid infrastructure is planned and implemented accordingly. As these are only recommendations, DSOs can implement customised grid planning rules to better use the ±10% voltage bandwidth for their specific conditions. Furthermore, the recommended allocation of the voltage bandwidth becomes obsolete when voltage-regulated distribution transformers are installed; they allow voltage deviations to be reset to the desired level. Thus, DSOs have a higher voltage bandwidth in the LV grid and, when widely used, also a higher voltage bandwidth in the MV grid [47].

In islanded grid operation, compliance with the voltage bandwidth is achieved to a large extent through the design of the grids as well as through the provision of reactive power by generation plants or grid resources (see Section 3.3.4.1). Additionally, OLTCs, as an effective device to control voltage, are already installed in the

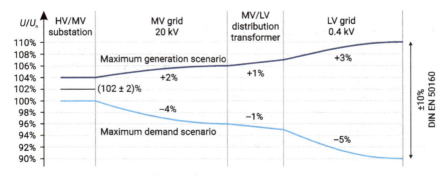

Figure 3.23   Allocation of the permissible voltage bandwidth in the low-voltage and medium-voltage grid

distribution network. Due to the usually small size of islanded grids, a possible voltage instability caused by the reactive power demand of long transmission lines (cf. Section 2.5) is not a cause for instability.

However, some devices for regulating the voltage and some grid features also pose challenges for island grid operation.

- Reactive current compensation systems are often installed at large industrial customers to provide symmetrical reactive power for the operation of motors. Reactive current compensation systems in consumer installations are either designed as controlled central systems or are switched on and off in a decentralised manner with the reactive power consumers. Depending on the operating mode and load situation, decentralised compensation systems may supply more reactive power than the load requires.
- The characteristic curve of a smart inverter with $Q(U)$ control is designed for interconnected grid operation and can lead to excessive reactive power feed-in and voltage increases during islanded grid operation.
- No active control option of the DSO in the low-voltage grid (unless a VRDT is installed).
- A stable operation of very weak grids requires additional compensation units or the provision of reactive power by oversized inverters.

In summary, voltage control in islanded distribution grids is primarily a challenge in weak grids. Nevertheless, the behaviour of compensation systems and smart inverters has to be considered in the design phase.

3.3.4 Requirements for generation units to support the operation of the power system

3.3.4.1 Static voltage support through the provision of reactive power from DGs

As already discussed in Section 3.3.3.1, in the low-load case, distributed generation units (DGs) contribute to a strong voltage increase at the end of the line and a possible violation of the voltage band. In the high load case, it would be desirable if the DGs could support the static voltage control. Therefore, DGs must be able to take part in the steady-state voltage control by providing or absorbing reactive power. With reactive power control, it is possible to increase or decrease the voltage U_{PCC} at the point of common coupling (PCC) as illustrated in Figure 3.24. However, the feed-in of electrical energy with a power factor $\cos\varphi = 1.0$ increases the voltage U_{PCC} and could lead to problems due to overvoltage in the grid.

Therefore, DGs must be able to provide or absorb reactive power to support static voltage control in the electric supply grid. The operating ranges are often designated according to the excitation-dependent reactive power behaviour of synchronous generators:

- Overexcited operation: reactive power provision (acts like a capacitance)
- Underexcited operation: reactive power consumption (acts like an inductance)

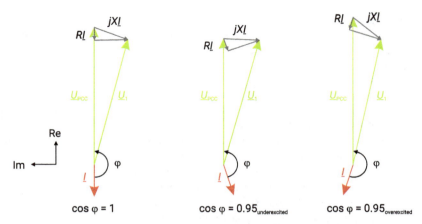

Figure 3.24   Schematic voltage profile with feed-in of a DG for different power factors

The following options for setting the reactive power value are defined in the various national standards (e.g. [29,30,48]), depending on the voltage level, location and size of the power-generating module:

- Fixed power factor $\cos\varphi$
- Active power-dependent power factor $\cos\varphi(P)$
- Fixed reactive power value Q
- Voltage-dependent reactive power control $Q(U)$

For further information, see also Chapter 4.

3.3.4.2   Short-circuit current and dynamic voltage support

In the event of a short circuit, currents that are many times greater than the rated currents flow to the short-circuit point. Nevertheless, grid stability must be maintained, voltage drops must be limited locally and the fault must be safely cleared by the corresponding protective devices. The short-circuit current provided by the corresponding sources causes voltage drops at the network impedances up to the short-circuit location, so that the characteristic funnel-shaped voltage profile with the short-circuit location as its centre (Figure 3.25) is formed. The spatial propagation of the voltage funnel depends on the physical properties of the generators (generator type and amplitude of the short-circuit current) and the electrical distance to the fault location (resulting impedance) [49]. Voltage drops of more than 10% are also referred to as brownouts.

Definition 1 (Brownout). *During a brownout, there is no complete power outage but rather a temporary, slight voltage drop in the power grid. Electricity continues to flow to the end user, but the voltage drops noticeably (flickering and dimming of light bulbs). Usually, short-term voltage drops do not cause any severe damage, but electric devices may respond differently to such a brief voltage drop. Usually, a brownout occurs together with a network overload caused by an unexpectedly high*

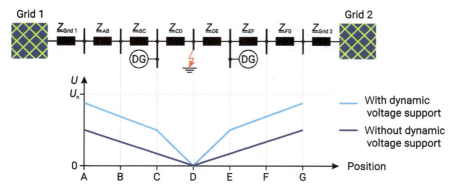

Figure 3.25 Schematic illustration of the voltage drop after a fault at fault location D

demand for electricity. This temporary network instability can be caused by excessive electricity demand, severe weather events or technical malfunctions.

The (steady-state) short-circuit power S_k is used as a measure to which extent a network is able to maintain the voltage in the event of a short-circuit fault.

$$S_k = \sqrt{3} \cdot U_n \cdot I_k \tag{3.15}$$

Since the voltage collapses during a short circuit (Figure 3.25) and becomes zero at the short-circuit point, the short-circuit power is a fictitious quantity.

The final stationary value of the short-circuit current is superimposed by transient and subtransient processes. While the steady-state short-circuit current is usually used for the thermal dimensioning of systems, the largest instantaneous value of the short-circuit current is decisive for the mechanical design. This occurs approx. 5 to 10 ms after the short-circuit occurs and is referred to as the peak short-circuit current i_p. On the other hand, the magnitude of the initial symmetrical short-circuit current I_k'' is decisive for the function of the protective devices. Ultimately, the various short-circuit quantities can be determined from the initial symmetrical short-circuit current according to [50].

The level of short-circuit currents depends on numerous parameters. Among the most important are

- the topology of the network,
- the short-circuit impedances of the equipment,
- the type and location of the grid connection of generation plants, and
- the voltage regulation of the DG unit during the short circuit.

The last two points will now be considered in more detail. In the past, the electrical energy supply was dominated by large conventional power plants with synchronous generators in the transmission grid and short electric distances to the load centres. However, these plants are increasingly being replaced by decentralised plants with inverters in the distribution grid, which results in time-dependent fluctuation at all grid levels and varying electric distances between the DGs and load centres.

100 *Intended and unintended islanding of distribution grids*

However, due to grid reinforcement measures, the grid impedance is reduced, and the coupling of the voltage levels is improved.

Conventional power plants' large synchronous machines have significantly higher short-circuit currents than wind power or photovoltaic systems. This is due to the principle of maintaining the magnetic flux in rotating electrical machines during faults or the reactances of the machine during transient processes. As a result, the short-circuit power provided by a synchronous generator can exceed its rated power several times.

The short-circuit current of the synchronous machine has a DC component decaying to zero and an AC component fading to a steady-state value. It is characteristic that the short-circuit current has a high initial peak ($\approx 6 \ldots 8$ times the rated current). In synchronous machines, the transient and the subtransient reactance determine the balancing process in the event of a terminal short circuit and, thus, the short-circuit current.

The short-circuit behaviour of power-electronically coupled systems with inverters or full converters is determined by electronic control and electronic protective functions that prevent an impermissible high energy input into the DC link capacitor and the switching power semiconductor module as well as its destruction [51]. The maximum short-circuit current is between 1.0 and 1.5 times the rated current [51–55] and depends on the control speed and the current-carrying capacity of the power semiconductors. After the fault occurs, the inverter continues to feed in an almost sinusoidal current. According to EN 60909-0, DGs with inverters are simulated in the short-circuit current calculation by a current source with an infinite impedance in the positive sequence.

With the doubly fed induction generator (DFIG), a different behaviour can be observed: Initially, current peaks (of a similar magnitude to the synchronous generator) caused by the machine dominate, which quickly decay and then are replaced by a short-circuit current set by the inverter control [53,56].

As already discussed in Chapter 1, the available short-circuit current at a grid node can also be characterised by the short-circuit ratio (SCR):

$$SCR = \frac{S_{\text{SCC}}}{P_{\text{r}}} \tag{3.16}$$

The S_{SCC} is the amount of power flowing at a given point in case of a short circuit. It is mainly dependent on the rated voltage U_{G} and the absolute value of grid impedance Z_{grid}, which can be measured at this point. The grid impedance is the sum of impedances of many grid components (lines, transformers, etc.) and loads [57]. Since the maximum short-circuit power of an IBR is around 1.0, the SCR represents the share of the short-circuit current which is provided by the grid.

If DGs are now inserted at nodes C and E as in Figure 3.25, the voltage drop at these nodes is lower in the short-circuit case with all other grid parameters unchanged. If the short-circuit power at nodes A and G is also increased, either by connecting additional generation units or by reducing the grid impedance (strong grid), the voltage drop at nodes A and G becomes smaller and the 'voltage funnel' becomes narrower. If the short-circuit power of the feeding grid is reduced (weak grid), the voltage drop increases and the spatial expansion of the 'voltage funnel' increases.

In addition, the voltage level at the PCC can be actively influenced by controlling the reactive current fed in (Figure 3.24). Finally, it must be ensured that DGs do not immediately disconnect from the grid in the event of voltage dips, but continue to operate for a certain period of time. If, e.g., all DGs were to disconnect from the grid immediately in the event of a short-circuit fault, the voltage would collapse in a larger area of the grid and there would be a power deficit after the short-circuit shut-down. This can lead to a non-selective disconnection of larger sub-grids that are not directly affected by the triggering fault. However, if the voltage is dynamically supported, the power supply can be resumed immediately after the fault has been cleared and the voltage has been restored. This process is also known as fault ride-through (FRT) [58].

Inverter-coupled systems do not inherently react in a voltage-supporting manner to a short circuit due to their control mechanism. However, according to technical connection guidelines, they are required to provide voltage support and supply an appropriate reactive current as part of the FRT after detecting a fault. Nonetheless, the inverter's short-circuit current supply is limited by its nominal power. This can result in a disproportionately lower available voltage-supporting contribution, particularly during periods when a high proportion of power demand is met by DG units. For further information, see also Chapter 4.

In the case of islanding, the short-circuit power of the upstream grid or the transformers is no longer available for the function of the protective elements in the islanded grid. Therefore, either another protection concept must be used to ensure personal and system protection or the system must be adapted to the low short-circuit power, e.g. by using suitable inverters and additional energy storage, or inadmissible fault voltages must be avoided by source-side measures and suitable short-circuit protection must be provided.

3.4 Further requirements for stable islanded grid operation

To ensure high quality, reliability and safety in electricity transmission and distribution, the grid operators continuously work on keeping the frequency, voltage and load of the grid equipment within the permissible limits or returning it to the normal range after faults. These services, which are absolutely necessary for the smooth operation of the electric power system, are referred to as ancillary services (Figure 3.26). Their stable and reliable provision is the core task of the system operators (TSO and DSO).

A distinction is made between:

- Frequency control to compensate for imbalances between inputs and outputs and to keep the grid frequency at its set-point,
- Voltage control, to always keep the voltage within a predefined range,
- System restoration, to restart the power supply as quickly as possible after the unlikely event of a power outage, and
- Operational management, in which the grid operators monitor the correct grid operation and intervene in a controlling manner if necessary.

The ancillary services are predominantly provided by large generation plants or other grid elements. The TSOs are solely responsible for frequency stability and

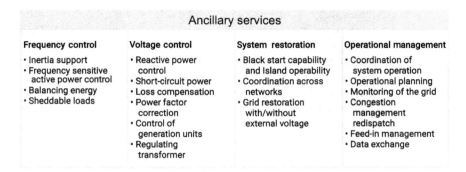

Figure 3.26 Ancillary services to guarantee a stable operation of the electric power system

network restoration (with the cooperation of the DSOs). DSOs and TSOs are each responsible for voltage control and power system operation (in coordination with each other).

In the case of intended and unintended islanded grid operation, the task of ensuring a secure and reliable grid operation is now transferred to the operator of the islanded grid. However, the crucial difference with an unintended islanded grid is that the grid operator is usually unaware of this; he assumes that the sub-grid has been de-energised. In addition, the operating parameters usually fluctuate much more in an intended islanded grid than in regular operation. Both factors increase the risk of personal injury or damage to equipment. In addition, the frequency and voltage control strategy is more advanced and sophisticated to ensure the stability and reliability of the islanded grid.

The focus of the previous sections in this chapter was on frequency and voltage control. Further requirements for stable islanded grid operation shall now be discussed.

3.4.1 Transition from interconnected to islanded grid operation and vice versa

In the following two sections, we will take a closer look at the transition from interconnected to intended islanded grid operation, the operation of this electrical island and resynchronisation with the interconnected grid. We will also discuss the differences and challenges of operating islanded sub-grids compared to interconnected grid operation.

3.4.1.1 Transition to islanded grid operation

As already discussed in Chapter 1 (Figure 1.11), the transition from interconnected to islanded grid operation can be realised through a supply interruption and subsequent grid restoration or without an interruption. Even though the uninterrupted transition is desirable from the customer's point of view, some questions and challenges arise in this case:

- What is the trigger signal for isolating the islanded grid from the main grid: Falling below or exceeding certain frequency/voltage thresholds? Here, a compromise must be found between the requirement to support the interconnected grid as long and as well as possible and the requirement to keep the effects on the islanded grid as low as possible.
- Which power plant(s) immediately take over the frequency and voltage control in the islanded grid? At least one power plant must be equipped with droop control or a communication-based control method must already be implemented.
- Can a possible power surplus or power deficit be compensated by the power plants in the islanded grid without triggering protection elements (over-/underfrequency or over-/undervoltage protection)?
- The function of the load dispatch centre and the system responsibility for the islanded grid must be taken over immediately, either by appropriately trained personnel, by a control system or by a decentralised self-regulating mechanism that acts like a control system (droop control and load self-regulating effect).

These are only the first obvious questions. As explained in Sections 3.1 to 3.3, the system parameters change fundamentally during the transition from interconnected to islanded grid operation and answers to all the points listed in Figure 3.26 must be ready.

A strategy for system restoration is also necessary. Even with an uninterrupted transition to islanded grid operation, it cannot be excluded that the operating parameters may be outside the permissible limits, and therefore, the islanded grid operation must be temporarily interrupted again.

Figure 3.27 illustrates the uninterrupted transition from interconnected to islanded grid operation. The system on the left consists of two subsystems (represented by two scales) that are rigidly connected to each other by green connecting rods. Generation and consumption are in equilibrium in the overall system (but not necessarily in each subsystem). Frequency deviations have an effect on both subsystems due to the rigid coupling. As a result of an event, e.g. falling below certain frequency thresholds, the grid breaks down into two sub-grids and the existing rigid connection breaks. The upper sub-grid could be stabilised, while the lower sub-grid went black. The lower sub-grid could be rebuilt, e.g. from bottom up and the balance is restored.

Figure 3.27 Uninterrupted transition from interconnected to islanded grid operation

104 *Intended and unintended islanding of distribution grids*

In the transition from interconnected to islanded grid operation with supply interruption and subsequent network reconstruction, the points described above are less challenging, since

- necessary operating parameters (e.g. droop control and protection settings) can be set without a time limit,
- any necessary operating personnel can be called in before the separated network is restarted in islanded grid mode and
- switching operations – switching on/off certain consumers/consumer groups, generation units, lines, etc.

can be carried out before the network is reconstructed.

At this point, however, it should not be concealed that the bottom-up network reconstruction is time-consuming and also poses technical challenges (e.g. transformer inrush currents during energising transformers).

3.4.1.2 Resynchronising

If the existing islanded grid shall be reconnected with the main grid, this can again be realised through a supply interruption or without interruption through synchronisation. In addition, two existing islanded grids can also be synchronised to increase the size of the system, use additional resources and improve the stability of the islanded grid.

First, the 'hard' reconnection of an unsupplied sub-grid with another grid shall be considered. This case is the simplest form of resynchronisation. As shown in Figure 3.9 and explained in Section 5.4.5, a sudden load increase results in a frequency drop. To keep the frequency drop within permissible ranges, maximum allowable ranges of power exchange and allowable switchable increments must be defined. Currently, the power grids are designed in such a way that each distribution grid can be resupplied top down by 'hard' switching from the transmission grid, either directly or in increments.

When synchronising two grids (in operation), three parameters – voltage amplitude, frequency deviation and phase angle difference – of the two grid areas must be compared with each other. The successful switching process takes place at the synchronisation point (measured with a synchroscope), the time at which the three parameters are within permissible small deviations (ideally none at all) from each other (Table 3.5). Correct synchronisation does not lead to any exchange of active and reactive power and results in significantly less 'disturbances' than a hard connection of loads or grid areas. Connecting two unsynchronized AC power systems likely results in high currents, which will severely damage any equipment. However, this is usually prevented by correctly configured parallel switching devices.

Depending on the load/generation characteristics and the control strategy (fixed or variable frequency) of the islanded grid, different methods are used for resynchronisation.

First, we consider an islanded grid supplied only by a diesel generator at a constant frequency of 50/60 Hz. By changing the set-points of the speed regulator and exciter of the diesel generator, the first two requirements (frequency deviation and

Table 3.5 Established limits for reconnection of an islanded grid to another power system in accordance with IEEE 1547 [48][1]

Aggregate rating of DG units (S in kVA)	Frequency difference (Δf in Hz)	Voltage difference (ΔU in %)	Phase angle difference ($\Delta \delta$ in °)
0–500	0.3	10	20
500–1500	0.2	5	15
>1500	0.1	3	10

[1] Other national regulations may deviate from these limit values.

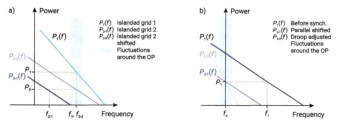

Figure 3.28 (a) Shifting of the droop characteristic for synchronisation of two islanded grids; (b) shifting of the droop characteristic for synchronisation the islanded grid with the interconnected grid

voltage amplitude) can be satisfied. If a small frequency deviation of, e.g. 0.05 Hz is added to the specific frequency set-point, the voltage angle of the islanded grid will change continuously and thus the second requirement for the voltage angle difference is fulfilled the sooner or later at some point.

If the frequency of the islanded grid is controlled via a droop control characteristic, this can be used advantageously for synchronisation with another islanded grid or with the interconnected grid. Figure 3.28(a) shows the droop control characteristics and current OPs of two islanded grids. The residual load P_1 of the larger islanded grid results in a frequency f_1. By shifting the $P(f)$-characteristic of the smaller islanded grid (from $P_{2-1}(f)$ to $P_{2-2}(f)$), both frequencies can be synchronised.

If the islanded grid shall now be synchronised with the interconnected grid, there is the challenge that a reduction of the islanded grid frequency to 50 Hz with simultaneous feed-in from DGs leads to a power increase in the islanded grid. This again bears the risk of an active power imbalance. A fixed frequency setting is, therefore, not possible for resynchronisation with the interconnected grid. To lower the frequency, the $P(f)$ characteristic of the grid-forming power plant is shifted parallel so that the island frequency is 50 Hz (see Figure 3.28(b)).

With this method, it is important to note that if the frequency is lowered by a parallel shift of a droop characteristic with low droop values ($P_{2-1}(f)$), there is

106 *Intended and unintended islanding of distribution grids*

a danger that strong underfrequencies can occur in a changing residual load situation. This danger exists, e.g., if the PV systems are shaded. Therefore, it can be advantageous to shift the characteristic curve parallel and increase its slope simultaneously.

3.4.2 Operation of the islanded grid

As already mentioned, in the case of intended islanded grid operation, the task of ensuring a secure and reliable grid operation is transferred to the operator of the islanded grid. Consequently, the operator of the islanded grid is responsible for providing all ancillary services. In the following paragraphs, several key aspects (without claiming to be exhaustive) are discussed, which are necessary for a safe and reliable operation of the islanded grid.

The same statements apply in principle to unintended isolated grid operation. However, the various aspects of stable islanded grid operation are not 'actively managed' but arise more by chance due to a 'favourable constellation' of the current operating parameters. This is also the reason why unintended isolated grids are undesirable or why they can be operated stably for a certain time and the sooner or later 'fall black'.

Islanded grid operation requires a balance between power generation and demand. It must, therefore, be ensured at all times that sufficient generation capacity is available to meet this demand. Suppose the system has a significant amount of renewable resources. In this case, special care should be taken to ensure that the system is able to meet the demands of the customers even when the renewable resources are not available (e.g. at night or in the case of low water). To achieve the desired power sharing between the generation units in operation and to avoid overloading, the control characteristics of the generation units (droop-based or communication-based) in the islanded grid operation mode will differ from that during interconnected operation.

Furthermore, it is important to have sufficient reserve capacity – in the case of a hydropower plant, this is the unused water flowing over the weir – to meet the demand in case of unexpected load or generation changes. If generation capacity cannot be guaranteed to follow demand, islanded grid operation requires effective load management to ensure that the system is able to meet the demands of the customers. This can be done by shedding non-critical loads, balancing the generation and loads, or using energy storage systems.

In addition to the power balance, the grid operator is also responsible for reactive power balance and voltage stability. In principle, the same applies here as for the power balance, with the difference that the voltage can usually be controlled more easily as no primary energy is required. Finally, the grid operator is also responsible for the safe and robust operation of the islanded grid. This includes suitable protection concepts, monitoring load flows and equipment utilisation.

This list makes clear that even in the quasi-stationary operation of an islanded grid, many different aspects have to be considered and coordinated, and should be taken into account in the conceptual design stage. Therefore, it is essential that operators are trained for this special operating case and supported by operational management tools and automation.

References

[1] Andersson G. Power System Analysis – Power Flow Analysis, Fault Analysis, Power System Dynamics and Stability [Lecture Notes]. ETH Zürich.: EEH Power Systems Laboratory; 2011.

[2] Wagner C, Greve M, Tretschock M, *et al.* Systemsicherheit 2050 – Systemdienstleistungen und Aspekte der Stabilität im zukünftigen Stromsystem [Expert's report]. Emil-Figge-Str. 76, D-44227 Dortmund: ef.Ruhr GmbH; 2020.

[3] Brunken E, Mischinger S, and Willke, J. Systemsicherheit 2050 [Study]. *Deutsche Energie-Agentur GmbH* (dena), Berlin: dena; 2020.

[4] Open data [Homepage on the Internet]. Fingrid; 2024 [updated daily; cited 2024 Feb 18]. Inertia of the Nordic Power System. Available from: https://ww w.fingrid.fi/en/electricity-market-information/InertiaofNordicpowersystem/.

[5] TEXASRE. 2021 Assessment of Reliability Performance [Summary Report]. Austin, TX 78744: Texas Reliability Entity, Inc.; 2022.

[6] Al Kez D, Foley AM, Ahmed F, *et al.* Overview of frequency control techniques in power systems with high inverter-based resources: Challenges and mitigation measures. IET Smart Grid;n/a(n/a):1–23. Available from: https://i etresearch.onlinelibrary.wiley.com/doi/abs/10.1049/stg2.12117.

[7] European Commission. Commission Regulation (EU) 2017/2195 – Establishing a guideline on electricity balancing [Commission Regulation]. Brussels, Belgium; 23 November 2017.

[8] Rezkalla M, Pertl M, and Marinelli M. Electric power system inertia: requirements, challenges and solutions. *Electrical Engineering.* 2018;100:2677–2693.

[9] IEEE. IEEE Standard for Synchrophasor Measurements for Power Systems. IEEE Std C371181-2011 (Revision of IEEE Std C37118-2005). 2011; pp. 1–61.

[10] Tuohy A and Kelly A. Meeting the Challenges of Declining System Inertia [Technical Update]. Palo Alto, CA 94304-1338, USA: EPRI, Electric Power Research Institute; 2019. Available from: https://www.epri.com/research/pro ducts/000000003002015131.

[11] IRENA. Grid Codes for Renewable Powered Systems [Report]. Abu Dhabi: International Renewable Energy Agency (IRENA); 2022.

[12] ENTSO-E. Inertia and Rate of Change of Frequency (RoCoF) [Report]. Brussels, Belgium: ENTSO-E AISBL; December 2020.

[13] Consentec. Description of the balancing process and the balancing markets in Germany – Explanatory document on behalf of the German transmission system operators. [Study]. Aachen, Germany: Consentec GmbH; 2022.

108 *Intended and unintended islanding of distribution grids*

[14] UCTE. *UCTE Operation Handbook*, Version 2.5 [Handbook]. Brussels: Union for the Co-ordination of Transmission of Electricity (UCTE); 2004.

[15] Hofmann L. *Elektrische Energieversorgung – Band 3 Systemverhalten und Berechnung von Drehstromsystemen*. Berlin, Boston: De Gruyter Oldenburg; 2019.

[16] Regelleistung.net [Homepage on the Internet]. Berlin, Pulheim, Bayreuth, Stuttgart: 50Hertz Transmission GmbH, Amprion GmbH, TenneT TSO GmbH, TransnetBW GmbH; [cited 2022 Nov 17]. Frequency Containment Reserve; Secondary Control Reserve; Minute Reserve; Prequalification for the provision and activation of balancing services.

[17] Beck M and Scherer M. Overview of ancillary services. [Study]. swissgrid; 2010.

[18] Henninger S. Netzdienliche Integration regenerativer Energiequellen über stromrichtergekoppelte Einspeisenetze mit integrierten Energiespeichern [Dissertation]. Friedrich-Alexander-Universität Erlangen-Nürnberg. FAU University Press Erlangen, Germany; 2019.

[19] Machowski J, Lubosny Z, Bialek JW, *et al*. *Power System Dynamics: Stability and Control*. The Atrium, Southern Gate, Chichester, West Sussex, UK: John Wiley & Sons Ltd; 2020.

[20] Schwab A. *Elektroenergiesysteme: Erzeugung, Übertragung und Verteilung elektrischer Energie*. Berlin: Springer; 2009.

[21] Schaefer H. Frequenz-Wirkleistungs-und Spannungs-Blindleistungs-Regelung [IfE Schriftenreihe, Heft 23]. TU München: Lehrstuhl für Energiewirtschaft und Kraftwerkstechnik; 1990.

[22] European Commission. Commission Regulation (EU) 2016/631 – Establishing a network code on requirements for grid connection of generators [Commission Regulation]. Brussels, Belgium; 14 April 2016.

[23] Sommerhuber A. Vergleich von Frequenzregelungsstrategien in Inselnetzen [Diploma thesis]. TU Graz: Institut für Elektrische Anlagen; 2019.

[24] Straußberger S. Erbringung von Systemdienstleistungen durch dezentrale Erzeugungsanlagen [Master thesis]. Technische Universität München: Lehrstuhl für Energiewirtschaft und Anwendungstechnik; 2014.

[25] ENTSO-E. Operation Handbook, Policy 1: Load-Frequency Control and Performance, Version 3.0 [Operation Handbook]. Brussels, Belgium: ENTSO-E; March 2009.

[26] Weber HW, Asal HP, and Grebe E. Characteristic Numbers of Primary Control in the UCPTE Power System and Future Requirements. *ETG '97 Summer Meeting*. 1997; pp. 395–405.

[27] Scherer M. Frequency Control in the European Power System Considering the Organisational Structure and Division of Responsibilities [Ph.D. Thesis]. Eidgenössische Technische Hochschule ETH. Zürich, Switzerland; 2016.

[28] Brundlinger R, Strasser T, Lauss G, *et al*. Lab tests: Verifying that smart grid power converters are truly smart. *IEEE Power and Energy Magazine*. 2015;13(2):30–42.

Control of electric power systems 109

[29] EN 50549-1. Requirements for generating plants to be connected in parallel with distribution networks – Part 1: Connection to a LV distribution network – Generating plants up to and including Type B [European Standard]. Rue de la Science 23, B-1040 Brussels: CENELEC; 2019.

[30] EN 50549-2. Requirements for generating plants to be connected in parallel with distribution networks – Part 2: Connection to a MV distribution network – Generating plants up to and including Type B [European Standard]. Rue de la Science 23, B-1040 Brussels: CENELEC; 2019.

[31] Han H, Hou X, Yang J, *et al.* Review of power sharing control strategies for islanding operation of AC microgrids. *IEEE Transactions on Smart Grid* 2016;7(1):200–215.

[32] Rokrok E, Shafie-khah M, and Catalão J. Review of primary voltage and frequency control methods for inverter-based islanded microgrids with distributed generation. *Renewable and Sustainable Energy Reviews.* 2018 02;82:3225–3235.

[33] Pinto J, Carvalho A, and Morais V. Power Sharing in Island Microgrids. Frontiers in Energy Research. 2021;8(Article 609218):1–14. Available from: http s://www.frontiersin.org/articles/10.3389/fenrg.2020.609218.

[34] Li C, Wu Y, Sun Y, *et al.* Continuous under-frequency load shedding scheme for power system adaptive frequency control. *IEEE Transactions on Power Systems.* 35(2):950–961.

[35] Delfino B, Massucco S, Morini A, *et al.* Implementation and comparison of different under frequency load-shedding schemes. In: *2001 Power Engineering Society Summer Meeting. Conference Proceedings* (Cat. No. 01CH37262). vol. 1; 2001. pp. 307–312.

[36] ENTSO-E. Technical Background and Recommendations for Defence Plans in the Continental Europe Synchronous Area [Report]. Brussels, Belgium: ENTSO-E; October 2010.

[37] Rudez U and Mihalic R. Monitoring the First Frequency Derivative to Improve Adaptive Underfrequency Load-Shedding Schemes. *IEEE Transactions on Power Systems.* 2011;26(2):839–846.

[38] Eremia M and Shahidehpour M. *Handbook of Electrical Power System Dynamics.* Piscataway, NJ 08854: IEEE Press; 2013.

[39] Schulze-Buxloh W. *Elektrische Energieverteilung.* Teil 2 [Hochschulskripten]. Essen: Verlag W. Girardet; 1981.

[40] Scheffler J. Bestimmung der maximal zulässigen Netzanschlussleistung photovoltaischer Energiewandlungsanlagen in Wohnsiedlungen [Dissertation]. Technische Universitat Chemnitz. Chemnitz, Germany; 2002. Available from: http://archiv.tu-chemnitz.de/pub/2002/0131/index.html.

[41] Kerber G. Aufnahmefähigkeit von Niederspannungsverteilnetzen für die Einspeisung aus Photovoltaikkleinanlagen [Dissertation]. Technische Universität München. Munich, Germany; 2011. Available from: https://mediatum.ub.tum .de/?id=998003.

[42] European Committee for Electrotechnical Standardization (CENELEC). DIN EN 50160: Voltage characteristics of electricity supplied by public distribution networks [Standard]. CENELEC; 2011.

110 *Intended and unintended islanding of distribution grids*

[43] Harnisch S, Steffens P, Thies HH, *et al.* Planung-und Betriebsgrundsätze für ländliche Verteilnetze – Leitfaden zur Ausrichtung der Netze an ihren zukünftigen Anforderungen. In: *Neue Energie aus Wuppertal*, Band 8. Wuppertal, Germany: Zdrallek, M.; 2016. pp. 1–210.

[44] Wintzek P, Ali SA, Monscheidt J, *et al.* Planungs-und Betriebsgrundsätze für städtische Verteilnetze – Leitfaden zur Ausrichtung der Netze an ihren zukünftigen Anforderungen. In: *Neue Energie aus Wuppertal*, Band 35. Wuppertal, Germany: Zdrallek, M.; 2021. pp. 1–287.

[45] VDE. Power Generating Plants in the Low Voltage Network (VDE-AR-N 4105) [Technical Connection Rule]. VDE Verlag GmbH, Berlin: Forum Netztechnik/Netzbetrieb im VDE (FNN); 2019.

[46] VDE. Technical Connection Rules for Medium-Voltage (VDE-AR-N 4110) [Technical Connection Rule]. VDE Verlag GmbH, Berlin: Forum Netztechnik/Netzbetrieb im VDE (FNN); 2018.

[47] Bayer B and Marian A. Innovative measures for integrating renewable energy in the German medium-voltage grids. *Energy Reports*. 2020;(6): 336–342.

[48] IEEE. IEEE Standard for Interconnection and Interoperability of Distributed Energy Resources with Associated Electric Power Systems Interfaces. IEEE Std 1547-2018 (Revision of IEEE Std 1547-2003). 2018; pp. 1–138.

[49] Altschäffl S. Einfluss zunehmender Einspeisung aus Wechselrichtersystemen auf Kurzschlussauswirkungen im deutschen Übertragungsnetz [Dissertation]. TU München. München, Germany; 2016.

[50] IEC 60909-0:2016. Short-circuit currents in three-phase a.c. systems – Part 0: Calculation of currents [Standard]. IEC; 2016.

[51] Plenz M, Grumm F, Meyer MF, *et al.* Szenariobasierte Analyse der Kurzschlussströme im deutschen Niederspannungsnetz unter Verwendung der CIGRE-Referenznetze. *E & I Elektrotechnik und Informationstechnik*. 2021;138:289–299.

[52] SMA. Short-Circuit Currents Information on Short-circuit Currents in SMA PV Inverters [Technical Information, Iscpv-TI-en-21, Version 2.1]. Sonnenallee 1, 34266 Niestetal, Germany: SMA Solar Technology AG; 2021.

[53] Schürhuber R. Die Kurzschlussnorm IEC 60909-0: 2016 – Neues und Änderungen. *E & I Elektrotechnik und Informationstechnik*. 2016 06;133:228–235.

[54] Li R, Booth C, Dyśko A, *et al.* A systematic evaluation of network protection responses in future converter-dominated power systems. In: *13th International Conference on Development in Power System Protection 2016 (DPSP)*; 2016. pp. 1–7.

[55] Fischer F. Netzanschluss von Erneuerbare-Energien-Anlagen. 3rd ed. Anlagentechnik für elektrische Verteilungsnetze. Berlin: VDE Verlag; 2022.

[56] Morren J and de Haan SWH. Short-Circuit Current of Wind Turbines With Doubly Fed Induction Generator. *IEEE Transactions on Energy Conversion*. 2007;22(1):174–180.

[57] Grunau S and Fuchs FW. *Effect of Wind-Energy Power Injection into Weak Grids* [Article]. Christian-Albrechts-University of Kiel, D-24143 Kiel, Germany: Institute for Power Electronics and Electrical Drives; 2012.

[58] Palm S and Schegner P. Static and transient load models taking account voltage and frequency dependence. In: *2016 Power Systems Computation Conference (PSCC)*; 2016. pp. 1–7.

Part II

Operational and planning issues

Chapter 4

Behaviour at grid connection point

Michael Finkel[1], Georg Kerber[2] and Herwig Renner[3]

The increasing integration of non-conventional sources in power systems – and the consequent reduction of conventional generation power plants – has changed the electricity mix of countries and the way electric power systems are operated. As a result, control issues become more complex, and due to the loss of inertia the time to react to frequency changes and disturbance events, such as power plant outages, becomes shorter [1]. This implies, among other things, that the technical requirements of new power plants that are going to be integrated into the grid have to be updated and redesigned so that stability and quality of electricity supply can be maintained.

In this context, countries have developed technical documents, commonly known as grid codes. Grid codes are governmental regulations or normative regulations made mandatory by law that define the connection and behaviour rules generators in power systems operating in parallel with the grid have to satisfy. The rules differ in each country, and the corresponding operator is responsible for establishing those conditions. The grid codes define not only the technical and operational requirements the generators must meet but also usually the compliance process for the compliance assessment.

These requirements for generation involve operating ranges under normal and adverse grid conditions, technical functionalities that these units must be able to fulfil automatically or by remote command, dynamic behaviour profiles, and restrictions, among other critical performance aspects. Power electronic inverters, which serve as the interface between renewable energy generation plants and the grid, play an important role in this issue, as technological advances in these elements allow DG units to provide advanced grid-supporting functions [2]. Besides grid codes for generation, power quality standards exist, ensuring electromagnetic compatibility and proper function of devices in the grid. Especially in islanded systems with reduced system strength, reduced power quality can be an issue.

This chapter presents a selection of important regulations of the grid codes for interconnected electrical systems and elaborates on the minimum technical requirements that generators in power systems must satisfy to be granted grid access. The

[1]Department of High Voltage Engineering and Electric Power Supply Systems, Augsburg Technical University of Applied Sciences, Germany
[2]Electrical Power Engineering and Networks, Munich University of Applied Sciences, Germany
[3]Graz University of Technology, Institute of Electrical Power Systems, Austria

116 *Intended and unintended islanding of distribution grids*

first section contains more general considerations about grid codes and explains why each country has its own technical standards. For this reason, only selected grid codes can be presented and discussed in the second section. The advantage here is that there are efforts to harmonise the individual grid codes to reduce the effort for manufacturers to adapt their products to the requirements of the individual countries.

In the case of islanded grids, the range of individual technical rules is more significant, as islanded grids have special requirements, especially concerning frequency and voltage control as well as fault ride-through (FRT) capabilities depending on their size, generation, and load structure. Furthermore, it makes a difference whether an islanded grid was intended as an islanded grid by design, e.g. due to its location on an island, or was disconnected from the superordinate interconnected grid due to an external event and is operated as an islanded grid. In the first case, special requirements for specific equipment can be taken into account during the construction and expansion phase of the islanded grid. In the second case, the components 'only' fulfil the requirements for interconnected system operation.

In order to limit the scope of this chapter, primarily the most important regulations of the superordinate Grid Code of the European Union (EU) are presented, and differences to other national regulations (e.g. Australia and South Africa) are discussed. In the last section, aspects regarding power quality parameters and the way they are affected by islanded grid operation, are presented. For additional information, please refer to the literature.

4.1 Introduction to grid codes

In a future, largely inverter-dominated electrical energy supply system, the instantaneous inherent properties of synchronous generators (e.g. inertia and voltage support in the event of a short circuit) will no longer be available to the same extent as today. This affects the way power systems are operated. In order to ensure reliable and stable grid operation, DGs must take over more and more system services. Since the conditions differ from country to country, the grid connection codes must be tailored to the requirements of the respective countries. The following explanations are mainly based on [3–5].

4.1.1 Variable renewable energy impacts the way power systems are operated

As already discussed in Chapter 1, the currently existing interconnected power systems have very often been formed by merging smaller sub-grids. The operating characteristics of these grids are primarily determined by the inherent characteristics of the used synchronous generators (cf. Chapters 2 and 3). The growth of variable renewable energy sources (RESs) has introduced challenges to power system operation, as these sources are variable and uncertain, and replacing conventional synchronous generators changes the dynamic behaviour of the system.

Furthermore, three trends being observed in power systems are decentralisation, digitalisation, and electrification of end users (cf. Figure 1.7), which are driving the

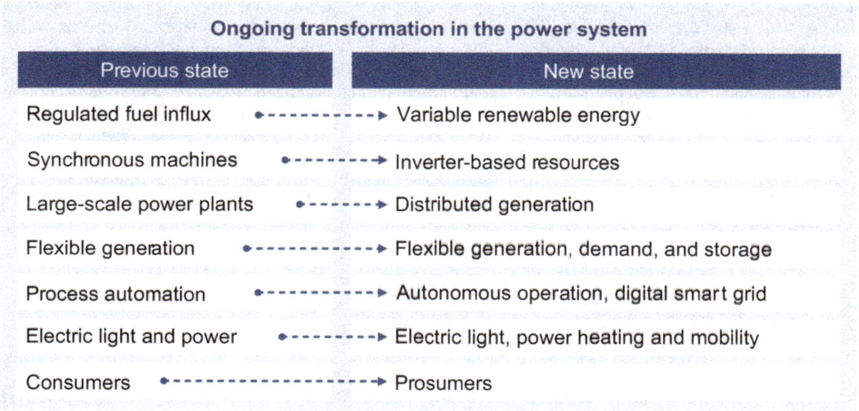

Ongoing transformation in the power system

Previous state	New state
Regulated fuel influx	Variable renewable energy
Synchronous machines	Inverter-based resources
Large-scale power plants	Distributed generation
Flexible generation	Flexible generation, demand, and storage
Process automation	Autonomous operation, digital smart grid
Electric light and power	Electric light, power heating and mobility
Consumers	Prosumers

Figure 4.1   Technological transformation trends in the power system [4]

growth of the power system in a new and different direction. All of this comes at a cost to the system. The system operator has to ensure that the system is both flexible – able to accommodate the frequent imbalances between demand and supply – and stable – able to recover in the event of any contingency [4].

As the power system changes and becomes more volatile on all grid-levels, there is a need for better monitoring and coordinated control of the different assets. Multiple stakeholders, including independent power producers, regulators, and planners, are involved in managing the system. Nevertheless, the system operator is still responsible for real-time monitoring, control, and power system operation. Coordination between different actors and different assets is only possible if credible regulations or principles governing their conduct, such as grid codes, are in place.

Grid codes set technical rules and behaviours for power system participants, including generators, loads, and storage assets, to ensure the security and reliability of the electricity supply. They facilitate the integration of distributed renewable energy generators and enable the private development of new plants. Grid codes serve multiple purposes, including coordination among different actors, increased transparency, and integration of DG units. They outline technical requirements for various aspects of the power system, such as planning, operation, and connection, to ensure smooth system functioning and build trust between power system actors.

Figure 4.1 lists the major paradigm changes transforming the power systems today. These trends have not only necessitated the inception of grid codes but also continue to influence their technical content.

4.1.2   Tailoring grid connection code requirements to system context

The grid connection code requirements vary with the share of synchronous generators and inverter-based resources (IBR) and the size of the power system. Figure 4.2 depicts the different technical requirements that are needed in a power system based on the expected penetration with RES.

118 *Intended and unintended islanding of distribution grids*

Always	Phase one	Phase two	Phase three	Phase four
- Protection systems - Power quality - Frequency and voltage ranges of operation - Visibility and control of large generators - Communication systems for larger generators	- Output reduction during high frequency events - Voltage control - FRT capability for large units	- FRT capability for smaller (distributed) units - Communication systems - Forecasting tool for RES feed-in	- Frequency/active power control - Reduced output operation module for reserve provision	- Integration of general frequency and voltage control schemes - Synthetic inertia - Stand-alone frequency and voltage control

Figure 4.2   Technical requirements for different phases of RES deployment [5]

Some technical requirements are always needed and are independent of the RES capacity. The increasing influence of the growing RES capacity on the power systems can be divided into the following four phases according to [5]:

- Phase one – RES not relevant at the all-system level
- Phase two – RES becomes noticeable
- Phase three – flexibility becomes a priority
- Phase four – power system stability becomes relevant

The phases in Figure 4.2 are indicative; there are no exact boundaries between the different phases. All listed requirements remain necessary when the next phase is reached.

In addition to the generation mix, the size of the electricity grid and the available interconnection capacity also influence the technical requirements in the different phases (cf. [4]). DGs of any size can noticeably impact system performance when connected to smaller systems since small systems tend to be less robust in the case of failures and generator outages. Therefore, facilities connected to smaller systems should be able to withstand wider frequency and voltage fluctuations. Additionally, controllability and FRT capabilities are needed even for small DGs on the low-voltage level at the initial RES integration stage [5].

It should also be taken into account that not all phases are always passed through. For example, in systems with very high shares of hydropower and low shares of fossil-fuel-based power plants to replace, hydropower often provides the needed storage capability and flexibility to accommodate RES fluctuations. However, hydropower plants have relatively low inertia. System operators hence have to either ensure that a certain minimum amount of inertia is always present in the system or provide alternative means to limit frequency deviation during low-inertia situations, like adding primary control reserve (FCR) or adding inertia [4].

4.1.3   The role of grid codes in electricity system regulation

Grid codes regulate different aspects of the power system, and they can take different names accordingly. For example, the European grid codes include network connection, operating and market codes. Because grid codes result from the stakeholders'

Behaviour at grid connection point 119

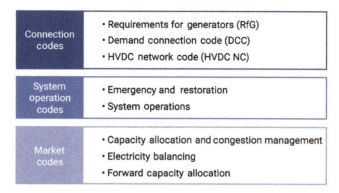

Figure 4.3 Current ENTSO-E grid codes

landscape and the power system organisation structure in place, each jurisdiction can have a different grid code structure and name. A prominent example of illustrating a system of grid code documents complementing each other is the EU Network Codes, a collection of documents that are classified as market codes, operation codes, and connection codes (Figure 4.3) [4].

This chapter focuses on the EU grid connection codes, particularly the provisions relevant to the connection of generators and IBR.

Grid connection codes specify the minimum technical requirements for DG units to be granted grid access. These requirements need to be appropriately designed to ensure system safety and stability with increasing shares of the corresponding generator technologies. If not appropriately designed, it can increase the risk of unplanned consumer supply interruptions and other grid incidents, causing unnecessary expenditures for grid and generator owners or preventing the system from reaching its renewable energy penetration targets. In addition, by providing appropriate rules for generation units, grid codes support the effectiveness of national and regional energy policies for renewable integration.

What is the purpose of grid codes?

Grid connection codes specify the minimum technical requirements all power plants need to meet to be granted grid access. They are developed to coordinate independent actors in power systems with different regulatory frameworks and are regularly updated to reflect technological and operational developments. Grid codes apply to all network users, regardless of whether the power system is operated by a specific operator or a vertically integrated utility, increasing transparency and enabling fair treatment. Establishing a grid code is an important step in opening up the power sector to new plant operators and enabling efficient integration of distributed generation units.

120 *Intended and unintended islanding of distribution grids*

4.2 The EU network code on requirements for generators

For Europe, the European Union has created Regulation (EU) 2016/631 [6] establishing a network code on requirements for grid connection of generators (NC RfG). This regulation provides a framework for the grid connection directives of the European countries. It was written by the TSOs with a clear focus on cross-border issues. Based on this regulation framework, the European countries had to specify the minimum technical requirements of the RfG and adopt their local and national guidelines on every voltage and power level down to 0.8 kW units, which were regarded as the smallest units of cross-border relevance. In Germany, for example, the following series of standards was created and set mandatory by law (Energy Industry Act (EnWG), Section 19, last amended in July 2023) for everyone (users and utilities):

- Technical Connection Rules for Extra High Voltage Network (VDE-AR-N 4130)
- Technical Connection Rules for High Voltage Network (VDE-AR-N 4120)
- Technical Connection Rules for Medium Voltage Network (VDE-AR-N 4110)
- Power-Generating Module Connected to the Low-voltage Network (VDE-AR-N 4105)

It should be noted that the NR RfG was revised and expanded during the writing of this book, particularly with regard to storage systems, e-mobility, and grid-forming inverters. The last aspect is of fundamental relevance to forming and operating intended islanded distribution grids. It is, therefore, recommended that, in addition to the explanations in this chapter, the requirements in the latest version of the RfG are also taken into account.

The NC RfG defines different types of power-generating modules (Table 4.1), which are either synchronous power-generating modules (SPGMs) or power park modules (PPM). A PPM is a unit or ensemble of units generating electricity, which

Table 4.1 Definition of generator types according to [6,7]

Gen. type	Capability from system perspective	Max. capacity threshold	Voltage level
D	Wide-scale network operation and stability, Ancillary services	10^1–75 MW^2	≥ 110 kV
C	Stable and controllable real-time dynamic response covering all operational network states	5–50 MW	<110 kV
B	Automated dynamic response, resilience to events, System operator control	0.1–1 MW	<110 kV
A	Basic capabilities to withstand wide-scale events, Limited automated response and control	0.8 kW	<110 kV

[1]Lower values: Ireland and Northern Ireland.
[2]Upper values: Continental Europe.

is either non-synchronously connected to the network or connected through power electronics, and that also has a single connection point to a transmission system, distribution system including a closed distribution system or HVDC system [6].

The range at the maximum capacity threshold takes into account the system impact of individual units in the different power systems. Thus, a 10-MW wind power plant in the smaller Irish synchronous system has to fulfil the same requirements that only generators of 75 MW and above would have to fulfil when connected to the much larger Continental synchronous system.

One of the fundamental principles of the NC RfG is to balance European-wide system needs and local specifics. For this purpose, only requirements relevant to European-wide system stability are defined exhaustively ('exhaustive requirements'). All others are defined as 'non-exhaustive', meaning further specifications and details are to be defined through national regulatory authorities. In detail, this means that [8–10]:

- Exhaustive requirements define the capabilities of generators by function or principle and by specified parameters. No further specifications are to be made on a national level.
- Non-exhaustive requirements only define basic capabilities without further specifying parameters and settings. Instead, the parameters and settings must be detailed on the national level, mainly to consider national or regional system characteristics to finally become 'exhaustive'.

In addition, there are also 'non-mandatory' requirements, where it is the subject of the national bodies to introduce obligatory rules.

An overview of the specified technical requirements is given in Figure 4.4. These requirements are technology neutral according to the system significance of the generators, depending on the maximum capacity of the generator and voltage level at the grid connection point. The sub-items with blue background colour are explained in more detail in the following sections.

The regulations of the EN 50549 series of standards 'Requirements for generating plants to be connected in parallel with distribution networks' [11,12] are also taken into account. As the RfG does not provide for harmonised standards, EN 50549 is intended to serve as a technical reference for the definition of national requirements in those areas where the requirements of the RfG allow for flexible implementation. EN 50549 intends to include all capabilities of generating plants necessary to operate generating plants in parallel to distribution grids and therefore goes beyond the scope of RfG, such as:

- Voltage operation range,
- Reactive power capabilities and control modes,
- Interface protection and anti-islanding function,
- Generation curtailment,
- Requirements to electrical energy storage systems (EESS).

122 *Intended and unintended islanding of distribution grids*

Aspect	Requirement	RfG Article	A	B	C/D
Frequency stability	Operating frequency ranges	13(1)(a)	x	x	x
	RoCoF withstand capability	13(1)(b)	x	x	x
	Limited frequency sensitive mode – over-frequency (LFSM-O)	13(2)	x	x	x
	Limited frequency sensitive mode – under-frequency (LFSM-U)	15(2)(c)			x
	Limitation of active power reduction at under-frequency	13(4)	x	x	x
	Automatic connection	13(7)	x	x	x
	Provision of synthetic inertia	21(2)			x
Robustness	Fault ride through (FRT)	14(3), 16(3)		x	x
	Post-fault active power recovery	20(2)(c)		x	x
Voltage stability	Voltage ranges	16(2)			x
	Reactive power capability	20(2)(a)		x	x
	Fast fault current injection	20(2)(b)		x	x
System restoration	Black-start capability	15(5)(a)		x	x
	Take part in isolated network operation	15(5)(b)		x	x
General system management	Control schemes and settings	14(5)(a)		x	x
	Electrical protection and control schemes and settings	14(5)(b)		x	x
	Priority ranking of protection and control	14(5)(c)		x	x
	Information exchange	14(5)(d)		x	x

Figure 4.4 Overview of system aspects and technical requirements addressed by the NC RfG (adapted from [7,10])

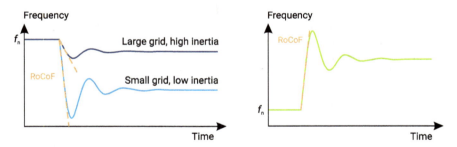

Figure 4.5 Frequency changes due to connections/disconnections on the load or generation side

4.2.1 Frequency stability

As shown in Figure 4.5, connections/disconnections on the load or generation side lead to transients. The amplitude of the overshoots and the steady-state end value depend on the inertia and the control behaviour of the involved generating modules. This results in requirements for the connected power-generating modules. It must be ensured that the power-generating modules stay connected to the network, operate in the expected operating frequency range and operate up to the specified values of the frequency change rates.

To cope with and compensate for such frequency deviations, FCR are deployed by generators running in frequency-sensitive mode (FSM). Suppose system

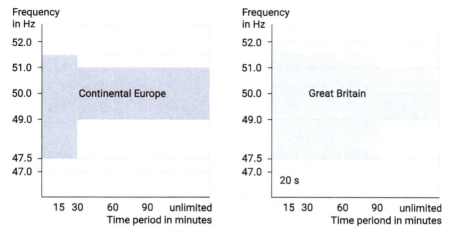

Figure 4.6   Time periods for operation in under- and over-frequency situations. Left: Minimum requirement for Continental Europe; Right: Most stringent requirement (Great Britain)

frequency deviation cannot be stabilised by the FCR resources only. In that case, the limited frequency-sensitive mode (LFSM) adjusts the output of the power generation modules in response to changes in system frequency to support the stabilisation of the system. LFSM has two modes: LFSM-U (under-frequency) and LFSM-O (over-frequency).

4.2.1.1   Operating frequency ranges

Maintaining the functionality of power-generating modules despite deviations in system frequency from its nominal value is critical for ensuring system security. Major disturbances to the system can cause significant deviations, which result in the separation of synchronously interconnected areas due to imbalances between generation and demand. If there is a generation surplus, the frequency will increase, while a lack of generation will cause a drop in frequency. The magnitude of the frequency deviation depends on various factors such as the amount of imbalance, generation profile, system inertia, FCR, and frequency response speed [13].

Figure 4.6 shows the minimum frequency range requirements in Continental Europe on the left side. The most stringent requirements (Great Britain) are also shown on the right side.

4.2.1.2   RoCoF withstand capability

One of the main new specifications introduced with the NC RfG is the requirement for generators to withstand frequency variations. The requirement aims at ensuring that generation modules that are connected to the network will remain connected after severe system incidents (e.g. system splits or loss of large generator in a smaller system) [14]. So that the necessary activation time for control reserves can be

124　*Intended and unintended islanding of distribution grids*

bridged without exceeding the frequency limit values, the RoCoF withstand capability has to be specified [15]. Two important aspects that can be defined for the RoCoF requirement are:

- Maximum threshold for RoCoF, above which tripping is permitted.
- Sliding frequency measurement window

Most of the continental European countries opted for a minimum of 2 Hz/s measured over a period of 500 ms. As already explained in Chapter 2, system inertia and RoCoF are indirectly proportional. System inertia is typically lower for small synchronous areas with typically a higher share of non-synchronous generation to total generation in operation, such as for Ireland or GB, where a single loss of a generator or HVDC interconnector can result in a change in system frequency that is markedly greater than what could be in Continental Europe synchronous area [16]. Therefore, it is not surprising that a lower value of 1 Hz/s (for 500 ms) is required in GB and Ireland. The RoCoF indirectly defines the speed for automatic last measures, such as under-frequency load shedding (Section 3.2.1). The higher the RoCoF, the faster the load-shedding relays have to react to keep the frequency in the permissible band.

4.2.1.3　Limited frequency-sensitive mode

FSM is a power system operating mode in which the power output of a generator is adjusted in response to changes in system frequency. In FSM, the generator output is decreased when the system frequency increases above a certain threshold and increases when the frequency decreases below a certain threshold. This operation mode helps to stabilise the system frequency and maintain the power balance.

Limited frequency-sensitive mode – over-frequency
LFSM-O is to be activated when the system is in an emergency state after a severe disturbance, which has resulted in a major generation surplus, and the frequency deviation cannot be mitigated by the FCR resources only. In such cases, FCR resources are fully deployed, but system frequency cannot be stabilised and increases further. The slow activation of FCR resources (due to high RoCoF) can also contribute to high frequencies [16]. The NC requires from each relevant TSO the specification of the actual frequency threshold f_1 (between and including 50.2 and 50.5 Hz) and droop settings (in a range of $s = 2\%–12\%$)* for its control area to provide the active power frequency response according to (4.1) and Figure 4.7.

$$\Delta P = \frac{1}{s} \cdot \frac{f_1 - f}{f_n} \cdot P_{ref} \tag{4.1}$$

The active power decrease of all power generation modules according to the LFSM-O specifications shall be as fast as technically feasible with an initial delay that shall be as short as possible (less than 2 s) [14]. In the case of LFSM-O activation and increasing frequency, power-generating modules shall continuously decrease the active power towards the minimum regulating level according to the selected

---

*The active power droop relative to the reference power might also be defined as an active power gradient relative to the reference power. A droop of 2% to 12% represents a gradient of 100% to 16.7% P_{ref}/Hz.

Behaviour at grid connection point 125

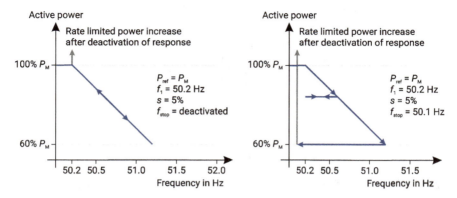

Figure 4.7 Example of active power frequency response to over-frequency without (left) and with (right) configured deactivation threshold [11,12][†]

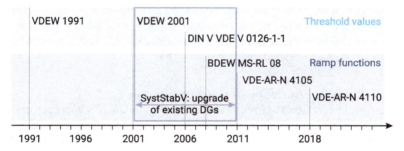

Figure 4.8 Overview of German standards for type A and B generators in LV- and MV-distribution grids

droop. For type A generators, the TSO may allow alternatively that such behaviour is emulated by disconnection at randomised frequencies [16].

The reduction of the feed-in power in the form of a ramp function shown in Figure 4.7 represents the latest development for the required behaviour at over-frequency. However, DGs and flexible loads have different $P(f)$ behaviours depending on their commissioning date. Figure 4.8 shows an overview of German standards for generation units in distribution grids. Initially, the DGs had to reduce the active power output at defined threshold values. Due to a fixed shutdown value of approx. 25 GW of small-scale plants at 50.2 Hz in Germany and their risk for the European interconnected grid (50.2 Hz problem), ca. 300,000 existing DGs had to

---

[†]Extracts from DIN EN 50549-2 (VDE 0124-549-2), edition October 2020, are reproduced for the notified and limited circulation with permission 132.024 of DIN German Institute of Standardisation and VDE Association for Electrical, Electronic & Information Technologies. For further reproductions or editions, separate permission is necessary. The additional comments reflect the authors' opinions. Decisive for the usage of standards are the editions with the latest date of issue. These most recent editions can be ordered from VDE VERLAG GMBH, Bismarckstr. 33, 10625 Berlin, Germany, www.vde-verlag.de.

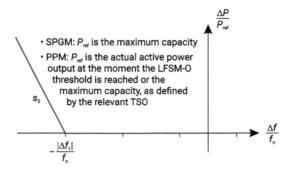

Figure 4.9   Limited frequency-sensitive mode – under-frequency (LFSM-U) [6,14]

be retrofitted as of 2012 (SysStabV – Systemstabilitätsverordnung) and active power reduction in the form of the ramp functions shown in Figure 4.7 was prescribed for new plants. Thus, the behaviour of the sum of all DGs installed in a grid has grown generically over time and changes continuously, since the grid operator does not have to be notified of an age-related replacement of DGs.

Limited frequency-sensitive mode – under-frequency
The objective of the LFSM-U requirement is to automatically increase the active power output of power-generating modules in case of under frequency, to increase the frequency back towards its target value (usually 50.0 Hz) and prevent load shedding [14]. This may occur in cases of major disturbances to the system, such as the loss of a large generation or, in more extreme cases, from a system split. Therefore, the NC requires from each TSO the specification of the actual frequency threshold (between and including 49.8 Hz and 49.5 Hz) and droop settings (in a range of 2 to 12%), to provide the active power frequency response according to Figure 4.9. Identical to the LFSM-O the active power frequency response shall be activated as fast as technically feasible with an initial delay that shall be as short as possible (less than 2 s).

Such reserves can be provided by storage systems and power-generating modules, which are operating at partial load and hence still have the possibility to increase generation proportionally to the deviation of frequency from its nominal value. But also fast controllable loads such as EV-charging systems can activate a similar behaviour as mentioned in [17] but only required in VDE-AR-N 4110. Also, the two standards EN 50549-1/-2 define parameters for the LFSM-U mode for both type A and B modules. However, it needs to be emphasised that the information given within these standards serves as recommendations and is not an obligation to operate a power-generating module at a reduced load level. According to the RfG, the responsible system operator has the right to specify many parameters for its control area in coordination with the TSOs of the same synchronous area. It is, therefore, not surprising that in some countries the LFSM-U requirement has also been implemented for type A modules as well as storage systems [10].

4.2.1.4 Limitation of active power reduction at under-frequency

The requirement to reduce active power output at under-frequency may appear detrimental to system security. Therefore, each generator shall reduce active power output as little as technically feasible in such a situation. However, the ability to maintain maximum active power output with falling frequency differs for different generation technologies. While many typologies of power-generating modules do not have specific technology limitations, some SPGMs, such as gas turbines or combustion motors, have technological limitations to maintain their maximum active power at low frequencies/rotation speed due to a lower mass flow rate.

Gas turbines commonly operated in the power system include a shaft-driven air compressor at the turbine inlet. When the combined cycle gas turbines (CCGT) are synchronously connected to the grid, any disturbance in the system resulting in the decrease of frequency will cause the compressor to slow down. This results in a reduction of the mass flow of air through the turbine and a reduction of the active power output of the CCGT. This effect is much stronger at high ambient temperatures. Thus, due to this physical phenomenon, gas turbine output drops significantly with falling frequency. This phenomenon of decreased maximum power capability at low frequencies can also be observed for other technologies used in the power sector especially if the low frequency is combined with low voltage [18].

A permissible reduction in the maximum active power output with falling frequency for power-generating modules of types A to D is only accepted if the power-generating module has inherent technology constraints. Active power reduction is defined as a reduction gradient that shall lie within predefined limits. The lower limit corresponds to a reduction by 2% of the maximum capacity (P_{max}) per frequency drop of 1 Hz below a frequency of 49 Hz. The upper limit corresponds to a reduction by 10% of the maximum capacity per frequency drop of 1 Hz below a frequency of 49.5 Hz [10].

4.2.2 Voltage stability and robustness

Identical to frequency stability, voltage stability can also be distinguished between three-time ranges: the normal operating voltage range, the time range immediately after the occurrence of a fault and the range of voltage stabilisation and recovery. Suppose a voltage dip is caused by the disconnection of an element in the system because of a short circuit. The magnitude of the voltage dip is determined by the fault location and fault type whereas the duration is dictated by protection technologies [19]. The recovery slope likely depends on the strength of the interconnection and reactive power support of the power-generating modules [20].

This results in the following requirements, which power generation units must fulfil:

- Remain connected to the grid within the defined operating voltage ranges
- Injection or absorption of reactive power to maintain the voltage level within the specified range

128　*Intended and unintended islanding of distribution grids*

Table 4.2　Minimum voltage withstand capabilities for the Continental Europe synchronous area [6]

Voltage base	Voltage range	Time period for operation
110–300 kV	0.85–0.90 p.u.	60 min
	0.90–1.118 p.u.	Unlimited
	1.118–1.15 p.u.	To be specified by each TSO, but not less than 20 min and not more than 60 min
300–400 kV	0.85–0.90 p.u.	60 min
	0.90–1.05 p.u.	Unlimited
	1.05–1.10 p.u.	To be specified by each TSO, but not less than 20 min and not more than 60 min

- Ride through and maintain synchronism during transient events such as grid faults and load steps in the system
- Providing fast fault current

4.2.2.1　Voltage ranges

Voltage withstand capabilities are only mentioned in NC RfG for type D power generation modules (Table 4.2).

The two European standards EN 50549-1/-2 define ranges with respect to voltage and time periods for this requirement. For EN 50549-1, an unlimited operation is required between voltage values of 0.85 – 1.1 per unit EN 50549-2 also requires unlimited operation; however, the voltage range is from 0.9 to 1.1 per unit.

4.2.2.2　Fault ride through

The objective is to limit the potential loss of generation after a fault on the distribution or transmission system to avoid more severe disturbances, i.e. frequency collapse in a synchronous area causing demand tripping and unexpected power flows, system splitting, load shedding, and even blackouts [14]. In the case of a fault on the transmission system level, a voltage drop will propagate across large geographical areas around the point of the fault during the period of the fault (cf. Chapter 3). The ability of a power-generating module to withstand and recover from such abnormal conditions or events is referred to as robustness.

To ensure robustness and limit the potential loss of generation after a fault, RfG requires all generators of types B, C, and D to provide FRT capabilities. A generator must be capable of remaining connected to the grid if the voltage is above the lower FRT limit curve and below the upper FRT limit curve[‡]. The shape of the profile of the lower FRT limit curve (Figure 4.10) is given through various voltage/time parameters as illustrated in Figure 4.10. The NC RfG provides areas in which the FRT parameters are to be defined. The most stringent requirements according to the NC RfG are

---

[‡]A high voltage ride-through (HVRT) capability is not addressed within the NC RfG. The two standards EN 50549-1/-2 define HVRT requirements, which PGMs must fulfil.

Behaviour at grid connection point 129

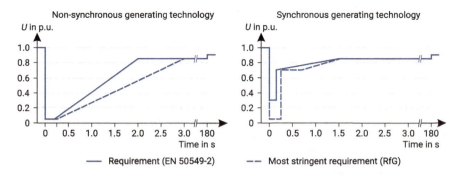

Figure 4.10 Low-voltage ride-through (LVRT) capability for type B non-synchronous generating technology (left) and synchronous generating technology (right) [12]†

illustrated in Figure 4.10 with dashed lines. Additionally, the requirements for type B modules according to EN 50549-2 are drawn. In addition, the LVRT requirements for two-phase, three-phase faults and consecutive faults were defined in Germany and requirements for over-voltages (HVRT) were introduced.

4.2.2.3 Reactive power capability

Voltage control of a grid is essentially determined by the reactive power balance (at least for a grid based on synchronous generators). As described in Chapter 3, the desired voltage bandwidth can only be guaranteed through local measures such as transformer tap changers, power factor correction, reactive power provision by generation units and changes in the network topology.

Therefore, the NC RfG defines requirements for the reactive power capabilities of generators intending to keep slow (quasi-stationary) voltage changes in the grid within tolerable limits. These capabilities have to be provided by all generators from type B upwards upon specification from the relevant TSO/national authority [7]. The RfG defines only the boundaries in the U–Q-space (Figure 4.11) within which the generation unit shall be capable of providing reactive power at its maximum capacity P_{max}. The power-generating module shall be capable of moving to any operating point within its U–Q/P_{max} profile in appropriate timescales to target values requested by the relevant system operator.

In addition, the generating plant shall be capable of meeting the requirements specified below over the entire operating frequency and voltage range. Outside these ranges, the generating plant shall follow the requirements as well as technically feasible. The required accuracy is defined in the local grid codes.

As the detailed conditions and specifications in the national proposals differ for the generator types (B, C, and D) and technologies (SPGM and PPM), Figure 4.11 shows only the requirements for SPGMs and some selected countries.

The location, size, and shape of the U–Q diagram differ significantly from one member state to another. This is due to the fact that the NC RfG enables free or partially free design, and even more than one U–Q diagrams may be defined. This depends, however, on the generator type (B, C, and D), technology (SPGM and PPM)

130 *Intended and unintended islanding of distribution grids*

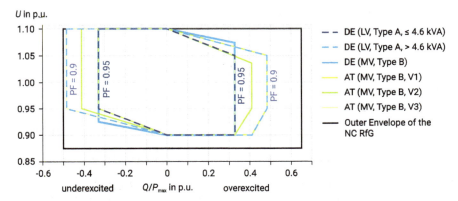

Figure 4.11   U–Q diagram: Different requirements for reactive power capacity with maximum capacity for SPGMs for selected countries [10][¶]

and the voltage level of the grid connection point. Therefore, Figure 4.11 shows only the requirements for SPGMs and some selected countries.

Even if the definition range, the shape, and the position of the individual curves in the U–Q space differ significantly, all curves have in common that in the case of low-voltage values, a voltage support by the provision of reactive power (overexcited operation) is always required (cf. Section 3.3.4). On the other hand, in the case of high voltage values a voltage-reducing behaviour by absorbing reactive energy (underexcited operation) is required.

In addition to the requirements for the provision of reactive power at the operating point P_{max} of the generating module, there are also requirements when operating at an active power output below the maximum capacity ($P < P_{max}$).

For PPMs, there are many differences between the individual member states. As with Q-U diagrams, the diagrams depend on the voltage level or the system type. For example, Figure 4.12 shows selected active power-reactive power capabilities requested for types A and B generators in the Austrian and German grid codes. In the Austrian grid code for type B generators [21], the grid operator can choose between three capability versions for PPM interconnections (Standard: V2). In the German grid codes [22,23], a reactive power capability is defined for type B generators, and a minimum power factor capability is defined for type A generators [24].

So far, only the operating ranges in which the power-generating module shall be capable of absorbing/providing reactive energy have been defined. The actual control mode to be implemented is specified by the grid operator. The following options for setting the reactive power value are defined in the various national standards (e.g. [11,12,25]), depending on the voltage level, location, and size of the generating plant:

[¶]Figure reprinted with permission from CIGRE, Technical Brochure 906: Distributed Energy Resource Benchmark Models for Quasi-Static Time-Series Power Flow Simulations © 2023.

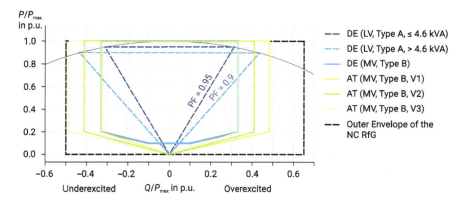

Figure 4.12 P–Q diagram: Different requirements for reactive power capacity below the maximum capacity for PPMs for selected countries [10,24]¶

- Fixed reactive power value Q
- Voltage-dependent reactive power control $Q(U)$
- Active power-dependent reactive power control $Q(P)$
- Fixed power factor $\cos\varphi$
- Active power-dependent power factor $\cos\varphi(P)$

4.2.2.4 Fast fault current injection

In case of faults in the network, in addition to the requirements of Section 4.2.2.3, generating plants shall have the capability to provide additional reactive current up to the current limitation of the generating plant.

In addition to FRT requirements, national regulations must also outline the specifics for providing fast fault current injection. The NC RfG offers a fundamental definition for the injection capability of fast fault current, while the detailed implementation is left to the national level. Therefore, the current documents vary significantly regarding the level of detail provided, depending on the country. Some nations offer comprehensive specifications for implementation, while others only provide a basic definition [7].

The required additional reactive current is usually defined as proportional to voltage deviation from the nominal system voltage. Therefore, the additional reactive current can be described with the proportional k-factor [26]. The magnitude of the additional reactive current is proportional to the k-factor and the voltage during (U) and before the fault (U_0). The k-factor can be described as

$$\frac{\Delta I_Q}{I_r} = k \cdot \frac{\Delta U}{U_n} = k \cdot \frac{U - U_0}{U_n} \qquad (4.2)$$

and is illustrated in Figure 4.13.

Tuning the k-factor can lead to better voltage support during a fault. The required k-factor in the different national standards ranges between 0 and 10 [7]. Studies [27,28] have shown that a high k-factor leads to a lower voltage drop during a fault.

132 *Intended and unintended islanding of distribution grids*

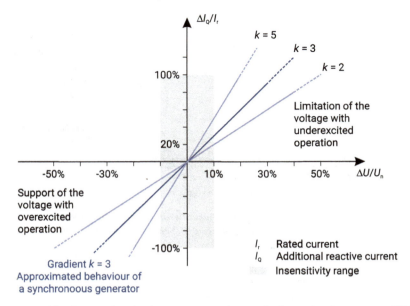

Figure 4.13 Principle of voltage support during faults and voltage steps [12][†]

However, a too-high k-factor will sometimes not result in better voltage support than a lower k-factor because of the current limiter of the inverter. However, the overcurrent capability of the inverter is limited (see Section 3.3.4). Hence, a too-high k-factor will not necessarily result in better voltage support.

4.2.3 System restoration

The 'system restoration' category in the European NC RfG addresses the technical requirements that power-generating modules should meet to support the restoration process of the electrical grid – as specified in the Emergency and Restoration Code [29] – after a blackout or other major system disturbance.

System restoration refers to the process of bringing the electrical grid back to regular operation after a major disturbance or blackout. To support system restoration, the RfG requires power-generating modules to have capabilities to synchronise with the electrical grid and provide power to the grid as it is being restored.

In particular, the 'Black-start capability' and the 'Capability to take part in isolated network operation' have been included as optional requirements in the ENTSO-E network codes (RfG and HVDC), where the relevant TSO is allowed to request these functions. Moreover, the RfG requires power-generating modules to be able to provide reactive power support, which is essential for voltage control and stability during the restoration process. Reactive power is required to stabilise the grid and restore the voltage level after a disturbance.

4.2.3.1 Black-start capability

Although critical for power systems, the black-start capability is a non-mandatory requirement for all generation units as it is only required for those that kickstart the restoration process. Once these units are connected, and the system is stabilised, subsequent generation units do not need black-start capability. The number of generation plants required to initiate the restoration process depends on the topology and characteristics of the local electric system, and thus, the specific details must be determined by the relevant network operator [14].

System characteristics like network topology, demands, and generation mix significantly impact the need for black-start capability of a particular power-generating module. The main system characteristics are [14]:

- System topology: After a blackout, it is necessary to proceed with the restoration process as soon as possible. Restoration processes are very complex and depend on the topology and circumstances of the electrical system at the moment of the blackout. The use of a far-located power generation module to restore an area is often not possible due to over-voltages and electrical connection transients. The need for a black-start capability from power generation modules is more related to a specific area than a whole system needs.
- Loads situation and characteristics: The final purpose of a restoration process is to supply all-system loads, but during the restoration process, the loads are used to stabilise the island, reduce over-voltages, and maintain the island frequency. The characteristics of the loads (power factor, connection peak, etc.) must be taken into account when deciding the need for a black-start capability from a particular power generation facility.
- Electrical protection system: The electrical protection system must be capable of tripping an element in case of an electrical fault. Sometimes it is necessary to connect a power-generating module to have the short-circuit power needed by the electrical protection system.

Black-start capability is closely tied to the technology of the power-generating module. Hydro plants, for instance, are traditionally employed for this service if they are available in the power system. They have a reliable primary energy source, a wide operation range in active and reactive power and a very low need for auxiliary power during operation. However, for power generation modules based on RESs, such capability will only assist in the restoration process when the primary energy source, such as wind or sun, is available.

4.2.3.2 Capability to take part in isolated network operation

This provision of NC RfG aims to set requirements for power-generating modules to enable them to operate in an isolated network after its disconnection from the interconnected system and control frequency and voltage in this isolated network special regulations in selected grid codes for islands [14]. The primary goal is to ensure the availability of generating units to operate under these specific conditions, maintain a continuous power supply in the isolated network and rapidly reconnect it to the interconnected system. Therefore, the defined requirements consider frequency and

134   *Intended and unintended islanding of distribution grids*

voltage ranges and control features, e.g. the plant shall be able to operate in specific frequency and voltage ranges and adjust its active power output automatically according to the actual frequency in the isolated network [14].

The network code requires that operation in an isolated network shall be possible within the defined frequency and voltage limits. If required, the plant shall be able to operate in FSM mode. The power-generating module shall be able to reduce the active power output in case of power surplus from its previous operating point to any new operating point within the $P-Q$-capability diagram as much as inherently technically feasible, but at least an active power output reduction to 55% of its maximum capacity shall be possible [14].

Load characteristics and their allocation in the system, together with the network topology and the availability of generating units, have an impact on the network's ability to be operated when isolated from the interconnected system.

When selecting power-generating modules to take part in island operation for system restoration, the relevant network operator should consider the flexibility of different generation technologies [14]. For example, hydro units are more flexible than thermal units due to differences in their regulation speed and active power output range between the minimum regulating level and maximum capacity. It is important to note that the speed/frequency regulation (i.e. (L)FSM) of the units determines their availability and functionality for isolated network operation.

Thus FSM and accumulatively LFSM with high-quality performance should be ensured by power-generating modules to increase the probability of operating an isolated network stably. Furthermore, active power frequency response within LFSM should also be as fast as possible. Another important function of the power-generating modules to be considered is its ability of house load operation[§], which further supports the operation of an isolated network [14].

4.3   Comparison of selected grid codes

Since the underlying legal framework differs from country to country, the individual grid codes also differ. Nevertheless, most grid codes set regulations for the following topics:

- Requirements for connecting to the grid, including voltage quality and the provision of active and reactive power.
- System behaviour during power grid failures, including considerations for islanding and black-start capability.
- Grid protection in case of technical issues with the connected generator.
- Technical requirements for generators to enable remote control (and power curtailment) during grid disturbance.
- Specifications for the connected power generator to support for ancillary services.
- Responsibilities of the network operator concerning grid capacity and architecture.

[§]Operation of a power-generating module to supply power only to its own electric loads.

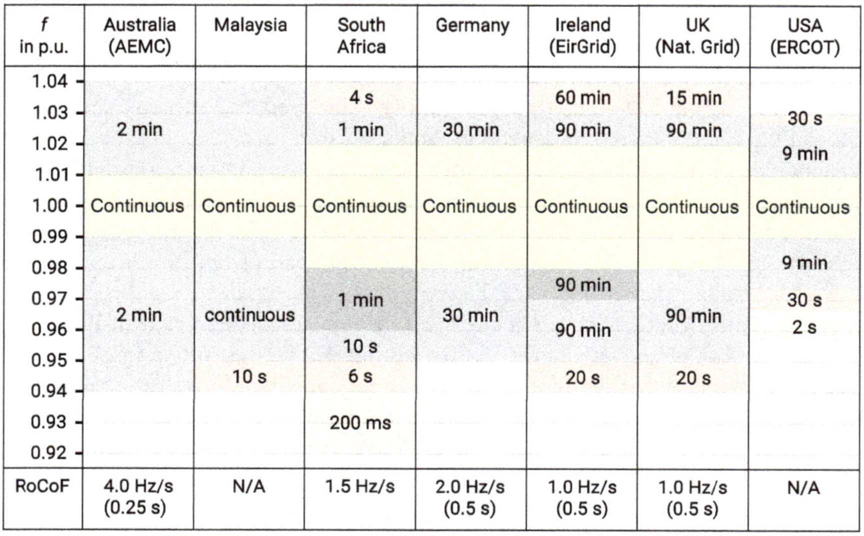

f in p.u.	Australia (AEMC)	Malaysia	South Africa	Germany	Ireland (EirGrid)	UK (Nat. Grid)	USA (ERCOT)
1.04							
1.03	2 min		4 s		60 min	15 min	30 s
1.02			1 min	30 min	90 min	90 min	
1.01							9 min
1.00	Continuous	Continuous	Continuous	Continuous	Continuous	Continuous	Continuous
0.99							
0.98							9 min
0.97	2 min	continuous	1 min	30 min	90 min	90 min	30 s
0.96			10 s		90 min		2 s
0.95		10 s	6 s		20 s	20 s	
0.94							
0.93			200 ms				
0.92							
RoCoF	4.0 Hz/s (0.25 s)	N/A	1.5 Hz/s	2.0 Hz/s (0.5 s)	1.0 Hz/s (0.5 s)	1.0 Hz/s (0.5 s)	N/A

Figure 4.14 Operating frequency ranges required in some selected grid codes (data sources: [23,30–37])

- Organisational aspects of daily operations, such as forecasting power generation, metering and billing of electricity, scheduling maintenance, and handling resulting downtime.

Grid codes often differ only in how the requirements are parametrised rather than differing substantially in the types of requirements made. This has historic origins. Countries at the forefront of RES integration have often been the first to test and develop new requirements. Other countries have then used these as a reference for their own grid codes. Thus, network operators have been able to share experience in setting RES generator requirements and avoid the mistakes made by early adopters of RES technology. One example is the 50.2 Hz problem in Germany. In addition, a degree of commonality between grid codes is in every country's interest because it makes manufacturer compliance easier and therefore reduces costs [3].

After presenting the superordinate European NC RfG in the previous section, this section will now consider the grid codes of other selected countries. This selection covers a wide spectrum of geographical locations, size, RES integration level, and electrical interconnection with other countries. It should be noted that the national regulatory frameworks are subject to continuous changes and revisions.

4.3.1 Frequency stability

Figure 4.14 shows the frequency tolerance bands specified in the different grid codes. Most grid codes require time-unlimited (continuous) operation in the range of ±0.02 per unit. This requirement can be traced back directly to the IEEE C50.13 and IEC60034-3 standards.

136 *Intended and unintended islanding of distribution grids*

In serious contingency (emergency) critical situations, the frequency may be outside the range of the normal operating conditions and is within a typical range of $-0.06/+0.04$ per unit. In this range, operation is required for a prolonged time period. This period is longer for synchronously independent systems such as Ireland and the UK compared to large interconnected systems to account for the higher frequency sensitivity of smaller systems. Australia has a narrow continuous operation range but extended time-limited ranges owing to the fact that the Australian system is prone to system splits, which can cause large short-term frequency deviations [4]. Outside of these specified ranges, disconnection from the system may be allowed.

As derived from theory in Chapters 2 and 3 and also illustrated in Figure 4.5, the resulting RoCoF and frequency deviations depend on the system inertia and the control characteristic of the power-generating modules involved. The combination of RoCoF and frequency-limits defines the requirements for additional functions such as automatic under-frequency load shedding and limits for the maximum delay of protection relays. Therefore, in addition to the specified frequency ranges also values for the RoCoF are specified that generation modules should be able to tolerate without disconnecting from the grid. As can be seen in Figure 4.14, the RoCoF requirements in the various grids differ significantly. In large interconnected systems, RoCoF is not (yet) an issue unless there is a system split and smaller islands are created, but it is very clearly visible in smaller, synchronously independent systems.

Historically, the RoCoF threshold in Ireland was 0.5 Hz/s over 500 ms, while, in GB, it was 0.125 Hz/s with no time delay. This was in the context of systems with high inertia and many synchronous generators. However, with diminishing inertia, operating the system under the 0.125 Hz/s threshold has become increasingly difficult [38]. Therefore, the RoCoF parameters in Ireland and GB have recently been relaxed to 1 Hz/s (500 ms). Also Australia and Texas have had to undertake measures to limit RoCoF. The various reasons can be summarised as follows [4]:

- A high RoCoF is generally undesirable as it reduces the time window for both the primary control (FCR) to limit the frequency deviation and for the automatic under-frequency load shedding (UFLS, see Section 3.2.1). This means that the UFLS must become faster and more sensitive, which increases the risk of false tripping.
- Synchronous generating units are mostly not designed to withstand high RoCoF events, which physically strain the generator, drivetrain, and prime mover.
- RoCoF is used as an indicator in some anti-islanding protection schemes, potentially resulting in disconnecting DGs at high RoCoF events, based on the assumption that fast and steep frequency changes can only be observed in unintentionally disconnected grid segments.

RoCoF issues can be addressed in different dimensions. The first and most obvious one is a requirement for a certain minimum amount of inertia that must always be present to limit the RoCoF to an acceptable value. Ireland is a good example in terms of reducing either the amount of necessary system inertia or the amount of conventional generation that needs to be online to achieve the necessary inertia [4].

The issue has been tackled on multiple levels in parallel (see DS3 programme [39]), in particular at:

- Generators are incentivised to reduce their minimum stable active power output, since the inertia of a generator is independent of the power output, as long as the generator remains online ('inertial floors').
- Generators and DGs have been requested to retrofit or confirm withstand capability for RoCoF values above the initially required 0.5 Hz/s.
- Fast frequency response (FFR) (see Section 3.1.2) has been introduced as an additional ancillary service to quickly limit frequency deviations, especially in high RoCoF or under-frequency events within 0.5–2 s. This has incentivised investments in resources that can respond faster, e.g. battery storage systems.

Besides Ireland, also Great Britain, the Nordic synchronous system, and the Electric Reliability Council of Texas (ERCOT) have also introduced 'inertia floors'. However, this measure is only a short-term solution, as it inevitably limits non-synchronous penetration and, thus, the contribution of RES. The implemented measures are very similar and differ only in minor technical details. However, inertia floors and a reduction of the minimum stable output still rely on synchronous generators. Therefore, changing the system equipment and grid codes to tolerate higher RoCoF values is also in focus.

EirGrid and SONI, the TSOs for Ireland and Northern Ireland, have changed the RoCoF requirements for controllable PPMs to facilitate the delivery of the renewable targets while maintaining operational security on the power system. Controllable PPMs are required to withstand a RoCoF event of 1 Hz/s over 500 ms (EirGrid Code Version 12, 2023) instead of 0.5 Hz/s (EirGrid Code Version 6.0, 2015). Although the Danish power system is a large interconnected system with high synchronous inertia, it has a RoCoF standard of 2.5 Hz/s rolling over 80 ms following the loss of a large generator. Currently, Germany and the USA have relatively large electricity demand networks with rigid AC power interconnections. Therefore, they are unlikely to exhibit issues related to RoCoF until a considerably high penetration of IBR levels is achieved [40].

In Australia, there is currently a RoCoF limitation only in Western Australia (0.25 Hz/s, 500 ms). All generators have to be compliant with National Electricity Rules (NER) that impose 4.0 Hz/s over 250 ms timeframe to meet the automatic access standard and 1.0 Hz/s measured over a window of 1.0 s to meet the minimum access standard [40,41].

While inertia floors and reduction of minimum stable output still rely on synchronous generators, recently, FFR has been the subject of heightened interest. This is because FFR addresses the inertia-related RoCoF issue directly, and FFR procurement is technology neutral. In Ireland, for example, the new FFR system service requires an active power response within 2 s, to be sustained for at least 8 s, with faster responses also incentivised. Similar products have been introduced in other countries and are summarised in Table 4.3. They all have a very fast response speed in common. The duration of the service ranges from a few seconds to an hour.

138 Intended and unintended islanding of distribution grids

Table 4.3 Comparison of FFR services (Database: [42–46])

Region	TSO	Response speed	Duration
Australia	AEMO	0.5–1 s	6 s
Chile	CEN	1 s	>5 min
Canada	Hydro-Québec	1.5 s	9 s
USA	ERCOT[1]	0.25 s	15 min
		0.5 s	60 min
Ireland Island	EirGrid and SONI	≤2 s	>8 s
Great Britain	National Grid	1 s	15 min
Finland	Fingrid	0.7–1.3 s	>5 s

[1] ERCOT has two FFR products.

In Canada, the provision of synthetic inertia by wind turbines has been already required by the TSO Hydro-Québec since 2005. It should be noted that the grid structure of Hydro-Québec is very different from that of central Europe: long 750 kV transmission lines, large distances between generation and consumption, comparatively 'poor distribution of inertia', which generally leads to large line angles and stability problems. The Hydro-Québec transmission connection requirement stipulates in detail that wind power plants must be equipped with an inertia emulation system. Canadian experience showed that synthetic inertia from wind power plants can contribute significantly to system stability but also revealed some issues with the post-event output power dip and compliance testing for the functionality [4,47].

4.3.2 Robustness: FRT envelopes

Large synchronous generators have, for many decades, been able to ride through faults in the system. However, it is critically important that other generators which are affected by a temporary reduction in voltage, until the fault is cleared, do not also trip. FRT requires DGs to stay online for a certain time during voltage dips to prevent an excessive loss of generation and a subsequent drop in system frequency. As already presented in Figure 4.2, the FRT capability of the generation units, therefore, plays an important role in increasing the RES share. It is, therefore, hardly surprising that FRT requirements can be found in all transmission system grid codes worldwide and that this requirement has also very often found its way into the connection rules of DGs in the distribution grid.

Figure 4.15 shows comparative graphs with the LVRT requirement profiles for some selected countries. For reasons of clarity, the HVRT requirements are not shown. The typical shapes of the FRT profiles for synchronous and non-synchronous generating technologies presented in Figure 4.10 can also be found in Figure 4.15.

However, both the depth (0%–30%) and duration of the voltage drop to be passed through and the time it takes to reach the 85% or 90% voltage levels again vary considerably from country to country. The shape and parameters of the LVRT envelope depend on the system's response to a grid fault and on the corresponding

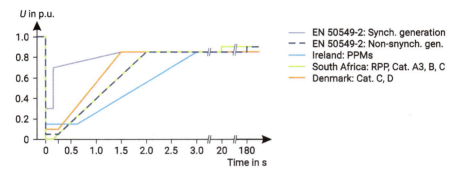

Figure 4.15 Comparison of various countries FRT requirements (data sources: [12,32,33,35,48])

needs of further system components. This response is directly dependent on the fault type, the protection scheme, and the capability of the connected generators and loads to remain connected and recover to normal operation after a fault. The protection scheme, in turn, depends on the critical fault clearing time of the system, determined by the rotor angle stability of the connected synchronous machines [4].

Although only envelopes corresponding to three-phase faults have been represented in Figure 4.15, some grid codes require separate characteristics depending on the short-circuit type (e.g. Germany).

4.3.3 Black-start capability

As already described in Section 4.2.3 the 'Black-start capability' and the 'Capability to take part in isolated network operation' are optional requirements in the ENTSO-E network codes (RfG and HVDC) and the relevant TSO is allowed to request these functions. In Europe, very few TSOs have yet made the provision of black-start a mandatory capability to comply with [49]. Nevertheless, there are many investigations worldwide on how the supply can be restored in the event of a blackout with the help of the current, but in future, a significantly lower number of conventional black-start capable generation units in combination with IBR. Black-start capability has been demonstrated for a range of IBR:

- Voltage source converter (VSC-HVDC) in Ireland [50] and Denmark [51]
- The use of large battery energy storage systems (BESS) with grid-forming inverters [52]
- Several wind turbine, solar inverter, and battery inverter manufacturers have successfully demonstrated black-start capability, often requiring the use of diesel generators or BESS [53–55] (for further examples, see also Chapters 10 and 14)

Even if IBRs will play an important role in grid reconstruction in the future, current standardisation activities are aimed more towards larger units connected to the transmission grid. Although it is no standard, a report of BC Hydro [56] describes the

140 *Intended and unintended islanding of distribution grids*

technical attributes of power generators for planned islanding in the integrated distribution system at 35 kV and below. This report focuses on new PG interconnections but could be applied to the retrofit of an existing PG for planned islanding capability.

4.4 Power quality aspects in islanded grids

Power quality is defined as the 'characteristics of the electricity at a given point on an electrical system, evaluated against a set of reference technical parameters' [57]. It encompasses parameters such as voltage, frequency, harmonics, and waveform distortion. It describes the deviation of the voltage at a specified point of evaluation from its ideal sinusoidal waveform with constant amplitude and frequency. Deviations can be small or large, the latter called 'events'. In general, these deviations are caused by 'non-ideal' currents, including distorted currents (harmonics), unbalanced currents, fluctuating currents or excessive high currents (fault currents), leading to a voltage drop across the respective grid impedance [58].

An important power quality parameter is the main frequency. Frequency deviations are caused by an imbalance in generation and load. The magnitude of frequency deviations will depend on the rotating energy (inertia) of the system and the control dynamics of generation units, participating in frequency control (see Chapters 2 and 3).

Thus, power quality is strongly connected to system strength. System strength refers to the ability of an electrical power system to maintain stable voltage and frequency under various operating conditions, including disturbances and faults. It is a measure of the robustness of the grid infrastructure and its capacity to withstand sudden changes in load or generation, closely connected to the grid impedance.

Strong systems are characterised by their ability to recover from disturbances and maintain stable operations quickly. As discussed in Chapter 2, the definition of system strength is ambiguous in grids with a high share of inverter-coupled generation. In normal operation with the system's voltage close to the nominal voltage, the equivalent grid impedance (Thevenin impedance) will be determined by the inverter's controller behaviour. As long as the inverter's current limits are not violated, the devices will be capable of maintaining the voltage, thus behaving like a strong grid with low grid impedance. In case of faults, the inverter will run into current limitation, which is equivalent to a weak grid with rather high impedance.

This means, that the representation of the grid at the point of evaluation will differ according to the power quality phenomena that are under investigation. The same grid can behave like a strong grid for small deviations and at the same time as a weak grid for large deviations (voltage dips). In islanded grids, it can be assumed that the system strength is rather low compared to interconnected systems, especially for medium and high voltage levels.

Examples of relevant power quality standards in distribution systems are EN 50160 [59], IEC 61000-x series [60,61], IEEE 519 [62], and IEEE 1547 [25].

In the following, a brief description of relevant power quality parameters is given.

Voltage dip

A voltage dip (or voltage sag) is a voltage variation with a remaining voltage below 90% and a duration from half cycle up to 1 min. In strong grids, voltage dips are typically caused by fault currents. However, in weak grids starting currents of equipment can also be a source of voltage dips. Furthermore, the area affected by the dip is larger than in strong grids. Voltage dips can disturb the operation of electronic devices and other sensitive equipment connected to the grid.

Voltage fluctuation (flicker)

Small voltage fluctuations can lead to irritating light flicker. It is caused by fluctuating loads like welding machines, gang saws, or electric arc furnaces. The irritating effect depends on the magnitude and frequency of the fluctuation as well as the lighting source. Modern LED lamps are less sensitive than incandescent lamps.

The magnitude is proportional to the grid impedance, respectively inverse to the system strength. In weak grids, larger voltage fluctuations will occur. In this case, system strength refers to the equivalent Thevenin impedance for normal operation with a voltage close to nominal voltage.

Voltage unbalance

Voltage unbalance (or voltage asymmetry) refers to an uneven distribution of voltage magnitudes or phase angles. It is typically induced by unbalanced loads and, in rare cases also by unbalanced grid impedance. Voltage unbalance causes extra losses and mechanical stress (vibration) in rotating machines.

The magnitude is proportional to the grid impedance, respectively inverse to the system strength. As for voltage fluctuations, system strength refers to the equivalent Thevenin impedance for normal operation with a voltage close to the nominal voltage.

Harmonics

Harmonics is a way to describe the deviation from an ideal sinusoidal waveform (distortion) using the Fourier transform. Classical harmonics cover the frequency range from DC to 2.5 kHz, while supra harmonics range up to 150 kHz. Harmonic voltages are caused by harmonic currents, originating from power electronic equipment. Excessive harmonic distortion can lead to overheating, increased losses, and malfunction of connected devices.

The harmonic grid impedance for low-order harmonics up to a few 100 Hz will depend on the system strength. However, for higher frequencies, the harmonic impedance will be dominated by parallel and serial resonance, formed by inductance and capacitance of the grid. Thus, system strength loses its significance with higher frequency.

Frequency

Frequency stability only depends on system inertia (rotating energy, see Chapter 2) and frequency control characteristics (primary frequency control, frequency containment reserve, see Chapter 3) and is therefore independent of the grid impedance. In interconnected systems, frequency deviations usually play a minor role due to large inertia.

142   *Intended and unintended islanding of distribution grids*

Table 4.4   Frequency limits with respect to power quality

Operation mode	Frequency range	Time frame
Synchronous connection to	50 Hz ± 1%	During 99.5% of a year
a transmission system	50 Hz + 4%/−6%	During 100% of the time
No synchronous connection to	50 Hz ± 2%	During 95% of a week
a transmission system	50 Hz ± 15%	During 100% of the time

Difference between interconnected and islanded systems

The main difference between interconnected and islanded systems will be the ability to control and maintain frequency. Islanded systems have significantly lower rotating energy related to a possible imbalance between generation and load, leading to a higher rate of change of frequency (RoCoF) and larger frequency deviations. Thus, grid codes for islanded systems usually allow a wider frequency band for normal operation (Figure 4.14). This is also considered in the European standard EN 50160 [59], which explicitly sets wider margins in the case of islanded operation.

For other power quality parameters, no extra limits are given for islanded operation, although it is very likely that system strength will be reduced in small islanded systems. Only for (slow) voltage variations islanded, the reference [59] states that in cases of electricity supplies in networks not interconnected with transmission systems, higher voltage variations are allowed.

Due to the limited short-circuit capacity, faults in the islanded system will cause harmful dips with low remaining voltage and a huge part of the system will be affected. The same will be true for fast voltage fluctuations (flicker), voltage unbalance, and low-order harmonics. In the case of high-order harmonics, the effect of islanding is harder to predict, since the harmonic impedance is mainly affected by resonance effects. However, in weak systems resonance frequencies will be shifted generally towards lower frequencies.

Poor power quality can lead to issues such as increased losses, equipment malfunction and damage, and it is clear that reduced system strength in islanded systems generally degrades power quality. To address these challenges, islanded grids often employ advanced control systems, energy storage technologies (such as batteries or flywheels) and intelligent grid management algorithms. Additionally, proper maintenance and regular monitoring of grid parameters are essential to ensuring sufficient power quality in islanded grids.

Although the focus in islanded systems is put on an available and stable supply, one should not lose sight of appropriate power quality.

References

[1] ENTSO-E. Frequency Stability in Long-Term Scenarios and Relevant Requirements [Report]. Brussels, Belgium: ENTSO-E; December 2021.

[2] Beires PP, Moreira CL, Lopes JP, *et al.* Defining connection requirements for autonomous power systems. *IET Renewable Power Generation.*

2020;14(1):3–12. Available from: https://ietresearch.onlinelibrary.wiley.com/doi/abs/10.1049/iet-rpg.2019.0430.

[3] IRENA. Scaling up Variable Renewable Power: The Role of Grid Cides [Report]. International Renewable Energy Agency (IRENA); 2016.

[4] IRENA. Grid Codes for Renewable Powered Systems [Report]. Abu Dhabi: International Renewable Energy Agency (IRENA); 2022.

[5] IEA. System Integration of Renewables – An Update on Best Practice, License: CC BY 4.0 [Report]. Paris: International Energy Agency (IEA); 2018. Available from: https://www.iea.org/reports/system-integration-of-renewables.

[6] European Commission. Commission Regulation (EU) 2016/631 – Establishing a network code on requirements for grid connection of generators [Commission Regulation]. Brussels, Belgium; 14 April 2016.

[7] Bründlinger R. Grid Codes in Europe – Overview on the current requirements in European codes and national interconnection standards [Presentation at NEDO / IEA PVPS Task 14 Grid Code and RfG Workshop, Tokyo]. Vienna, Austria: AIT Austrian Institute of Technology; 2019.

[8] Bründlinger R, Schaupp T, Graditi G, *et al.* Implementation of the European Network Code on Requirements for Generators on the European national level; 2018. p. 1–9.

[9] ENTSO-E. Parameters of non-exhaustive requirements – ENTSO-E guidance document for national implementation of non-exhaustive requirements [Implementation Guidance]. Brussels, Belgium: ENTSO-E; April 2013.

[10] European Commission. Implementation of the Network Code on Requirements for Grid Connection of Generators [Study]. Brussels, Belgium: Directorate-General for Energy, Internal Energy Market; February 2021.

[11] EN 50549-1. Requirements for generating plants to be connected in parallel with distribution networks – Part 1: Connection to a LV distribution network – Generating plants up to and including Type B [European Standard]. Rue de la Science 23, B-1040 Brussels: CENELEC; 2019.

[12] EN 50549-2. Requirements for generating plants to be connected in parallel with distribution networks – Part 2: Connection to a MV distribution network – Generating plants up to and including Type B [European Standard]. Rue de la Science 23, B-1040 Brussels: CENELEC; 2019.

[13] ENTSO-E. Network Code for Requirements for Grid Connection Applicable to all Generators – Requirements in the Context of Present Practices [Working Document]. Brussels, Belgium: ENTSO-E; June 2012.

[14] ENTSO-E. Implementation Guideline for Network Code "Requirements for Grid Connection Applicable to all Generators" [Implementation Guidance]. Brussels, Belgium: ENTSO-E; October 2013.

[15] Nuschke M. Frequenzstabilität im umrichterdominierten Verbundnetz [Dissertation]. Fraunhofer IEE Kassel and TU Braunschweig. Nobelstraße 12, 70569 Stuttgart; 2021.

[16] ENTSO-E. Limited frequency sensitive mode – ENTSO-E guidance document for national implementation for network codes on grid connection [Implementation Guidance]. Brussels, Belgium: ENTSO-E; January 2018.

144 *Intended and unintended islanding of distribution grids*

[17] Lehner J and Kerber G. Necessary Contribution of Electric Vehicles to Limited Frequency Sensitive Mode. *1st E-Mobility Power System Integration Symposium*, Berlin, Germany. 2017; p. 1–4. Available from: https://mobilityintegrationsymposium.org/wp-content/uploads/sites/7/2017/11/3A_1_EMob17_084_paper_Lehner_Joachim.pdf.

[18] ENTSO-E. Maximum Admissible active power reduction at low frequencies–ENTSO-E guidance document for national implementation of conditions for maximum admissible active power reduction at low frequencies [Implementation Guidance]. Brussels, Belgium: ENTSO-E; April 2021.

[19] Patne NR and Thakre KL. Factor affecting characteristic of voltage sag due to fault in the power system. *Serbian Journal of Electrical Engineering*. 2008 01;5:171–182.

[20] Mali S, James S, and Tank I. Improving low voltage ride-through capabilities for grid connected wind turbine generator. *Energy Procedia*. 2014 12;54:530–540.

[21] E-Control. TOR Erzeuger: Anschluss und Parallelbetrieb von Stromerzeugungsanlagen des Typs B [Technical Connection Rule]. Rudolfsplatz 13a, 1010 Wien: Energie-Control Austria für die Regulierung der Elektrizitäts-und Erdgaswirtschaft (E-Control); 2019.

[22] VDE. Power Generating Plants in the Low Voltage Network (VDE-AR-N 4105) [Technical Connection Rule]. VDE Verlag GmbH, Bismarckstraße 33, 10625 Berlin: Forum Netztechnik/Netzbetrieb im VDE (FNN); 2019.

[23] VDE. Technical Connection Rules for Medium-Voltage (VDE-AR-N 4110) [Technical Connection Rule]. VDE Verlag GmbH, Bismarckstraße 33, 10625 Berlin: Forum Netztechnik/Netzbetrieb im VDE (FNN); 2018.

[24] CIGRE. Technical Brochure no. 906: Distributed Energy Resource Benchmark Models for Quasi-Static Time-Series Power Flow Simulations [Technical Brochure]. CIGRE, Paris, France: CIGRE WG C6.36; 2023.

[25] IEEE. IEEE Standard for Interconnection and Interoperability of Distributed Energy Resources with Associated Electric Power Systems Interfaces. IEEE Std 1547-2018 (Revision of IEEE Std 1547-2003). 2018; pp. 1–138.

[26] Thengius S. Fault current injection from power electronic interfaced devices [Master's thesis]. KTH Royal Institute of Technology, School of Electrical Engineering and Computer Science; 2020.

[27] Weise B. Impact of K-factor and active current reduction during fault-ride-through of generating units connected via voltage-sourced converters on power system stability. *IET Renewable Power Generation*. 2015;9(1):25–36. Available from: https://ietresearch.onlinelibrary.wiley.com/doi/abs/10.1049/iet-rpg.2014.0116.

[28] Ndreko M, Popov M, Boemer JC, *et al.* Sensitivity analysis on short-circuit current contribution from VSC-HVDC systems connecting far and large offshore wind power plants. In: *IEEE PES Innovative Smart Grid Technologies*, Europe; 2014. p. 1–6.

[29] European Commission. Commission Regulation (EU) 2017/2196 – Establishing a network code on electricity emergency and restoration [Commission Regulation]. Brussels, Belgium; 24 November 2017.

Behaviour at grid connection point   145

[30]   AEMC Reliability Panel. Frequency Operating Standard [Standard]. Level 15, 60 Castlereagh St, Sydney, NSW, 2000: AEMC Reliability Panel; 2020.

[31]   AS/NZS 4777 2:2020. Grid connection of energy systems via inverters, Part 2: Inverter requirements [Standard]. Standards Australia Office, 20 Bridge Street, Sydney: EL-042 (Renewable Energy Power Supply Systems & Equipment); 2020.

[32]   Suruhanjaya Tenaga Energy Commission. Electricity Supply Act (Act 447) – Grid Code for Peninsular Malaysia (Amendments) 2020 – KOD/ST/No. 2/2010 (Pindaan 2020) [Grid Code]. Putrajaya: Suruhanjaya Tenaga Energy Commission; 2020.

[33]   nersa. Grid Connection Code for Renewable Power Plants (RPPs) connected to the electricity transmission system (TS) or the distribution system (DS) in South Africa [Grid code]. Pretoria: National Energy Regulator of South Africa (nersa); 2022. Available from: https://www.sseg.org.za/wp-content/uploads/2014/07/SAGC-Requirements-for-Renewable-Power-Plants-Rev-3.1.pdf.

[34]   VDE. Technical Connection Rules for High-Voltage (VDE-AR-N 4120) [Technical Connection Rule]. VDE Verlag GmbH, Berlin: Forum Netztechnik/Netzbetrieb im VDE (FNN); 2018.

[35]   EirGrid. EirGrid Grid Code, Version 12 [Grid Code]. Dublin: EirGrid Plc.; 2023. Available from: https://www.eirgridgroup.com/site-files/library/EirGrid/GridCode.pdf.

[36]   National Grid. The Grid Code, Issue 6, Revision 16 [Grid Code]. Warwick, CV34 6DA: National Grid Electricity System Operator Limited; 2023. Available from: https://www.nationalgrideso.com/document/162271/download.

[37]   NERC. Standard PRC-024-2 – Generator Frequency and Voltage Protective Relay Settings [Standard]. Atlanta, Washington: North American Electric Reliability Corporation (NERC); 2023. Available from: https://www.nerc.com/pa/Stand/ReliabilityStandards/PRC-024-2.pdf.

[38]   Al Kez D, Foley AM, Ahmed F, *et al.* Overview of frequency control techniques in power systems with high inverter-based resources: Challenges and mitigation measures. *IET Smart Grid.* 2023;6(5):447–469. Available from: https://ietresearch.onlinelibrary.wiley.com/doi/abs/10.1049/stg2.12117.

[39]   DS3 Programme [Homepage on the Internet]. Dublin: EirGrid Plc.; 2023 [cited 2023 Aug 25]. What is the DS3 Programme?. Available from: https://www.eirgridgroup.com/how-the-grid-works/ds3-programme/.

[40]   Al kez D, Foley AM, McIlwaine N, *et al.* A critical evaluation of grid stability and codes, energy storage and smart loads in power systems with wind generation. *Energy.* 2020;205:117671. Available from: https://www.sciencedirect.com/science/article/pii/S0360544220307787.

[41]   GHD. Advice for the 2022 Frequency Operating Standard review – System Rate of Change of Frequency | A GHD survey of international views [Report]. Australia: GHD Pty Ltd; 18 November 2022.

146   *Intended and unintended islanding of distribution grids*

[42] AEMO. Market Ancillary Service Specification Consultation [Issues paper]. Melbourne VIC 3000: Australian Energy Market Operator Ltd (AEMO); May 2022.

[43] CEN. Estudio de control de frecuencia y determinacón de reservas – Parte 1 informe final [Study]. Las Condes, Santiago: Coordinador Eléctrico Nacional (CEN); June 2020.

[44] Hydro Québec. Decision D-2022-088: Technical Requirements for the Connection of Generating Stations to the Hydro-Québec Transmission System (ETRC) [Technical Requirement]. Montréal (Québec) H2Z 1A4, Canada: Hydro Québec; July 2022.

[45] New Dynamic Services (DC/DM/DR) [Homepage on the Internet]. Warwick, CV34 6DA: National Grid Electricity System Operator Limited; 2023 [cited 2023 Dec 04]. Find out more about Dynamic FFR. Available from: https://www.nationalgrideso.com/industry-information/balancing-serv ices/frequency-response-services/new-dynamic-services-dcdmdr.

[46] Reserves and balancing power [Homepage on the Internet]. Finland: Fingrid; 2023 [cited 2023 Dec 04]. Fast Frequency Reserve (FFR). Available from: https://www.fingrid.fi/en/electricity-market/reserves_and_balancing/f ast-frequency-reserve/#procurement.

[47] Rezkalla MMN, Pertl MG, and Marinelli M. Electric power system inertia: requirements, challenges and solutions. *Electrical Engineering.* 2018;100(4):2677–2693.

[48] ENERGIENET. Technical regulation 3.2.2 for PV power plants above 11 kW [Technical regulation]. Denmark: ENERGIENET; 2016. Available from: https://en.energinet.dk/media/evsijtqt/technical-regulation-3_2_2-for-pv-power-plants-above-11-kw.pdf.

[49] ACER. Implementation Monitoring Report of the Network Code on Requirements for Grid Connection of Generators – Third edition [Report]. Slovenia: ACER – European Union Agency for the Cooperation of Energy Regulators; 2020.

[50] MacLeod N, Cowton N, and Egan J. System restoration using the "black" start capability of the 500 MW EIRGRID East-West VSC-HVDC interconnector. In: *IET International Conference on Resilience of Transmission and Distribution Networks (RTDN)* 2015; 2015. p. 1–5.

[51] Sørensen TB, Kwon JB, Jørgensen JM, *et al.* A live black start test of a HVAC network using soft start capability of a voltage source HVDC converter. In: *CIGRE Aalborg Symposium*, June 2019. CIGRE; 2019.

[52] Zhao Y, Zhang T, Sun L, *et al.* Energy storage for black start services: a review. *International Journal of Minerals, Metallurgy and Materials.* 2022 04;29:691–704.

[53] AEMO. Electricity Rule Change Proposal – Future System Restart Capability [Proposal]. Melbourne VIC 3000: Australian Energy Market Operator Ltd (AEMO); 2019.

[54] Pagnani D, Kocewiak u, Hjerrild J, *et al.* Integrating black start capabilities into offshore wind farms by grid-forming batteries; 2023. Available from: https://ietresearch.onlinelibrary.wiley.com/doi/abs/10.1049/rpg2.12667.

[55] Mahieux C and Oudalov A. Microgrids – The mainstreaming of microgrids using ABB technologies. ABB review 4|14. 2014; pp. 54–60. Available from: https://library.e.abb.com/public/b5d0b09ecd79799c83257e03004d91cf/ABBReview4-2014_72dpi.pdf.

[56] BC hydro. Distribution Power Generator Islanding Guidelines [Report]. British Columbia, Canada: BC Hydro; June 2006.

[57] IEC 61000-4-30. Electromagnetic compatibility (EMC) – Part 4-30: Testing and measurement techniques – Power quality measurement methods [Standard]. IEC; 2015.

[58] Bollen MHJ. *Understanding Power Quality Problems: Voltage Sags and Interruptions*. Wiley-IEEE Press; 2000.

[59] EN 50160. Voltage characteristics of electricity supplied by public electricity networks [European Standard]; 2022.

[60] IEC 61000-2-2. Electromagnetic compatibility (EMC) – Environment – Compatibility levels for low-frequency conducted disturbances and signalling in public low-voltage power supply systems [Standard]. IEC; 2002.

[61] IEC 61000-2-12. Electromagnetic compatibility (EMC) – Part 2-12: Environment – Compatibility levels for low-frequency conducted disturbances and signalling in public medium-voltage power supply systems [Standard]. IEC; 2003.

[62] IEEE. IEEE Standard for Harmonic Control in Electric Power Systems. IEEE Std 519-2022 (Revision of IEEE Std 519-2014). 2022;p. 1–31.

Chapter 5

Power system restoration

Holger Becker[1] and Christian Hachmann[1]

5.1 Motivation for power system restoration and types of outages

5.1.1 General

Modern interconnected electric power systems rarely fail on a huge scale and the reliability of power supply is on a very high level. However, errors, failures, extreme natural events, and sophisticated deliberate attacks (physical and cyber) can occur, and the technical design and implementation for an uninterrupted operation covering all eventualities are not economical and almost not feasible. Therefore, when necessary, restoration must take place as quickly as possible to limit the impact of electric power outages on society and economy in case of supply interruption. Since many partners with different responsibilities are involved, a concerted and coordinated action is necessary. To this end, it is imperative to develop the procedures in advance and to coordinate them with the relevant parties. Restoration strategies and measures are necessary to reduce downtime.

5.1.2 Historical blackouts

Outages can differ in geographical extent, the affected grid levels, the availability of resources, and generation facilities, as well as the number of affected partners among other categories. A corresponding categorisation enables a speedy and effective action. Table 5.1 gives an example of the classification of outages [1], names very briefly the restoration procedure, and shows some historical examples.

The challenge tends to increase with the size of the affected area. However, the causes and effects are so diverse that a strict evaluation is difficult and will not be made here. Major disruptions can vary so much in terms of severity and impact that a general assessment is difficult, as the following two examples show. In the snow chaos in Münsterland (Germany and Netherlands, relative small area) in 2005, damage to overhead power lines caused by a combination of weather-related ice formation and the use of old poles made of a brittle type of steel led to an interruption

---

[1]Fraunhofer IEE; University of Kassel, Germany

Table 5.1 *Classification of outages [1] and historical examples*

Type	Characteristics	Restoration	Examples
Regional outage	• Outage regionally limited • Transmission system essentially not affected • No relevant imbalance of active power in the transmission system	• Reconnection	• Münsterland snow chaos 2005 [2] • Berlin 2019 • London 2022 • Paris 2022
Subnetwork formation	• Transmission system separated into several subnetworks (system split) Imbalance of active power in the interconnected network Stabilised operation after load shedding or generation reduction. One or more islands can collapse	• Stabilise subnetworks, resynchronisation	• ENTSO-E 2006 (three subnetworks) [3] • ENTSO-E 2021 (two subnetworks) [4]
Blackout with power provision from neighbour TSO	• Supraregional outage Intact neighbour network exists Restoration from neighbour network possible 'Top-Down' according to [5]	• Voltage provision (agree max. P and Q), start of generation units, reconnection of loads	• Italy 2003 [6,7] • Sweden 2003 [7,8] • Greece 2004 [9] • India 2012 [10] • Turkey 2015 [11]
Blackout without power provision from neighbour TSO	• Supraregional outage No intact neighbour network exists Restoration from inside 'Bottom-Up' according to [5]	• Open network islands with blackstart units, start further generation units, expand and synchronise islands	• USA/Canada 1965 [6,12] • Sweden 2003 [7,8]

Power system restoration 151

in the power supply to around 250,000 end customers in the winter season. Due to the massive unavailability of infrastructure, mobile emergency power supplies were set up to provide rapid assistance, but only a fraction of those affected benefited from it. Even after 4 days, not all customers were connected to the grid again. In contrast, the 2006 system split affected the entire European interconnected grid and thus all continental European transmission system operators (TSOs). To stabilise the underfrequency areas, more than 15 million people were temporarily cut off from the power supply. After less than 2 hours, the interconnected grid could be synchronised again and the customers who were cut off could be resupplied.

5.1.3 Restoration strategies

With regard to the restoration strategy in the event of supraregional outages, depending on the circumstances, mixed forms of top-down and bottom-up can also be used to speed up restoration. For example, during the blackout in Sweden, the failed grid in southern Sweden and Denmark was energised from the intact grid in northern Scandinavia. Denmark should have been reenergised using two blackstart-capable power plants. But this failed, so at the end, Denmark was reenergised from Sweden. One unit of the Ringhals nuclear power plant was able to switch to houseload-island operation so that the grid island could be expanded from there to accelerate restoration. It is always easier and more reliable to extend a possibly large grid island and to switch voltage into the failed grid.

Islands may form in a planned manner or unintentionally (see also Chapter 1) and can offer useful resources to restoration but also pose additional challenges, especially when small and uncontrolled islands form within the distribution system.

The smallest possible intact grid island is a power plant that managed to switch to houseload-island operation, as in the 1965 USA/Canada outage. It is not guaranteed that a stable grid island or a power plant in houseload-island operation will be available in the event of an outage, so blackstart-capable power plants must be kept in reserve to be prepared for reliable restoration under all circumstances.

5.1.4 Techno-economic trade-off

Procuring and holding resources for network restoration like blackstart capabilities, training of operators and redundancy measures mean expenditures that do not generate any income. So there is a trade-off between reducing the cost of preparation for restoration and the ability for fast and reliable restoration. While there can be guidance from an engineering perspective, it is a societal issue, and fundamental decisions are ultimately to be made by politicians and regulators. Furthermore, the likelihood of blackout and restoration scenarios cannot be calculated based on power system data alone but hinges on probability estimates from numerous other fields, such as geology (for earthquakes), meteorology and climatology (for extreme weather), social and political science (for deliberate attacks), computer since (for Information and Communication Technology (ICT) vulnerability) and so on.

Methods for determining the costs for different outages provide assistance in risk assessment. For example, Sullivan *et al.* [13] provide methods for outages that last

152 *Intended and unintended islanding of distribution grids*

around 1 hour. Calculation methods for outages that last several hours are given in [14]. Mills and Jones [15] provide analysis for several outages and give assumptions up to 16 hours.

5.1.5 Emergency backup supply

Assumptions about the maximal duration of restoration also guide the requirements for backup power supply of critical infrastructures. Again, the severity of the most extreme events that must be reckoned with is subject to political judgement. Depending on the criticality of the respective facilities, recommendations range from 'provide a commercially reasonable chance of maintaining power for at least 3 days under all-hazards' to 'power should be sustained with no unplanned downtime under all hazards in excess of 30 days' [16].

5.2 Restoration strategies

Power system restoration procedures are highly dependent on the specifics of the individual outage. Therefore, power system operators typically make restoration plans for several scenarios to cover a broad range of outage events. Necessary resources are procured in advance and training exercises – often in collaboration with multiple grid operators as well as power plant operators – are conducted regularly to ensure actionability. There are two fundamentally different strategies for network restoration in the event of supraregional failures, which are selected depending on the boundary conditions and which differ in their basic procedure.

5.2.1 Top-down

If an outage encompasses only a fraction of an interconnected power system, restoration entails the stepwise reconnection of the unsupplied area from the operational parts of the system. For this purpose, operators often pre-determine compensated network sections that can be reconnected in one step, so-called 'start-up grids'. Neighbouring TSOs must agree on the permissible exchange of active and reactive power as well as the permissible switchable load. Usually, power plants are started up first to enable generation capacity.

If there are islanded sections within the unsupplied area, these can be resynchronised or – if this is not feasible due to lack of operational preparations like a synchronisation concept or missing synchronisation devices (see also Chapter 3.4) – must first be de-energised.

In general, top-down restoration is faster and less challenging than bottom-up restoration, which is part of the following section.

5.2.2 Bottom-up

If there is no neighbouring network being powerful enough to provide voltage and power to energise parts of the grid and to start further generating units, restoration has to begin from scratch [17]. The starting point can either be a power station that could manage to switch to household-island operation and that is able to energise parts of the grid, or a blackstart-capable unit (BSU). These units are able to start operation without energy taken from the public grid, to energise parts of the public grid, usually a predetermined start-up grid and to supply first loads. Ramping up the voltage instead of reconnection avoids inrush currents caused by inductive devices. It is common to determine several BSUs within a control area to ensure redundancy and to be able to start different island networks in parallel, to speed up restoration. As BSUs often have limited energy storage, the aim is to start up further generating units as soon as possible and to obtain generating power and ancillary services. To aggregate ancillary services, it is advantageous to synchronise different islands before starting bulk resupply. The number and cuts of start-up grids depends on several factors as availability of BSU, topology, organisational resources, involved partners, and so forth. The classic way is the build-down strategy [18], whereas blackstart is conducted at the transmission level and voltage is forwarded to the distribution level. As generation shifts from transmission to distribution level, a build-together strategy gains an advantage: When blackstart activities in the distribution level take place · in parallel with those in the transmission, load supply can start earlier. In this case, too, a fast synchronisation of transmission and distribution grids should be aimed for (build together) [18].

5.3 Ancillary services and secondary technology during power system restoration

5.3.1 Blackstart

A generating unit has blackstart capability if it can start operation without drawing power from the public grid, regulate a voltage with a specific amplitude and frequency at the generator or plant terminals, and energise a de-energised busbar or a sub-grid. As a prerequisite for the start of operation of a BSU, the energy required for this must be stored in sufficient quantity or otherwise available. Typically, plants are chosen that do not need much power for their internal processes and are equipped with a small generator or battery system. A common example are hydroelectric plants which require neither heating nor pumping but only control of some valves to start generating electricity and supplying a stable voltage. They are also often able to resume operation within a few minutes. Other candidates for a BSUs are gas turbine plants which require only moderate cranking power and starting times. Huge steam turbines, on the other hand, are less suited as BSUs since they usually need huge amounts of time and energy to start operating.

To be able to regulate the voltage with amplitude and frequency, the grid coupling of the unit must have grid-forming (meaning voltage-regulating) properties as a prerequisite. This is the case with directly coupled synchronous machines, where the magnetic field of the rotor induces a voltage in the stator windings, which is why the generators have a voltage source behaviour. Inverter-coupled systems are divided into grid-forming and grid-following or grid-supporting systems, with regard to the design of the control (see also Chapter 2.1). In the former, the voltage at the converter terminals is controlled with the hardware-related, inner control loop, which fulfils an essential requirement of a BSU. In current decentralised generation units, the grid coupling is usually carried out by grid-following or grid-supporting inverters with hardware-based current control. These inverters are therefore not able to energise a de-energised grid. They require a live grid to which they can connect, to start feed-in operation.

Due to the voltage source behaviour of a BSU, the fed-in active and reactive power are essentially determined by the load on the grid side. To avoid overloading a BSU, the grid operator must pay attention to the operating point in the generator power diagram. The steady-state frequency of a BSU is determined by the system control, both with directly coupled synchronous machines and with converter coupling. Common control types are 'fixed frequency' (isochronous speed control) or power control with frequency static. The latter makes it possible to divide the active power following load changes when operating in parallel with other generating units.

The possible active power feed-in depends on the available primary energy. Today, pumped storage power plants with a limited amount of energy are often used as BSU. For these, timely energetic replacement by another power plant or a generating unit must be planned for the restoration process. If distributed generators with volatile energy sources are used, the performance of the plant depends on the current weather situation. To determine availabilities, historical weather data can be statistically evaluated in the planning. For operational use and operational planning, corresponding forecasts are needed to determine the available feed-in power.

5.3.2 Voltage control

The contribution of a generating unit to voltage control is directly related to the behaviour of the reactive power fed in. Both the **reactive power capability** of a system and its **reactive power mode** are relevant and must be considered in a differentiated manner. The capacity describes the possible reactive power setting range, which is usually dependent on the feed-in active power and the terminal voltage. The reactive power mode determines the feed-in of reactive power in relation to several technical values.

The **reactive power capability** of a generating unit is a design parameter and is specified in the generator power diagram. The current grid codes (GCs) define the range as the minimum requirement at the point of common coupling (PCC). In principle, inverter-coupled systems can also feed in reactive power without active power (so-called Static Synchronous Compensator (STATCOM) function). This requires technical measures on the plant side to provide auxiliary power from the public grid even during plant standstill and therefore leads to increased active power consumption. For this reason, this function is not usually installed as standard in plants and is available as an option from many manufacturers. Figure 5.1 shows common forms of reactive power requirements.

The **reactive power mode** describes the reactive power feed-in of a system depending on other variables such as voltage, active power and, if applicable, specified setpoints. The basic function of a plant is determined by the type of grid coupling. Current-controlled units, which include almost all current distributed generators, actively feed in a reactive current or reactive power. Voltage-controlled units, such as synchronous generators or grid-forming inverters, control a voltage at their terminals, which, in combination with the load in the grid, results in the reactive power fed in. By means of a superimposed control, voltage control can also be

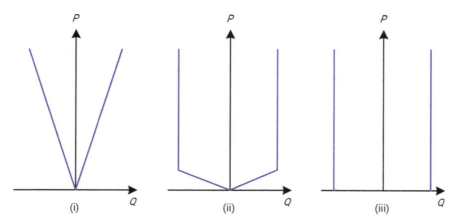

Figure 5.1 Reactive power requirements: (i) power factor, (ii) reactive power, and (iii) reactive power with STATCOM-functionality

implemented with current-controlled units and reactive power control with voltage-controlled units. The additional control loop reduces the dynamics, which is why such solutions are not used as standard applications.

If necessary, setpoints are received during operation directly via the network operators' control system or via third-party communication technology. The connection network operator specifies the desired reactive power mode. Depending on the operating mode required by the network operator, characteristic curves or fixed set points are implemented in the plant controller or in the individual unit controllers and selectable via parameters.

At the beginning of the network restoration, there is a need to compensate for long, unloaded or weakly loaded lines and cables. This can be covered by inductors or by selective reactive power feed-in of inverter-coupled systems. Likewise, a voltage increase caused by the Ferranti effect can be reduced in this way. Inductors have the advantage that they can be connected to a start-up grid in a de-energised state and thus already compensate the sub-network during voltage ramp-up. Inverter-coupled systems can only connect to the grid at the minimum voltage required by the system (typically 90% of the nominal voltage) and then contribute to compensation.

It is advantageous to use systems with constant reactive power capability (Figure 5.1(ii)) or, even better, with STATCOM-functionality (Figure 5.1(iii)), as the reactive power capability does not depend on the feed-in active power and, ideally, is even possible without active power feed-in. In this way, wind and photovoltaics (PV) systems, for example, can contribute to voltage support regardless of the weather.

Actual typical reactive power modes are as follows:

1. Reactive power voltage characteristic (Q/U curve). This control is used for voltage support at the PCC and is usually designed either as a limiting control or as a continuous control. The latter is defined by the slope and the point of intersection with the voltage axis (x-axis). As a rule, the slope is set as a parameter in the controller, and the point of intersection with the x-axis is usually specified as a voltage setpoint which can either be fixed or can be provided by

the network operator via remote control, which is often the case for larger units. This reactive power control mode must not be confused with 'real voltage control', where there is no fixed relationship between voltage and reactive power value, and where the voltage at the PCC is directly controlled within the reactive power capability of the individual units and depending on the controller with a permanent control error (when using a P controller) or without (when using a proportional-integral (PI) controller).

2. Reactive power in Mvar. Constant value for the reactive power to be fed in, either adjusted as a fixed value in the system controller, or provided as a changable value by the network operator via remote control.
3. Reactive power characteristic curve as a function of active power: Q/P curve. With this method, a characteristic curve is stored in the system controller in which the reactive power to be fed in is specified as a function of the active power.
4. Power factor cos φ. Constant value for the power factor, either adjusted as a fixed value in the system controller, or provided as a changable value by the network operator via remote control.
Power factor characteristic curve as a function of active power: $\cos(\varphi)/P$ curve. A characteristic curve is stored in the system controller in which the reactive power to be fed in is specified as a function $\cos\varphi\ (P)$.

Figure 5.2 shows graphics that represent the different reactive power modes. Whereas reactive power voltage characteristics are helpful for keeping voltages

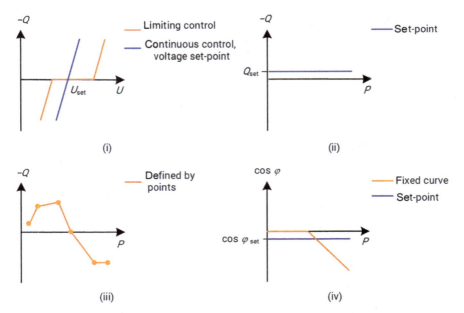

Figure 5.2 Exemplary reactive power modes: Q/U curve (i), fixed Q (ii), Q(P) curve (iii), and cosφ (iv)

within the allowable range, they inadvertently also change the operating points of other generators in the grid. Especially in the early stages of restoration, close coordination between plant and network operators is necessary to avoid instability or generator tripping due to the violation of over or under excitation limits.

5.3.3 Frequency control

Frequency control is a particular challenge during the early stage of network restoration, since in contrast to normal operation, there are relatively few active power reserves available, and the reconnection of formerly unsupplied grid sections goes along with relatively huge active power changes. Active power demand upon reconnection after a long and unplanned outage is also hard to predict since there is very little operational experience and insufficient data on load behaviour.

The availability of active power reserves within a network island determines the connectable load and thus has a significant influence on the speed of restoration. As long as conventional power plants dominate the generation fleet, the rotational energy of directly coupled generators is converted into electrical energy to supply the load immediately after reconnection. This leads to a speed reduction and thus also to a reduction of the frequency until power/frequency control takes over to stabilise frequency on a steady-state value.

It is the responsibility of the system operator to ensure that frequency does not exceed operational limits. The main task is to prevent the frequency from exceeding limits that lead to the shedding of load or generation power. While load shedding is organised centrally and thresholds are usually equal within an interconnected system, frequency-tripping values of distributed generators are determined with the valid grid code on the day of commissioning. It is crucial for system operation to know how much generation power is tripped at which frequency value. In Germany, the majority of distributed generators do not trip at a frequency between 47.5 and 51.5 Hz, but the first stage of load shedding happens if the frequency falls below 49.0 Hz. As a prerequisite for resupply of load, sufficient active power reserves must be available for its supply, and the frequency must not fall below the limit value for load shedding, even dynamically. Figure 5.3 shows in principle the activation of active power reserves as well as the frequency curve after a load recovery.

The additional power demand of the reconnected load is fed by the momentary reserve of the direct-coupled synchronous generators, leading to a decrease in speed. Speed control of a power plant does react on this and increases primary active power according to the power curve by opening the admission valve and adjusting the combustion process if necessary. Frequency deviation remains according to the static curve. By replacing the primary power with the secondary power reserve, the frequency is brought back to the target value. As long as only a small part of the transmission grid within a control area is energised or different island networks exist, it is impractical to use the central secondary controller, as with this a one-to-one connection to individual power plants is often not possible (see 5.3.5). Therefore, power plant commitment in the early restoration phase is often carried out by direct instruction from the grid operator control centre.

158   *Intended and unintended islanding of distribution grids*

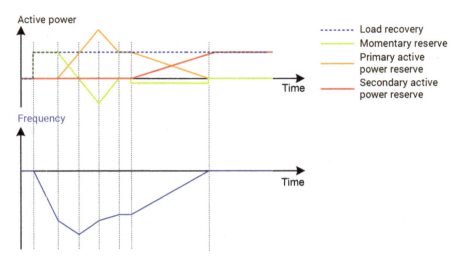

Figure 5.3   Active power reserve activation and frequency following a load recovery

The minimum frequency value following a load connection (nadir) is determined by the load size, the availability of momentary reserve and the dynamic of the primary power control. Experience shows that 5% of the nominal active power of synchronised generators can be reconnected as loads without taking the risk of a breakdown of the island [19]. This consideration is based on the assumption that the rotational parts of the power plants have a start-up time constant of $T_s = 10\,\text{s}$ (see (5.1)) and participate in the primary power-frequency control with a static (s) of 5% (see [19] and (5.2)). A safety margin of 100% to the load-shedding frequency of 49.0 Hz is taken into account.

$$T_s = 2H = \frac{J\omega_n^2}{S_n} \tag{5.1}$$

T_s: start-up time in s
H: inertia constant in s
J: mechanical inertia in kg m^2
ω_n: nominal circular frequency in s$^{(-1)}$
S_n: nominal apparent power of generator in MVA

$$s = -\frac{\Delta f}{\Delta P}\frac{P_n}{f_0}100\% \tag{5.2}$$

s: static in %
P_n: nominal active power of generator in MW
f_0: nominal frequency in Hz
f_{set}: frequency setpoint in Hz

In the early stage of system restoration, the secondary control is usually deactivated and unit commitment is done on direct instruction of the network operator control room. Figure 5.4 shows the power plant's behaviour following a load recovery. First, the power plant is loaded with the same value given as a setpoint (I, P_{set1}). The frequency is controlled to the setpoint. With a load recovery, the active power generated by the plant is increased and frequency follows the curve to operation point II (value below frequency setpoint f_{set}). To bring the frequency back to the setpoint, the active power setpoint has to be changed (P_{set2}). With this, the curve is moved in parallel, and therefore, the operation point moves to III, with which frequency is brought back to the setpoint.

As classical inverter-based distributed generators in power-optimised operation do not provide active power reserves, they do not contribute to frequency support after load connection and thus do not mean an advantage in speeding up load restoration (the stator of a wind turbine with a doubly fed induction generator is directly connected to the grid, which is why it does provide momentary reserve). When a distributed generator is operated at reduced power, active power reserves are available, and it can support the frequency after load recovery. The dynamic with which active power is activated depends on the generator's design and might be adjusted by parameters in the control system. In general, wind- and PV-generating units can adjust active power much quicker than conventional power plants with a relatively slow primary process can do, which is why they can support the frequency more strongly than conventional power plants with the same output [20]. When activation happens too fast, the island can lose stability [21], which is why a detailed stability assessment considering the dynamic behaviour of all generators needs to be conducted (see also Chapter 2).

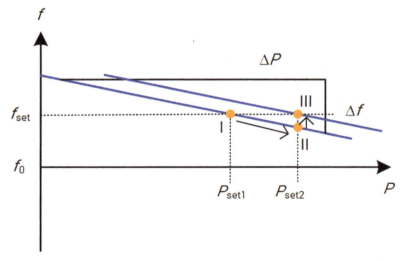

Figure 5.4 *Power/frequency static curve of a power station*

160   *Intended and unintended islanding of distribution grids*

5.3.4   Protection

In normal operation, power system protection is required to selectively and sensitively disconnect parts of the grid that contain short circuits. Furthermore, it is required that there are backup systems that work in case of a failure of a single protection device and still disconnect the faulty part along with only a reasonably small section of the grid.

Many common protection systems rely either exclusively (in the case of overcurrent protection) or partially on the detection of abnormally high current. Since the available short-circuit current is limited in the early phases of power system restoration, these systems typically do not reliably work in such situations.

Typically, operators do not design additional protection systems or settings for each stage of power system restoration. In case of a short circuit occurring during early restoration or of a faulty line being reenergised, generator protection might trip and consequentially, the island system can collapse. In this way, generator protection serves as a backup system that prevents severe equipment damage or excessive hazard to personnel.

Therefore, during network restoration, compromises have to be made on the issue of protection, and functionality is limited compared to normal operation. This is often accepted since the effort of tailoring protection systems for every combination of grid configuration and fault that can occur during restoration would go along with disproportionate effort.

5.3.5   Information and communication technology

During power system restoration, the responsible network operator needs to communicate with several parties. Coordination with up- and downstream as well as neighbouring network operators and generators is of particular importance. While in normal operation, the power plant commitment is taken over by aggregators and dispatch centres that do manage a portfolio of several power plants, in the case of restoration it must be possible to instruct individual power plants directly from the grid operator control centre.

Communication channels range from phone calls to directly sending and receiving measurements and set point values. Typically, components connected to the networks of other operators are not accessed directly but in an aggregated manner via the respective operator.

5.3.5.1   Awareness

Information is needed especially about the following items:

- Available generation resources along with their flexibilities and possible restrictions.
- Connectable load and the granularity in which it can be connected.
- Status of grid sections and the active and reactive power exchange expected at reconnection.
- Status of neighbouring island systems and possible synchronisation conditions.
- Possible exchange capacity of active and reactive power.

5.3.5.2 Capacity to act

To successfully achieve power system restoration, the responsible operator needs the following means of control:

- Switching of lines, substations, transformers, and so forth.
- Control of active power feed in (setpoints and possibly mode of operation).
- Connection of load in reasonably defined increments (typically by connecting individual feeders).
- Control of reactive power contribution by generators as well as dedicated compensation equipment.

5.3.5.3 Scope of backup supply

Since telecommunication hinges on electric power supply, there are some specific challenges during power system restoration.

Some communication systems will only become available after the resupply of the respective grid sections.

In some cases, at least 'blackout-robust' communication provisions are necessary. This means that the respective systems only start significant interaction with the grid (e.g. feed-in or consume relevant amounts of active or reactive power) after communication with the operator has been reestablished. This applies to many small- to medium-sized distributed generators and some larger residential or commercial loads.

For the most critical components, an autonomous backup power supply to the communication system is necessary. This includes operator control centres and substations of the transmission network as well as blackstart units and other important generators. The capacity of the backup supply needs to be large enough to cover the duration of most conceivable interruptions. Typically, these components are directly connected to the operator's dedicated control system since public third-party infrastructure (such as mobile radio and internet systems) do not meet the necessary requirements for sustained reliable operation in blackout situations.

5.4 Different phases of power system restoration

Power system restoration happens over several roughly consecutive phases. Whereas each phase is defined by a certain status of the grid and distinct priorities of the grid operator, the order is not strictly fixed, and the stages may overlap. There are different classifications in the literature that follow a common structure and only differ in individual points. Figure 5.5 shows a structure based on [22–25]. A jointly coordinated structure of the process with different consecutive phases facilitates prioritisation depending on external influences and enables coordinated action between different partners.

In the following, the individual phases of the restoration are explained: Section 5.4.1 introduces the planning phase, followed by an explanation of the grid preparation phase in Section 5.4.2, while the blackstart is the subject of Section 5.4.3. The system and network restoration phase is part of Section 5.4.4, while Section

Figure 5.5   Different phases during power system restoration

5.4.5 deals with load restoration, and finally the resume to normal operation is part of Section 5.4.6.

5.4.1 Planning phase

The type and extension of the outage influence the restoration strategy (see Section 5.2) and the necessary action. Therefore, a picture of the situation that is as precise as possible must first be drawn up, on the basis of which further action can be planned and agreed with the other actors. Since each network operator can only observe a section of the overall system, consultation with neighbouring as well as subordinate and superordinate network operators plus the operators of power plants and generation facilities is crucial. Today, this is done directly between the control rooms of the grid and power plant operators via specially secured voice connections using dedicated communication lines or satellite telephones. Of particular importance are resilience and authenticity of information. Measured values can be manipulated, as happened, for example, in the attack on the Ukrainian electricity system in 2015 [26]. Also, voice communication can be manipulated by AI-based voice bots. For this reason, the checking and verification of information, for example through particularly high data security or redundant systems, is becoming increasingly important.

An important decision criterion is whether a stable grid area with sufficient power reserves is available to transfer voltage and power to neighbouring grid areas. Furthermore, the availability of controllable generation capacity as well as ancillary services is important. Non-controllable, fluctuating generation capacity requires additional power reserves during restoration, which must be kept available accordingly. Equipment outages are also relevant. It is extremely advantageous if grid operators who are able to energise or maintain at least one network island, do coordinate possible synchronisation nodes in advance. Grid operators should also exchange information in advance about the size and characteristics of the loads and generation capacity that can be connected.

Based on the current situation and taking into account the previously identified and agreed boundary conditions, each grid operator determines the strategy for restoration and the required procedure. Grid operators without any BSU in their own grid need to wait for voltage provision from a neighbouring grid operator. The determined procedure is to be continually compared with the current situation and adjusted accordingly if new information becomes available (e.g. new unavailability of operating resources, new forecasts).

5.4.2 Grid preparation phase

For restoration, the network must be segmented into suitable parts. For this purpose, connections to neighbouring, subordinate and superordinate networks are first opened to limit the effects of faulty actions. Connections to other network operators should only remain closed or be closed after a direct bidirectional agreement. Further segmentations are to be oriented to the planned procedure for restoration and are ideally already defined in advance and validated by means of simulation calculations and network restoration training.

To reduce reactive power steps caused by connecting lines in the early phase of network restoration, compensated network sections should be prepared. These can be energised at once without causing significant load changes. This is particularly important for connections between BSU and large power plants or generation facilities (see also Section 5.4.3). Lightly loaded grids have only low damping, which is why switching over voltages and resonant oscillations can be problematic and should be checked in advance by simulations to be able to provide remedial measures if necessary [27].

Depending on the technical characteristics and requirements of the power plants and generation facilities, a minimum load may be required at an early stage. For example, steam power plants must be loaded with their minimum thermal output or, for stability reasons, a minimum load must be ensured for individual power plants. For this purpose, network sections need to be identified in advance and prepared accordingly, with which the necessary loads can be connected in steps of non-critical size. With the progressive expansion of renewable generation plants in the distribution grids, pure load areas will decrease in number and capacity, which is why loads can only be connected to an increasing extent in combination with decentralised generation. The power character of a grid area (load/generation ratio) must be provided for in the plans and taken into account when carrying out network restoration.

5.4.3 Blackstart phase

If the bottom-up strategy was chosen, in the preparation phase the BSU was instructed to prepare for the start-up process and a start-up network that can connect the BSU to at least one power plant or generation unit was separated from the interconnected network. To avoid inrush currents when inductive equipment is connected, the network should be energised by ramping up voltage. If the BSU is able to do so, the grid is connected at the lowest possible terminal voltage of the generator or the inverter, and then the voltage is ramped up to the target value. This is typically done within a timeframe of approximately 1 minute. Suitable inductances are to be connected to the mains already in the de-energised state to ensure sufficient compensation already at voltages below the nominal value. During voltage ramp-up and when switching on loads or power plants, the permissible reactive power value as well as the current values of the BSU must be observed.

During voltage ramp-up, special attention must be paid to the protection. In the case of voltage protection devices, it must be ensured that the release current of the protection device is not exceeded during the time in which the voltage is

164   *Intended and unintended islanding of distribution grids*

below the lower tripping threshold. Otherwise, appropriate remedial measures must be taken. Also, due to the low short-circuit power, the overcurrent protection will not function reliably in the event of a mains fault. Therefore, special attention must be paid to the generator currents in the early phase and, if necessary, a mains fault must be manually released.

Many BSUs are operated in island mode with isochronous speed control as standard. This is also possible in parallel operation with other generating units and power plants. It must be ensured that not more than one power plant in isochronous speed control or fixed frequency control is connected to a network because a defined state cannot be realised then. When operating a BSU in fixed frequency control in a grid island synchronised with another power plant that is in power control with superimposed frequency control (static), the other power plant cannot automatically participate in load absorption. In this case, special attention must be paid to the operating point of the BSU in order not to violate minimum or maximum load requirements. Ideally, the BSU should be started immediately in power control with superimposed frequency control.

As soon as the target voltage value is reached, the BSU can provide start-up power for one or more power plants. The duration of the start-up process depends on the type and status of the plant and can last up to several hours. After synchronisation with the grid, a power plant needs to be loaded with minimum capacity by reducing the power of the BSU if possible or by connecting further loads. In the case of a storage plant, the BSU can switch to storing mode to establish further reserves for possible further blackstarts if needed. Whether this can be done steplessly depends on the local conditions. Switching operations and waiting times may be necessary to change from generator operation to pump operation. The active- and reactive power of the BSU needs to be permanently observed to be able to initiate remedial measures if necessary.

Converter stations of HVDC connections can also act as BSU if the control mode is grid forming and sufficient power is available elsewhere in the DC network.

5.4.4   System and network restoration phase

The aim of this phase is to secure the grid operators' ability to act and to stabilise the restored system. For this purpose, the blackstart islands are extended by successively energising lines and cables. In this way, the emergency power supply of substations and power plants is replaced. The distribution network must also be taken into account. Before each switching operation, the reactive power demand of the sub-grids to be connected and the voltage profile in the grid must be taken into account. Due to the Ferranti effect, voltage increases occur over long, idle lines and cables, which must not reach critical values. Loads are mainly connected for system stabilisation, for example, to ensure the minimum load for power plants or to set them to a favourable operating point or to influence voltage and reactive power consumption.

By connecting controllable generation units and reactive power sources, available reserves are increased and robustness is enhanced. With the synchronisation of grid islands, active power reserves are combined and robustness is also increased. For this purpose, it is helpful to plan synchronisation nodes for neighbouring grid islands in advance. The synchronisation of grid islands in the transmission grid increases the transport capacity. With the increase in weather-dependent generation capacity, depending on the time of the outage during reconstruction, there can be different requirements for supraregional transport capacity. To counteract this variability in planning, it is advantageous to rapidly restore a backbone network that extends over wide areas. As the size of the restored network increases, properties such as redundancy (e.g. $(n-1)$ security), underfrequency load shedding and protective selectivity become more important and increasingly available.

5.4.5 Load restoration phase

As soon as the bulk of the transmission system is reenergised along with sufficient generating capacity and instantaneous reserve to simultaneously connect multiple distribution feeders, the operator's goal shifts towards resupplying the remaining customers and generators. For this purpose, permissible ranges of power exchange and allowable switchable increments are assigned to the underlying distribution system operators (DSOs). Frequency is continuously monitored and adjustments are made if necessary.

As soon as sufficient instantaneous reserve and generation capacity are available so that several load areas can be connected simultaneously without taking the risk that the frequency can reach critical values, load absorption can begin on a larger scale. The frequency curve must be permanently monitored and, if necessary, the load must be taken up in smaller parts or more slowly. The frequency can be brought back to the target value by adjusting the target power given to the power plants. A grid restoration controller can also be used here. This functions similarly to a secondary controller, except that it directly controls the power plants synchronised with the grid island and the selection of the power plants can be adjusted as the restoration progresses.

As the reconnection of loads progresses, the individual stages of the underfrequency load shedding are also reactivated. It is advantageous, if possible, to reactivate the individual stages in approximately the same order of magnitude.

5.4.6 Resume normal operation

After all costumers are – at least provisionally – resupplied, the transition to the normal stage of grid operation must be coordinated. In power systems that achieve the balance of generation and consumption via market mechanisms, the responsible parties need to check whether auction results and other power delivery contracts that have been settled before the blackout are still feasible. Market activity is resumed and power plant dispatch is successively transferred from the grid operators to the market players. If the disturbance went along with prolonged unavailability of grid components or generators, interim solutions need to be found.

166 *Intended and unintended islanding of distribution grids*

5.5   Outlook: external impact due to renewable energy transition

5.5.1   Decentralisation

Due to the ongoing decarbonisation – in many places accompanied by decentralisation – of the power system, a higher degree of coordination among the responsible TSOs and their underlying DSOs becomes necessary. Information about and control of small- and medium-sized generators located in the distribution system needs to be aggregated to facilitate effective power system restoration by the TSOs. Manually transmitting set points to individual generators will become less feasible.

As generating units are more and more installed within parts of the grid that have been pure load regions before, it will be less possible just to connect the pure load and system restoration phase and the load restoration phase will successively melt together.

5.5.2   Weather dependency

The shift towards renewable generators – often depending on the weather – also increases the number of different generation scenarios. Furthermore, the electrification of heating and transportation goes along with changes in the seasonal and weather-dependent load profiles. Therefore, the availability of forecasts for individual generating units or grid areas in the responsible operator's control centres is of increasing importance.

5.5.3   Changing load behaviour

There is a trend towards connecting more and more appliances to the internet and making load behaviour dependent on various ICT systems, ranging from dynamic tariffs to weather forecasts. Since these systems can be unavailable for some time even after the loads have been reconnected, their behaviour can significantly deviate from normal operation, introducing additional uncertainty.

Furthermore, heating systems as well as vehicle charging systems also exhibit high simultaneity in the aftermath of prolonged power interruptions due to cold load pickup. Since the penetration of these loads is increasing, this must be taken into account in the restoration plans and possibly also in demand connection codes.

5.5.4   Start-up times

Renewable generators such as wind turbines and PV plants typically only need a few minutes to restart and reach their full power output after reconnection to a grid and even these delays are mostly the result of network codes prescribing monitoring of connection conditions and limited ramp-up rates. This is much faster than the start-up procedures of most conventional power plants – which is several hours for typical thermal steam units. This can potentially speed up restoration considerably.

However, care must be taken to ensure that the plant's auxiliary power is also resupplied. In the case of high-voltage connected plants, this can also be provided from a local low-voltage grid.

Power system restoration 167

5.5.5 Inverter-based generation and load

Due to the increased penetration of inverter-coupled generators, new stability aspects need to be considered. Grid-forming control of inverters can potentially improve stability and allow contributions to system inertia by generators as well as loads [28]. However, there is a considerable delay between the understanding of possible favourable contributions and their actual implementation in the field due to the regulatory process, the product pipelines of the manufacturers and the life cycle of the components. Therefore, the requirements of future power system restoration need to be considered in the design of GCs decades before the full potential can be realised.

Furthermore, inverter-coupled generation or storage systems can increasingly also be used as BSUs. Requirements for the capabilities of BSUs that are currently expressed in terms referring to the rotating mass, inertia or short-circuit power of a synchronous machine will in some cases need adjustment for application to inverter-coupled generation. First units have been in operation for several years [29,30].

5.5.6 High-voltage direct-current systems

There is a trend towards an increasing number of HVDC systems as interconnectors between synchronous areas as well as embedded links within a synchronous area.

Conventional thyristor-based HVDC systems with DC current links need a considerable amount of short-circuit power for their operation and, therefore, can only be activated in advanced stages of power system restoration. This is also the case with current-controlled VSC HVDC systems, but less short-circuit power is required here.

On the other hand, a new HVDC systems are increasingly designed in a way that allow them to operate in a grid-forming manner, provide short-circuit current and inertia themselves, and can even be used as a blackstart unit if there is a sufficiently large and stable power system at their opposing end.

5.5.7 Decreasing distinction between system and load restoration

With more and more generation as well as ancillary services located at the lower voltage levels, it might no longer be feasible to energise a network skeleton in the transmission system before large-scale resupply of distribution-connected customers. Therefore, system restoration and load restoration – while already not strictly distinct – might overlap even more in the future.

5.5.8 Potential for distribution system islands

The trend towards decentralisation creates opportunities for island operation within the distribution system. Combined heat and power (CHP) plants are typically connected to the distribution system and can be combined with local renewables. Furthermore, some of these units, such as waste-to-energy plants, are already equipped with backup supplies for their auxiliary system for safety reasons, lowering the requirement for additional investment in blackstart capabilities [31].

168 *Intended and unintended islanding of distribution grids*

5.5.9 *Artificial intelligence in power system operation*

It is increasingly proposed to augment the operation of distribution systems via the use of machine learning technology [32]. This can allow for an increased degree of automation and thus facilitate faster and more distributed restoration. However, the reliance on training data can make these systems less reliable in rare, extraordinary situations such as power system restoration. Furthermore, increased automation of normal daily grid operations might go along with a decreased number of operators and personnel being less accustomed to manually performing grid operations. This might make it more difficult to manually take the necessary action in the aftermath of a disturbance when normal operation paradigms do not apply.

References

[1] 50Hertz Transmission GmbH, Amprion GmbH, Tennet TSO GmbH, TransnetBW GmbH, Grundlagen der Netzwiederaufbaukonzepte der deutschen ÜNB (Basics of restoration concepts of German TSO), Projektbericht NETZ:KRAFT, April 2015, https://www.iee.fraunhofer.de/content/dam/iee/energiesystemtechnik/de/Dokumente/Projekte/NetzKraft-Grundlagen_NWA_Konzepte-Projektbericht.pdf.

[2] Bundesnetzagentur für Elektrizität, Gas, Telekommunikation, Post und Eisenbahnen, Ed., Untersuchungsbericht über die Versorgungsstörungen im Netzgebiet des RWE im Münsterland vom 25.11.2005, June 2006.

[3] Union for the Co-ordination of Transmission of Electricity (UCTE), System Disturbance on 4 November 2006, Final Report.

[4] ENTSO-E – System Separation in the Continental Europe Synchronous Area on 8 January 2021 – 2nd update, 26 January 2021, https://www.entsoe.eu/news/2021/01/26/system-separation-in-the-continental-europe-synchronous-area-on-8-january-2021-2nd-update/.

[5] The European Commission, Commission Regulation (EU) 2017/2196, Establishing a Network Code on Electricity Emergency and Restoration.

[6] Kundur P. Major Power Grid Blackouts in North America and Europe, 2003.

[7] VDE Verband der Elektrotechnik und Informationstechnik e.V., Stromversorgungsstörungen in den USA/Kanada, London, Schweden/Dänemark und Italien, Anlässe und Abläufe, Ursachen und Konsequenzen, 27 November 2003.

[8] Larsson S and Ek E. The Black-out in Southern Sweden and Eastern Denmark, September 23, 2003, Svenska Kraftnät, the Swedish TSO.

[9] Vournas CD, Nikolaidis VC, and Tassoulis A. Experience from the Athens blackout of July 12, 2004. *2005 IEEE Russia Power Tech*, St. Petersburg, Russia, 2005, pp. 1–7, doi: 10.1109/PTC.2005.4524490.

[10] Report of the Enquiry Committee on Grid Disturbance in Northern Region on 30 July 2012 and in Northern, Eastern and North-Eastern Region on 31 July 2012, New Delhi, 16 August 2012, https://powermin.gov.in/sites/default/files/uploads/GRID_ENQ_REP_16_8_12.pdf.

Power system restoration   169

[11] ENTSO-E: Report on Blackout in Turkey on 31 March 2015, Final Version 1.0, 21 September 2015.

[12] Vassell GS. Northeast Blackout of 1965. *IEEE Power Engineering Review*, 1991;11(1):4.

[13] Sullivan MJ, Collins MT, Schellenberg JA, and Larsen PH. *Estimating Power System Interruption Costs, A Guidebook for Electric Utilities*, July 2018, https://emp.lbl.gov/publications/estimating-power-system-interruption [accessed 9 May 2023].

[14] Ratha A, Iggland E, and Andersson G. Value of lost load: how much is supply security worth? *2013 IEEE Power & Energy Society General Meeting*. Vancouver, BC, Canada, 2013, pp. 1–5, doi: 10.1109/PESMG.2013.6672826.

[15] Mills E, and Jones R. An Insurance Perspective on U.S. Electric Grid Disruption Costs. *Geneva Pap Risk Insur Issues Pract.*, 2016;41:555–586, doi: 10.1057/gpp.2016.9.

[16] Resilient Power Fact Sheet, CISA [online]. https://www.cisa.gov/resources-tools/resources/shares-documents [Accessed 29 Mar 2023].

[17] Becker H, Schütt J, Spanel U, and Schürmann G. The "SysAnDUk" project: ancillary services provided by distributed generators to support network operators in critical situations and during system restoration. *19th Wind Integration Workshop 2020, Virtual Conference*.

[18] Guo Y, Torabi-Makhsos E, Rossa-Weber G, *et al.* Review on Network Restoration Strategies as Part of the RestoreGrid4RES Project, *EnInnov* 2018.

[19] Adibi MM and Martins N. Power system restoration dynamics issues. *2008 IEEE Power and Energy Society General Meeting*, Pittsburg, PA, USA, https://doi.org/10.1109/PES.2008.4596495.

[20] Becker H, Schütt J, Schürmann G, Spanel U, Holicki L, and Malekian K. Opportunities to support the restoration of electrical grids with little numbers of large power plants through converter-connected generation and storages. *IET Renew. Power Gener.*, 2023;17(14):3496–3506. https://doi.org/10.1049/rpg2.12552.

[21] Becker H, Valois-Rodriguez MF, Holicki L, Malekian K, and Gartmann P. Evaluation of wind power plants' control capabilities to provide primary frequency support during system restoration. *2021 International Conference on Smart Energy Systems and Technologies (SEST)*, Vaasa, Finland, 2021, pp. 1–6, doi: 10.1109/SEST50973.2021.9543369.

[22] Consentec: Netzwiederaufbaukonzepte vor dem Hintergrund der Energiewende, Bericht für die deutschen Übertragungsnetzbetreiber 50Hertz Transmission GmbH, Amprion GmbH, Tennet TSO GmbH, TransnetBW GmbH, Restoration concepts against the background of the energy transition, July 2020.

[23] Braun M, Brombach J, Hachmann C, *et al.* The future of power system restoration. *IEEE Power and Energy Magazine*, 2018;16(6):30–41.

[24] Schütt J, Becker H, Koch J, Fritz R, and You D. Graphical user interface of an aggregation system to control a multitude of distributed generation during

170 *Intended and unintended islanding of distribution grids*

power system restoration. *19th Wind Integration Workshop 2020, Digital Proceedings*, https://doi.org/10.24406/publica-fhg-411531.

[25] 50Hertz, Amprion, Tennet, TransnetBW, Batrachtungen zum Netz- und Versorgungswiederaufbau, Teil des Berichts der Deutsche Übertragungsnetzbetreiber gem. § 34 (1) KVBG, Report of German TSO to System Restoration Without the Use of Coal Fired Power Plants, https://www.netztransparenz.de/portals/1/20201222%204UeNB.

[26] Liang G, Weller SR, Zhao J, Luo G, and Dong Z Y. The 2015 Ukraine blackout: implications for false data injection attacks. *IEEE Trans. Power Syst.*, 2017;32(4):3317–3318.

[27] Cheng GH and Xu Z. Analysis and control of harmonic overvoltages during power system restoration. *IEEE/PES Transmission and Distribution Conference & Excibition: Asia and Pacific*, Dalian, China, 2005.

[28] Heid J, Schittek W, Hachmann C, and Braun M. Asymmetric Contributions to Instantaneous Reserve by Generation. *Loads, and Storage*, doi:10.17170/kobra-202202015687.

[29] Gutierrez I, Crolla P, Roscoe A, *et al.* Operator considerations for the implementation of testing enhanced grid forming services on an onshore wind park. *Virtual 19th Wind Integration Workshop*, November 2020.

[30] Gryning MPS, Berggren B, Kocewiak LH, and Svensson JR. Delivery of frequency support and black start services from wind power combined with battery energy storage. *Virtual 19th Wind Integration Workshop*, November 2020.

[31] Hachmann C, Becker H, Theimer F, Thiel P, and Braun M. Local power system restoration and islanded operation with combined heat and power plants and integration of wind power. *ETG Congress*, 2021.

[32] Dalle Ave G, Carvalho T, Chakravorty J, Schmitt S, and Subasic M. Grid reconfiguration for congestion management of distribution grids using deep learning. *ETG Congress 2023*, Kassel, Germany, 2023.

Chapter 6

Protection

Holger Kühn[1] and Peter Schegner[2]

There is a controversial discussion about whether increasing the infeed of power plants with an inverter interface – and thereby considerably reducing synchronous generation temporarily – will lead to lower short-circuit currents, a general change in the behaviour during short circuits and a possible malfunction of protective relays. This chapter is not intended to and cannot replace a specialised book on protection. Instead, it is intended to illustrate the physical context and summarise the main aspects of the current discussion.

6.1 Introduction

The development of electrical energy distribution was largely characterised by the generation of electrical energy using synchronous generators (so-called type 1 units[*]). These units primarily determined the short-circuit behaviour and therefore also the protection concepts of power plants and grids. With the emergence of renewable generation plants, so-called type 2 units became predominant. These are mainly distributed generation (DG) units feeding electric energy into the grid via inverters. They are increasingly determining the energy generation landscape. Thus, it is not surprising that their behaviour also has a major influence in the event of faults in the grid and must be considered especially when setting up island grids.

Compared to energy conversion in electrical machines, inverters have fundamentally different characteristics. On the one hand, the semiconductor components have a very low thermal overload capacity limiting the short-circuit current of inverters to approximately 1.1 times the rated current. On the other hand, inherent physical characteristics are missing, such as the mass inertia of a synchronous generator or its impressed internal generated synchronous voltage. The behaviour of inverters does not depend on their physical parameters but is determined exclusively by their control. In principle, their control behaviour is freely adjustable within wide limits.

[1] Kühn – Netz und Systemschutz, Hannover, Germany
[2] Institute for Electrical Power and High Voltage Engineering, TU Dresden, Germany

---

[*] Here, we use the distinction according to the VDN Transmission Code, 2007. At the European level (ENTSO), generator types are divided into four classes: types A, B, C and D (cf. Chapter 4) according to their maximum capacity.

172 *Intended and unintended islanding of distribution grids*

In the case of islanding, the following aspects are of particular interest:

- the behaviour of generation systems with inverters in the event of a short circuit,
- the resulting influence on the short-circuit power,
- the interaction of inverters and the grid in special islanding grid constellations, and
- the functionality of different grid protection concepts.

6.2 Short-circuit behaviour of inverters

In order to ensure that the electrical grids fulfil their supply task, the behaviour of generation plants, consumers and the grid needs to be coordinated. Therefore, the requirements in the event of grid disturbances to be realised in the inverter control are specified in the grid connection rules (see Chapter 4). The grid connection rules vary between countries and between grid levels. They are continuously improved. In general, a network disturbance is characterised by frequency and/or voltage not being within predefined limits. From the inverter's point of view, there is a short circuit in the grid when one or more phase or phase-to-phase voltages at the inverter fall below a predefined threshold value of, for example, 80% or 85% of the nominal voltage U_n.

Basically, there are four possibilities of short-circuit behaviour for inverters, i.e. switch-off, zero-power mode (ZPM), feed-in of reactive current in the positive sequence, and feed-in of reactive current in the positive and negative sequence. Common to all procedures is that faults in the grid are detected due to the fact that the voltage at the inverters no longer is within the normal voltage band.

1. Switch-off mode (SOM):
    Under this procedure, the voltage monitoring at the inverter immediately and definitely switches the inverter off. Upon recovery of the grid voltage and expiry of a sufficient security period, the generation plants are allowed (but not obliged) to reconnect.
2. Zero-power mode (ZPM):
    In the event of a fault in the grid, the inverter is not allowed to feed currents into the grid but must not switch off either (or disconnect from the grid, respectively, e.g. by opening a switch). Instead, upon recovery of the voltage, it shall immediately feed active power into the grid.
3. Feed-in of reactive current in the positive sequence (positive sequence mode – PSM):
    In the event of a voltage drop, the inverter shall support the grid voltage by feeding in a (inductive) reactive current ΔI_1 in the positive sequence, a so-called 'fast fault current'. This proportion of reactive current shall be supplied in addition to the existing operational reactive current, its amount depending on the amount of the voltage dip in the positive sequence ΔU_1. Reactive current support has priority over the supply of any active current.
4. Feed-in of reactive current in the positive and negative sequence (positive- and negative-sequence mode – PNSM):

Under this procedure, in addition to the fast reactive current support in the positive sequence, an additional current ΔI_2 shall be supplied in the negative sequence, its amplitude depends on the voltage change of the negative-sequence voltage ΔU_2. This is another case where the reactive current support has priority over the supply of any active current.

Feeding in a zero-sequence current is not required because, if necessary, this is achieved by the neutral point treatment of the transformers.

6.2.1 Switch-off mode in the event of a fault

The advantage of this procedure was that immediately after the fault had occurred, all inverter feed-ins into the grid were switched off. Thus, the existing protective devices were able to identify the location of the fault and trip the breaker without intermediate feeds preventing the localisation of the faulty equipment. The switch-off mode (SOM) was typically used in networks with a small amount of power plants with an inverter interface and a powerful network connection to the superordinate voltage level. SOM was especially useful in radial grids where the protection system is designed to ensure that the short-circuit current is fed in from a single source, usually via a transformer from a superordinate voltage level.

Within the SOM is not specified when generation plants must reconnect to the grid. A short circuit in the transmission grid leads to very extensive voltage drops in subordinate grids. This means that SOM switches off a large number of generating units that are not directly affected by the grid fault. After the fault clearing, the power deficit had to be compensated by the primary control reserve until these generation plants had reconnected to the grid. When it became clear that by further expansion of renewable energy the loss of active power after a fault would reach the threshold of the primary control power of the interconnected grid, SOM could no longer be applied. As a consequence, grid connection rules for generation units with inverters were established requiring a grid-supportive behaviour in the event of a fault.

6.2.2 Zero-power mode

The grid-supportive behaviour is specified by voltage–time curves (fault ride-through curve, FRT). FRT curves define a fault in the grid as an area of voltage and time where a generation unit must not disconnect from the grid while a special grid-supporting behaviour is required.

Figure 6.1 shows such a voltage–time curve required in the event of voltage drops (low-voltage-fault ride-through) in low-voltage grids. In the shaded area, the power generating unit is not allowed to feed currents into the grid, but it has to stay connected to the grid. After a fault switch-off, the DG must, with an increase in voltage to more than 0.85% of the nominal system voltage, feed in the currents before the fault with as little delay as possible. With a voltage below the blue line, inverter infeeds are allowed to disconnect from the grid. We would like to point out that below this line, disconnecting from the grid is not mandatory. Mandatory for disconnection from the grid are the settings of the voltage relays (Table 6.1) required

174 Intended and unintended islanding of distribution grids

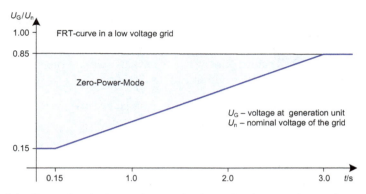

Figure 6.1 Example of a low-voltage ride-through curve in a low-voltage grid [1]

by the grid operator only. Zero-power mode (ZPM) is a standard in low-voltage grids and is often permitted in medium-voltage grids.

Due to the fact that conventional power plants with synchronous generators connected in high-voltage grids or extra-high-voltage grids are increasingly replaced by renewable generation plants with inverters, the short-circuit power decreases. The voltage drop increases and affects more and more customers and devices connected to the grid. This is another case where a power plant-like behaviour of the inverters, i.e. the feed-in of reactive current during a voltage drop, comes in useful. There are two procedures available for this purpose.

6.2.3 Feed-in of reactive current in the positive sequence

In the positive sequence mode (PSM), the amount of reactive current in the positive sequence is proportional to the voltage drop in the positive sequence. A voltage drop ΔU is defined as deviation of a voltage from a pre-fault-voltage U_{mean}, where U_{mean} may be averaged over a certain time period (e.g. 1 minute). U_1 shall be the voltage occurring in the positive sequence in the event of a fault. The reference voltage U_n is either the nominal voltage of the grid or a normal voltage, agreed upon in special grids. The grid shall by definition be symmetrical before the fault so that the pre-fault positive sequence voltage can be set to $U_{1n} = U_n$ and $U_{1\text{mean}} = U_{\text{mean}}$. Therefore the following applies to the positive sequence:

$$\Delta u_1 = \frac{U_1 - U_{1\text{mean}}}{U_n} \tag{6.1}$$

A voltage drop results always in a negative sign. The additional positive-sequence current to be supplied is determined from

$$\Delta i_1 = j \cdot k \cdot \Delta u_1 \tag{6.2}$$

with $\Delta i_1 = \frac{\Delta I_1}{I_r}$ and I_r are the rated current of the generation unit. Δi_1 becomes negative in the event of a voltage drop, which corresponds to the feed of an inductive reactive current into the grid or the feed-in of an overexcited synchronous machine, respectively. k is a dimensionless proportionality factor which can either be set to a standard value at a generation unit or which, especially in high-voltage and extra-high-voltage grids, can be set so that, taking account of transformers etc., a pre-set share to the short-circuit current results at the grid connection point. As a standard, the setting is $k = 2$ in medium voltage (MV) grids and approx. $k = 2.3$ in high voltage (HV) grids.

As a result of the feed-in of reactive current exclusively in the positive sequence, with phase-to-phase short circuits the voltage drop in the positive sequence is only half as high as with three-phase faults. Correspondingly, the grid-supporting reactive current also is only half as large as with three-phase faults. Therefore, it is unfavourable that the generation plant does not supply any negative-sequence current supporting the grid voltage.

6.2.4 Feed-in of reactive current in the positive and negative sequence

The requirement to feed in a short-circuit current adapted to the fault type (two- or three-phase) can be met by complementing the feed-in of a reactive current in the positive sequence by the feed-in of a reactive current in the negative sequence which shall be proportional to the voltage change in the negative sequence (Figure 6.2). In the symmetrical state of the grid before the fault occurs, $U_{2\text{mean}} = 0$, and

$$\Delta u_2 = \frac{U_2 - U_{2\text{mean}}}{U_n} = \frac{U_2}{U_n} \tag{6.3}$$

$$\Delta i_2 = j \cdot k \cdot \Delta u_2 \tag{6.4}$$

Δu_2 is always positive in the event of an asymmetrical fault.

Figure 6.3 shows a type 2 generation unit with symmetrical components whose inverter is designed as a current-feeding source in positive and negative sequence

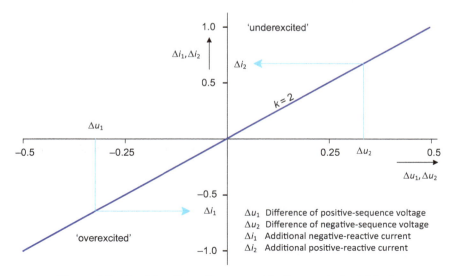

Figure 6.2 Feed-in of reactive current in the event of grid faults

176 *Intended and unintended islanding of distribution grids*

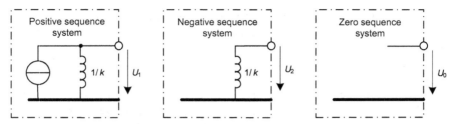

Figure 6.3   Generation unit with PNSM in symmetrical components

mode (PNSM). The current source is part of the positive-sequence system. This is controlled so that at normal operating voltage, the active current (corresponding to the supply of renewable energy) and the reactive current required by the grid operator are available at the terminal of the positive-sequence system. In addition, the power source generates a virtual inductive current that flows through the reactance $1/k$ when the operating voltage is applied. If the positive sequence voltage collapses due to a fault in the grid, the part of the additional reactive current that can no longer flow through the reactance due to the lower voltage flows into the grid and thus supports the grid voltage. This corresponds to an overexcited synchronous generator feeding an inductive current into the grid during voltage dips, i.e. having a capacitive effect itself.

The negative-sequence voltage is negligible in normal operation. It increases in the event of asymmetrical faults. As a result, an inductive current is fed from the grid into the generating unit in the negative-sequence system. In Figure 6.2, this can be seen from the different signs of Δi_1 and Δi_2. This different response in the positive- and negative-sequence system corresponds to that of a synchronous generator.

6.3   Behaviour of inverters in different grid constellations

6.3.1   Wind power plant – offshore grids

Offshore wind power plants (WPPs) connected to the transmission grid via converter stations and DC cables are an example of an island grid without rotating machines. Via line-commutated inverters, the wind turbines of an offshore WPP (left in Figure 6.4) feed into a 30 kV grid which shall be connected to an HVDC converter (High-Voltage Direct-Current) at sea (right) via a 30/150 kV transformer and a 150 kV cable. The nominal capacity of the converters and transformers shall be 300 MVA, the 150 kV cable shall be a single-core cable with a length of approximately 20 km and a capacitance of approximately 0.25 µF/km. The star points of the 150 kV transformers shall be grounded. The self-commutated rectifier of the HVDC converter shall be the grid-forming component; the phasing of the wind turbines current feed-in is geared to the grid voltage of the HVDC rectifier. The DC cable and the shore side are not shown here.

Both the control of the inverters of the wind turbines and the rectifier side of the HVDC converter shall be designed according to the specification [2] with PSM. In the event of a fault in the 30 kV or 150 kV grids, both the wind turbines and the

Protection 177

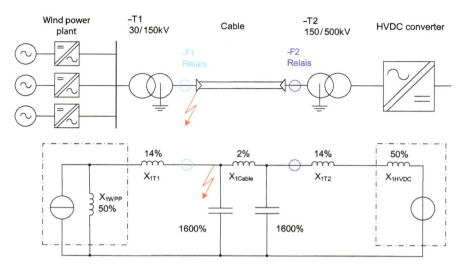

Figure 6.4 Connection of an offshore wind farm to an HVDC link

offshore converter station provide an additional reactive current, but exclusively in the positive sequence. This condition can be considered in a short-circuit current calculation by setting the internal source reactances X_{2WPP} and X_{2HVDC} in the negative sequence to infinity. Figure 6.5 shows the equivalent circuit diagram in symmetrical components for a phase-to-phase fault at the cable termination on the left-hand side of the 150 kV cable. In the following figures, an impedance of 100% corresponds to the value $\frac{U_n}{(\sqrt{3} \cdot I_r)}$ while $X_{1WPP} = X_{1HVDC} = 50\%$ corresponds to the factor $k = 2$.

In the negative sequence, as intended, due to the infinitely high internal reactance, the inverters with $I_{2WPP} = I_{2HVDC} = 0$ do not contribute to the short-circuit current.

In the positive sequence, the transversal path in the equivalent circuit diagram consists of the inverter inductance X_{1WPP} and of the cable capacitances. In the negative sequence, there is no internal reactance of the inverters because, as intended, they shall exclusively feed reactive current in the positive-sequence current. Thus, the negative sequence exclusively consists of the high-impedance cable capacitance. Since the equivalent circuit in the event of a phase-to-phase fault consists of a series circuit of the positive and negative sequence, the high-impedance cable capacitance determines the level of the short-circuit current which is nearly completely independent from the level of the nominal capacity of the wind farm and the HVDC system. In this example, the 'short-circuit current' would be only slightly higher than the cable charging current, making it unidentifiable even with a cable differential protection. However, with simple considerations like this, actual currents and voltages cannot be determined. The inverter controls detect the drop of a line-to-line voltage and determine a reactive current to be fed in, which, however, they are unable to supply due to the limitation of their intermediate circuit voltage and the grid parameters. In reality, distorted currents and voltages occur that cannot be reasonably evaluated

178 *Intended and unintended islanding of distribution grids*

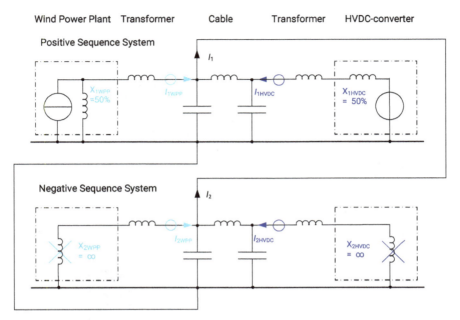

Figure 6.5   Phase-to-phase short circuit at the left cable termination

by protective devices. At this point, both modelling in symmetrical components and normal grid calculation programs fail, as limitations of the intermediate voltage of inverter and rectifier are not considered in these programs. Calculations of cases like this are very complex and only feasible if the original software of the inverter and rectifier control is integrated into the calculation program.

For completeness, we will consider a ground fault [3]. The cable capacity shall not be compensated, and both transformer neutral points are low-impedance grounded (see Figure 6.6). Nevertheless, in the event of a single-phase fault the positive sequence, the negative sequence and the zero sequence are connected in series, meaning that the fault current is essentially limited by the high impedance in the negative sequence.

Since in the event of grid faults in offshore AC grids with asymmetrical short circuits, only very low fault currents can occur due to the high impedance in the negative sequence system, a feed-in of reactive current exclusively in the positive sequence is not useful. In this case, a feed-in of reactive current in the positive and negative sequence would be imperative to ensure the function of the protective system.

As many cables are operated in compensation mode, we would like to consider briefly how the installation of compensation coils affects two- and single-phase short circuits in this grid. In the equivalent circuit diagrams, a compensation of the cable capacitance through compensation coils corresponds to a parallel circuit of compensation coil and cable capacitance both in the positive and negative sequence, and even in the zero sequence if the neutral point of the compensation coil is grounded.

Protection 179

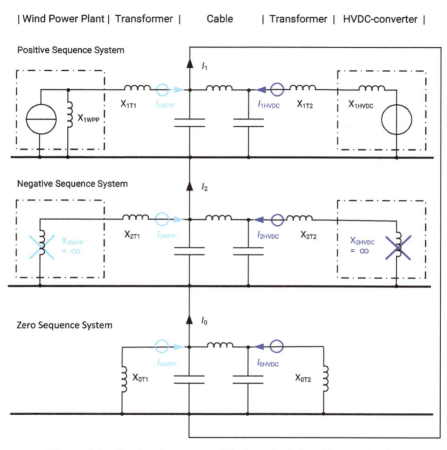

Figure 6.6 Single-phase ground fault at the left cable termination

With a theoretical degree of compensation of 100%, cable capacitance and compensation coil would form a parallel resonant circuit adapted to the mains frequency. This parallel resonant circuit has an infinite resistance, and there would be no current flow through the negative sequence. It is the same principle as in a compensated network, the only difference being that in our case, the negative sequence principally has an infinitely high impedance. Since, in practice, only approximately 80% of the cable capacitance is compensated, the short-circuit current would decrease again by a factor of approximately 5 versus a non-compensated cable.

6.3.2 Onshore grids

In onshore grids, the above-mentioned case of an offshore grid with an extremely high-impedance negative sequence usually does not occur. This is due to the loads generally having an impedance in the positive and negative sequence. One example of this is the sum of all single-phase loads in the low-voltage grids. In the event

180 *Intended and unintended islanding of distribution grids*

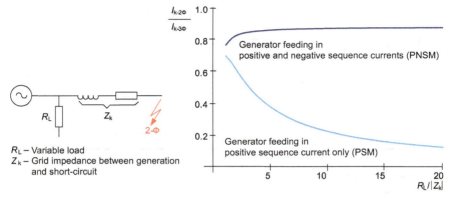

Figure 6.7 *Short circuit in the low-voltage grid with a load at the infeed point*

of asymmetrical faults, these occur as transverse elements in the negative sequence leading to a lower impedance and enabling the negative-sequence current to flow.

First, we will consider an island grid. Figure 6.7 shows a generator essentially powering a nearby load. There are no other loads in the grid. The grid has a phase-to-phase short circuit, the short-circuit impedance shall amount to 1 p.u., with an angle of 45°.

The diagram shows the phase-to-phase short-circuit current as a function of the load which is reduced from 1 p.u. to 1/20 p.u. The behaviour of a generator feeding a short circuit merely in the positive sequence is shown in light blue. The short-circuit reactance of the generator was assumed to be 0.5 p.u. which corresponds to the common set value of $k = 2$. The short-circuit current of a generator having a short-circuit reactance of 0.5 p.u. both in the positive- and in the negative sequence is shown in dark blue. In the case of the generator with a negative-sequence reactance, the short-circuit current is almost independent of the load, whereas the short-circuit current of the generator without a negative-sequence short-circuit reactance is extremely dependent on the load. With a low load – i.e. with a high load resistance – they merely supply a very low short-circuit current in the event of asymmetrical short circuits. This example shows that especially in island grids infeed of a (negative) inductive current during non-symmetrical faults is decisive for the fault current and therefore for the function of the protective system.

If no negative-sequence current flows in the generator supplying the short-circuit current, the negative-sequence current will have to flow via the loads in the grid, meaning that it is strongly load-dependent. The reason for this is easily discernible in Figure 6.8 (left) – the negative sequence current has only to flow via the loads, as the negative sequence, the load resistance and the short-circuit impedance are connected in series. Actually, the behaviour during asymmetrical short circuits strongly depends on the ratio of load to PSM generation. If the ratio is low, the short-circuit current will be low while overvoltage and voltage distortion in the healthy phase occur.

Normally, the grid with the short circuit is a part of an interconnected grid. In this case, the load resistance R_L in Figure 6.8 consists of all loads in the interconnected

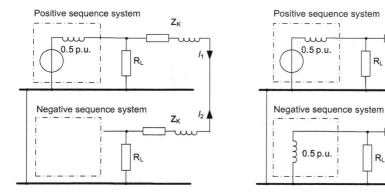

Figure 6.8 Equivalent circuit diagram in symmetrical components in the event of a phase-to-phase fault without (left) and with (right) a negative-sequence short-circuit reactance of the generator

grids, meaning that it is arbitrarily low. However, for the most part, these loads are located further away electrically, which is reflected in a series-connected grid reactance with the respective load resistance. In total, this results in a sufficiently low resistive inductive shunt impedance in the negative sequence so that even with asymmetrical short circuits, there will be a sufficient short-circuit current flow, as long as the ratio of PSM-controlled inverters and nearby loads is low enough. Otherwise, it is to be expected that the voltage drop will be spatially more extended than with inverters feeding short-circuit current in the positive and negative sequence (Figure 6.8 right). With these, the short-circuit current will always be higher and relatively independent from the load resistance due to the relatively lower negative-sequence reactance compared to the load resistance.

6.4 Development of short-circuit currents

Often there are concerns that the short-circuit power in grids with a high proportion of generation plants with inverters could decrease as the increase in renewable generation plants is accompanied by a decrease in short-circuit currents from synchronous generators. Before getting into detail about protection and islanding, this question should be clarified for the normal case of a transmission grid with its subordinate grid levels.

An estimation of the development of short-circuit power in the transmission grid was examined in a study [4] for Germany, with particular emphasis on the minimum short-circuit power, which is decisive for the function of grid protection.

Figure 6.9 shows the minimum short-circuit power of selected nodes in the German transmission grid in 2011 and the minimum short-circuit power predicted for the same nodes in 2033 [4]. The reference value for the data is the maximum short-circuit power of the strongest node in 2011. Interestingly, the strongest nodes in 2011 are also the strongest in 2033, albeit at a significantly lower level, while the

182 *Intended and unintended islanding of distribution grids*

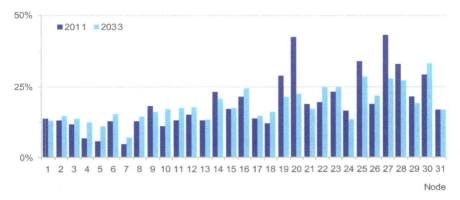

Figure 6.9 Minimum short-circuit power S_{kmin} in the German transmission grid [4]

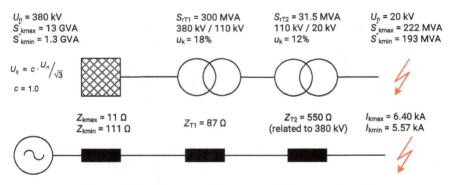

Figure 6.10 Short-circuit power at different grid levels

short-circuit power at the weaker nodes actually increases. In industrial regions with their high density of conventional power plants, the short-circuit power in the transmission grid tends to decrease due to the elimination of these power plants, while it increases in rural regions. The reason for this is the short-circuit current contribution of many renewable plants from subordinate rural grids, which in total replaces the short-circuit current contribution of conventional power plants. The short-circuit power will be higher at all nodes in 2033 than it was in 2011 at the weakest node. Therefore, the protective devices in the high-voltage and extra-high-voltage grids will not have any problems with reduced short-circuit power.

Using a small model consisting of an extra-high-voltage grid and two transformers in series (Figure 6.10) illustrates the influence of a changing short-circuit power in the superordinate grid on the short-circuit currents in the medium-voltage grid.

For the short-circuit power in the extra-high-voltage grid, it was assumed that it would decrease by a factor of 10 from 13 GVA to 1.3 GVA. This roughly corresponds to the values of nodes 20 and 7 in 2011 in Figure 6.9. Due to the large impedance of transformers, the short-circuit impedance of the transmission grid does not have a large influence on the short-circuit current in the medium-voltage grid, which is only reduced by approximately 13% from 6.40 kA to 5.57 kA. The concern occasionally expressed by medium-voltage grid operators that the short-circuit power could drop and possibly no longer be sufficient for the protection concepts currently used in medium-voltage grids fortunately does not apply as long as the medium-voltage grid is part of an interconnected high-voltage grid.

However, the provision of short-circuit power for the extra-high-voltage grid will change significantly. To date, short-circuit power has been mainly delivered by large power plants feeding directly into the extra-high-voltage grid. In the future, short-circuit power in the transmission grid will increasingly be provided from lower voltage levels. For this purpose, a sufficient number of renewable generation units with inverters need to actively contribute to supporting the grid voltage in the event of a short circuit by feeding in a suitable reactive current. Under these conditions, the existing structures of the medium-voltage grids, the protection concepts and the protection devices can be retained.

6.5 Grid protection concepts and short-circuit contributions of inverters

6.5.1 Extra-high-voltage grids and high-voltage grids

In high-voltage and extra-high-voltage grids, distance relays with impedance starter elements and signal comparisons as well as differential protection relays are predominantly used. Both types of relays clear any fault strictly selectively within a short period of time. Moreover, their function is largely independent of the level of the short-circuit current. A significant decrease in short-circuit currents at these voltage levels would not pose any problems for these protective devices, even if their level were to be significantly lower than the maximum load currents in the future.

High-voltage and extra-high-voltage grids are usually operated as meshed grids. This means that system and protection concepts are designed to detect and selectively disconnect short-circuit currents from all directions from which they can affect equipment in the event of a fault. In these networks, DG units that supply a short-circuit-like current in the event of a fault can be easily integrated and connected to a busbar. The connection of a DG with a T-feeder to an existing overhead line is also possible with a manageable amount for protection.

6.5.2 Medium-voltage grids

Medium-voltage networks are usually operated as radial grids. Since in these grids, short-circuit currents were to be expected exclusively from the feeding transformer, the system and protection concepts could be designed in a correspondingly simpler way. These concepts are often unsuitable for short-circuit currents from DG units and feedbacks from the low-voltage to the medium-voltage grid. Connecting DG that feeds in a short-circuit current is possible and is also carried out for hydropower plants, combined heat and power plants, etc., but requires a relatively high level of grid-related technical effort compared to the connected load.

184 *Intended and unintended islanding of distribution grids*

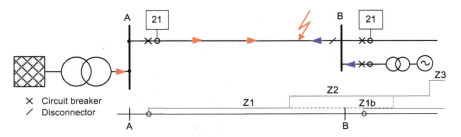

Figure 6.11 *Distance protection in a radial medium-voltage grid*

As an example, we consider a rural medium-voltage grid (Figure 6.11). Switchgear and protective devices of the network are designed for a short-circuit current exclusively from the feeding transformer (red). The overhead lines are protected by distance relays (21 in figure 6.11) which trigger an autoreclosure when a fault is detected in zone Z1b.

After the switch in A has been opened by the distance protection, the high short-circuit current from the transformer (red) is switched off. However, the short-circuit current (blue) of an inverter-based DG connected to B continues to feed the short-circuit during the dead time. Even if its short-circuit current only reaches the level of the load current of the DG, i.e. it can be very small, it is sufficient to prevent the arc from extinguishing.

On the other hand, the sometimes very small short-circuit currents of DGs are problematic because the protection concepts in medium-voltage systems are mainly based on the detection of currents that are greater than the highest load currents in the event of a fault. Examples of this are overcurrent relays with staggered tripping characteristics and distance relays. Distance relays in the medium-voltage network are usually equipped with overcurrent starters which also require short-circuit currents to detect a fault that is greater than the maximum load currents. Sensitive differential relays for cables, which are mainly used in urban networks, are an essential addition to the protection concept. Nevertheless, overcurrent protection is needed to cover faults on busbars, in the dead zones between the circuit breaker and current transformer and for backup protection. Thus, the basic statement is that in most medium-voltage networks, to ensure the function of the entire protection concept, short-circuit currents are required, which must be significantly greater than the respective maximum load currents. Therefore, especially medium-voltage grid operators often express concern that the short-circuit capacity and thus the level of the short-circuit currents could decrease since the increase in renewable generation plants is accompanied by a decrease in the short-circuit currents from synchronous generators. As we have seen in Chapter 6.4, this is no problem, as long as the medium-voltage grid is connected to a superordinate grid. But it will be a problem if the grid is islanded.

These considerations led to the fact that in Germany, depending on the voltage level, different behaviour is required in the event of a short circuit. The grid connection rules require that

Protection 185

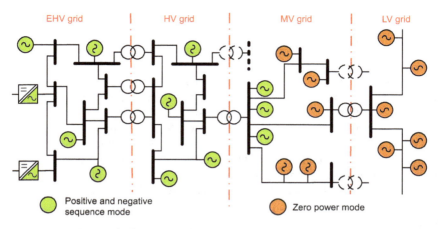

Figure 6.12 Network levels with PNS mode and ZP mode

- Generation units with a grid connection point in a high or extra-high-voltage grid must contribute to the short-circuit current. Therefore, the feed-in of reactive current in the positive and negative sequence is required [5,6].
- When connecting the DGs to a low-voltage grid, the 'zero-power mode' is prescribed [1].
- In medium-voltage grids, it is up to the grid operator to decide which of the two procedures he requires [7].

Figure 6.12 shows a frequently used compromise between the economic efficiency of grid expansion and the provision of short-circuit power. Converter stations with connections to offshore wind farms and DGs must always feed reactive power into the high and extra-high-voltage grid in the event of a short circuit; regardless of whether they are connected to a busbar or to a T-feeder. These DGs are shown in green in Figure 6.12.

In medium-voltage grids usually, the feed-in of a short-circuit current is required when the DG is connected either with a feeding transformer to the busbar of a transformer substation on the medium-voltage side or at grid connection points where the required grid infrastructure is available. DGs at all other grid connection points are operated in ZPM (shown in red in Figure 6.12). This allows for an easy connection of renewable generation units without having to carry out time-consuming and costly modifications to the protective devices in the grid. From the grid's perspective, ZPM is currently the only acceptable solution for low-voltage grids.

6.6 Protection of inverter-based generators

When protecting inverter-based generators, a distinction must be made between internal and external faults. Short-circuit protection for internal faults can simply be

186   *Intended and unintended islanding of distribution grids*

provided by fuses or conventional protective devices as there always is a relatively high short-circuit power on the grid side.

In the case of external faults, only the current from the DGs is available. For inverters, the maximum current is more or less limited to the nominal current of the system for reasons of economic dimensioning. This also applies to the short-circuit current due to the very low overload capacity of the semiconductor components, even in the short term, for reasons of self-protection. Therefore, the protective criterion 'overcurrent' cannot be applied to inverters to detect short circuits outside the own installation. The protective criterion 'impedance' is also largely ineffective for inverters. Since the current from the inverter is impressed by its control system, the impedance measurement criterion does not provide any other information about a short circuit outside the DG than measuring the voltage. Therefore, monitoring the voltage is the essential criterion for detecting a short circuit in the grid for inverter-based generators.

The voltage measurement criterion is used to activate the voltage-supporting behaviour of the system required by the grid connection rules in the event of a short circuit. If the voltage does not recover in time, the DG's protective devices will switch off the DGs after the time specified in the applicable FRT characteristic curve has elapsed. This is the case, for example, if the DG is located in the grid area that was switched off by the grid protection or if there is a large-scale fault with a long fault-clearing period in the grid.

Another current-independent measurement criterion is the frequency. In the event of major disturbances in the grid, over- and underfrequency relays disconnect systems with inverters from the grid when the permissible frequency limits are exceeded.

However, these protective devices are more important in the event of unintended islanding than in normal operation. Voltage and frequency relays ensure that generation systems with inverters are switched off before customer systems are damaged. Table 6.1 shows examples of the ranges of setting values and tripping times specified in grid connection rules. The values to be set in each case depend on the voltage level and the grid connection of the generation system. The tripping times of the system

Table 6.1   Setting values and tripping times of voltage and frequency relays in grid connection rules

Relay	Function	Setting values	Tripping time
Overvoltage	$U\!>\!>$	$1.2\text{–}1.25\ U_n$	0.1–0.2 s
Overvoltage	$U >$	$1.15\ U_n$	0.1–180 s
Undervoltage	$U <$	$0.8\text{–}0.85\ U_n$	1.0–3. 0 s acc. FRT curve
Undervoltage	$U\!<\!<$	$0.15\text{–}0.45\ U_n$	0.15–1.5 s acc. FRT curve
Overfrequency	$f >$	51.5–52.5 Hz	0.1–5.0 s
Underfrequency	$f <$	47.5 Hz	0.1–0.4 s

protection must be selected in such a way that the fault ride-through requirements are not undermined. For the undervoltage protection ($U<$) of a system – for which the FRT requirements have been set according to Figure 6.1, the tripping time should be at least 3 seconds. In general, the setting limits in the lower voltage levels are narrower and the tripping times shorter than in the higher voltage levels.

6.7 Protection for island grids

Figure 6.13 shows a part of a medium-voltage grid, which is usually supplied from the high-voltage grid via a 16 MVA transformer. The synchronous generator of a small hydropower plant, which is connected centrally at the feed-in point, is the grid-forming element in the case of an islanded grid.

The protective system and the setting of the relays are designed for a short circuit of about 3.8 kA from the superordinate grid on the main busbar, which results in a short-circuit current of some 580 A for faults at the end of the grid. Curve 2 (Figure 6.13) shows the voltage profile with supply from the transformer and hydroelectric power plant (HPP). The voltage at the main busbar reaches nearly 0.9 p. u., and the influence of a 2.3-MW WPP is low, even if it is – as assumed – operated in PNSM mode (curve 1).

Conversely, there will be an entirely different behaviour if there is no feed-in from the superordinate grid. As long as the short-circuit power in the island grid is provided by the synchronous generator of the HPP exclusively, the short-circuit

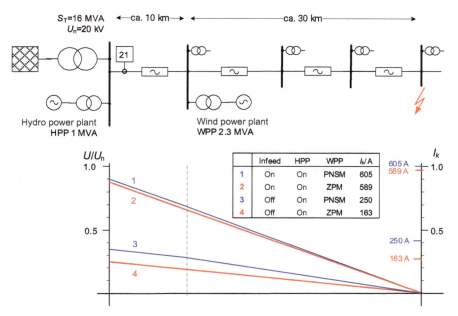

Figure 6.13   Medium-voltage grid, voltage profile during short circuits

188 *Intended and unintended islanding of distribution grids*

power at the central feed-in point decreases from 130 MVA to approximately 7.5 MVA. The short-circuit currents – and thus the behaviour of the protective devices – can be roughly estimated. Short-circuit current is quite low (166 A) for faults at the end of the grid and so is the voltage at the main bus (curve 4). These considerations show that protection concepts exclusively based on short-circuit currents quickly reach their limits in the case of island grid formations.

Short-circuit calculations become more complex if inverter-based generation has to be considered. The current contribution of DGs with ZPM is per definition zero and those of DGs with PSM cannot be calculated, as it depends on the load (for asymmetrical faults) – which is unknown, normally. So, we concentrate on DGs with PNSM. As a standard, short-circuit calculation programs simulate DGs as voltage sources with short-circuit impedance (see Figure 6.4, HVDC converter). Assuming a short-circuit reactance of 50% (corresponding to a short-circuit factor $k = 2$), a three-phase terminal short circuit will result in a calculated short-circuit current of $I_k = 2 \cdot I_r$; however, our inverter supplies nominal current only. In the design of electrical equipment, short-circuit currents that are too high certainly are on the safe side. Nevertheless, they are inappropriate when it comes to evaluating protective devices as this requires minimal short-circuit currents. To determine minimal short-circuit currents correctly; an iteration technique with three steps can be used:

1. Calculation of short-circuit currents with a source voltage $U_q = \frac{U_n}{\sqrt{3}}$
2. Replacement of all inverter feed-ins whose voltage is less than $\frac{U_q}{k}$ by a constant current source with $I = -j \cdot I_r$
3. Recalculation of the short-circuit currents

A comparison between the short-circuit currents and the voltage profiles 3 (WPP with PNSM) and 4 (WPP with ZPM) reveals a grid-supporting impact of the wind turbine with PNSM. Short-circuit current rises from 163 to 250 A, while the voltage significantly increases at the main busbar resulting in a reduced voltage drop (see Figure 3.24 in Chapter 3). On the other hand, the wind turbine's intermediate feed also falsifies the impedance measurement of the distance relay (21 in Figure 6.13). In case 1, the fault at the end of the line is measured at approximately 105% of the line length while in case 3 without a feed-in from the superordinate grid, it is even measured at 126%.

As a conclusion, it must be assumed that after an island grid formation, the protective concept in the medium-voltage grid does not work or only works to a limited extent. Particularly in the case of extended rural medium-voltage grids situations can occur where the short-circuit current at the end of a long overhead line is no longer sufficient to trigger an overcurrent protection or a distance protection with overcurrent excitation. In this case, the proper functioning of the grid protection will be jeopardized, and a selective disconnection of the short circuit will be impossible.

In the past, it could be assumed that in case of an island grid formation, the inevitable active power disequilibrium meant that the frequency limits were reached quickly, the under- and overfrequency relays of the generators were triggered and the island grid collapsed. With the introduction of the active power frequency response (see Chapter 3, Figure 3.15 and [7], Figure 17), especially when there is a high penetration with renewable energy generation systems, there is a high probability that the island grid stabilises with a slight overfrequency. The probability of a stable island grid operation increases even further when the DGs are operated with a reactive power/voltage characteristic $Q(U)$ ([7], Figure 8).

On the other hand, the failure of the grid protection concept is not necessarily a disaster either, because the lower short-circuit currents result from a higher short-circuit impedance. Therefore, in case of a fault, the voltage drop is larger, and the voltage funnel becomes more extensive. The generation units may then be disconnected one after the other by the voltage protection listed in Table 6.1. As a consequence, the island grid collapses. However, this 'emergency protection' should not be relied upon. If an island grid operation is to be set up, the short-circuit conditions and protection concepts and settings will have to be thoroughly examined in advance. Where appropriate, protection settings will have to be adapted or relays even replaced.

If precautions are considered in the case for unintended islanding, the unsatisfactory functioning of the grid protection can be improved if the distance relays with overcurrent excitation are replaced by those with impedance starters. In the case of digital relays, it is not necessary to replace the relays for this purpose, as a corresponding software adjustment is usually sufficient. In many MV grids, it should be sufficient to replace the distance protection devices at the bus bar of the central feed-in.

Regarding the protective function, the restoration of a collapsed medium-voltage island grid via the feeding transformer is unproblematic. If the medium-voltage grid is in a more or less stable island grid condition, the easiest solution is to force the island grid to collapse by shutting off generation plants and then restoring 'from above.' The reason for this is simple: generally, the medium-voltage grid will not be able to adapt its frequency to the frequency of the superimposed grid by controlling the active power generation; moreover, there will usually be no synchrocheck devices available for 'soft' switching in the MV and HV grid.

Of course, a medium-voltage grid designed for an island grid operation both in terms of the design of the protective concept as well as in terms of the design of the control of active power generation and voltage stability, can be restored starting with a grid-forming generation plant. Ideally, this plant should be installed in proximity to the central feed-in as this is where it can ensure its maximum impact on the grid. However, a complete 'bottom-up' grid restoration, i.e. feeding voltage into the high-voltage grid starting from medium-voltage grids, is likely to be difficult simply because of the charging capacity of the high-voltage grid, so that this is not supposed to be an option as a grid restoration strategy in the future.

References

[1] E VDE-AR-N 4105, 'Power Generating Plants in the Low Voltage Network,' VDE Verlag GmbH, Berlin, 2018.

190   *Intended and unintended islanding of distribution grids*

[2]   SDLWindV – Verordnung zu Systemdienstleistungen durch Windenergiean-lagen (https://www.gesetze-im-internet.de).

[3]   Kühn H and Quitmann E. The 'Old' Short Circuit In A 'New' Power System, *9th ETG-FNN-Tutorial Schutz-und Leittechnik*, Berlin, 2016.

[4]   dena-Studie Systemdienstleistungen 2033, 'Sicherheit und Zuverlässigkeit einer Stromversorgung mit hohem Anteil erneuerbaren Energien,' Deutsche Energie-Agentur GmbH (dena), Berlin, 2014.

[5]   E VDE-AR-N 4120, 'Technical Connection Rules for High-Voltage,' VDE Verlag GmbH, Berlin, 2018.

[6]   E VDE-AR-N 4130, 'Technical Connection Rules for Extra High-Voltage,' VDE Verlag GmbH, Berlin, 2018.

[7]   E VDE-AR-N 4110, 'Technical Requirements for the Connection and Operation of Customer Installations to the Medium Voltage Network (TAR medium voltage),' VDE Verlag GmbH, Berlin, 2018.

Chapter 7

Unintentional islanding detection

Sebastian Palm[1] and Peter Schegner[2]

Due to the significant growth of distributed generation (DG) units, the detection and shutdown of unintentional electrical islands is an important challenge for today's power supply system. Numerous islanding detection methods exist, which should be able to detect all kinds of unintentional island networks. However, real cases of unintentional electrical islands in distribution networks show that situations can occur in which unintentionally stable island networks are formed, and even many detection methods cannot cover every case. This chapter therefore:

- Explains the processes which lead to the forming of unintentional islands
- Introduces the wide range of islanding detection methods
- Shows limitations and presents criteria to evaluate the effectiveness of detection methods

Most of the investigations in this chapter were conducted in the course of the dissertation of the chapter author [1]. Other literature is explicitly referred to in the respective sections.

7.1 Occurrence of unintentional electrical islands

The increasing decentralisation of energy generation places high demands on both the resilience of the electrical grid and the provision of auxiliary services. One phenomenon associated with this change is the occurrence of unintentional electrical islands [2,3]. The term electrical island describes the system state in which a subgrid of the electrical supply system is separated from all upstream networks. However, it still has an electrical power balance and thus does not become de-energised. The disconnection from the upstream network can be caused, for example, by triggering protective devices, within the scope of automatic frequency relief or by switching actions of the operating personnel. An electrical island that occurs as a result of a switching action carried out with the aim of creating a de-energised state is called an unintentional electrical island.

[1]DIgSILENT GmbH, Gomaringen, Germany
[2]Institute for Electrical Power and High Voltage Engineering, TU Dresden, Germany

192   Intended and unintended islanding of distribution grids

To understand the occurrence of unintentional electrical islanding, it is necessary to look at the individual parts of the power system. The smaller the subgrids considered are, the greater the influence of the behaviour of DG units and electrical loads (ELs). In an unintentional island network, there is no centralised voltage and frequency control, so processes after a disconnection of the system result from the behaviour of these parts.

7.1.1   Phases of electrical islands

Once an electrical island has occurred, either a new steady state can be established in the islanded subgrid or the power supply collapses after a short period, e.g. as a result of severely unbalanced power levels. Depending on the duration that an unintentional electrical island exists, it is divided into the following categories according to [4]:

- Unstable island (<5 s)
- Quasi-stable island (5–60 s)
- Stable island (>60 s)

The controls effective in an electrical island (e.g. the current control of inverters) of DG units operate very quickly so that within the first 5 s, all decisive transition processes decay [5]. Unintentional islanding must be shut down within these 5 s according to the technical guidelines [6] in Germany. Between 5 and 60 s, mainly deliberately delayed processes such as power balancing via secondary controls, transformer stepping or reactive power statics react. After these processes, an electrical island that has not yet been shut down is called stable. Subsequent changes result mainly from stochastic switching of EL and the volatility of the dominant renewable primary energy sources, wind and solar. However, changes in the switching state within the islanded electrical network result in a completely new network topology and a new set of DG units and ELs, so transition processes occur again.

7.1.2   Behaviour of distributed generation units

DG units, which are connected to the electrical grid via power converters, generally have no grid-forming abilities and do not regulate voltage or frequency. Instead, an attempt is made to achieve the highest possible financial reimbursement by delivering the maximum possible active power, depending on the availability of the primary energy source. However, based on the applicable connection guidelines and the specifications of the network operator, additional auxiliary services must be performed.

7.1.2.1   Frequency-dependent active power reduction

If the frequency in the interconnected grid increases, this is usually an indication of a power surplus. The reason for this is the acceleration of large power plant generators, with which the surplus energy of the overall system is balanced. In order to counteract this frequency increase, DG units are required to reduce their fed-in active power above specified frequency limits with ΔP_{DG}. In Germany, for example,

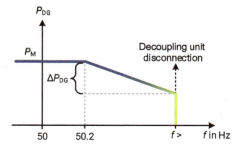

Figure 7.1 Schematic representation of active power reduction at over frequency

a power reduction of 40%/Hz is required above 50.2 Hz, based on the instantaneous power P_M when exceeding the threshold [6,7]. This requirement is described by (7.1) and visualised in Figure 7.1.

$$\Delta P_{DG} = \begin{cases} 0 & f \leq 50.2 \text{ Hz} \\ 20 \cdot P_M \cdot \frac{50.2 - f/\text{Hz}}{50} & f > 50.2 \text{ Hz} \end{cases} \qquad (7.1)$$

More detailed information about the frequency-dependant active power reduction can be read in Section 4.2.1.3.

7.1.2.2 Reactive power for static voltage support

With the decreasing number of large power plants, further operationally relevant auxiliary services must be taken over by DG units. Static voltage support in the electrical supply grid depends on a sufficiently large provision of reactive power. DG units must therefore be able to provide or absorb reactive power to support the voltage in the supply grid. It can and must therefore be assumed that the reactive power balance in an unintentional electrical grid can also be balanced by existing DG units.

More detailed information about the static voltage support can be read in Section 4.2.2.3.

7.1.2.3 Fault-ride-through

The behaviour of generating units to remain connected to the grid despite voltage drops is called 'Fault-ride-through' (FRT). The system rides through the fault instead of disconnecting from the grid. This behaviour must not jeopardise the detection of isolated grids. On the other hand, the islanding detection should not undermine the grid-supporting behaviour of generating units. Ideally, the systems should only switch off when there is an isolated grid. The required FRT behaviour can make it complicated to detect unintentional islanding, in particular, if the island is caused by a fault.

More detailed information about the FRT can be read in Section 4.2.2.2.

194 *Intended and unintended islanding of distribution grids*

7.1.3 *Behaviour of electrical loads*

Traditionally, the EL for investigating electrical islands is assumed to be an resistor, inductor and capacitor (RLC) parallel circuit. This is considered the worst-case scenario, as it can easily be used to set up a parallel resonant circuit. However, the behaviour of real loads often deviates from this and is no less critical in the case of unintentional islanding. Two different load models are explained in the following, which can be used to examine different islanding conditions.

7.1.3.1 RLC-parallel-load

In most studies on unintentional islanding and also islanding test methods from standards, the EL is assumed to be a parallel connection of the elements R, L and C, hereafter referred to as L_RLC. This arrangement behaves as a parallel resonant circuit in which R represents the damping and L and C represent the energy storage. The active and reactive power consumption of these three elements is calculated with (7.2)–(7.4).

$$P_R(U) = \frac{U^2}{R} \tag{7.2}$$

$$Q_L(U, f) = U^2 \cdot 2\pi \cdot f \cdot L \tag{7.3}$$

$$Q_C(U, f) = -\frac{U^2}{2\pi \cdot f \cdot C} \tag{7.4}$$

The power consumption of the whole parallel circuit L_RLC as a function of voltage and frequency can be represented analytically with (7.5) and (7.6). The reference values P_0 and Q_0 represent the power consumed by the load for the reference values f_0 and U_0.

$$P_{RLC}(U) = P_0 \cdot \left(\frac{U}{U_0}\right)^2 \tag{7.5}$$

$$Q_{RLC}(U, f) = Q_0 \cdot \left(\frac{U}{U_0}\right)^2 \cdot \frac{f_0 - 4\pi^2 \cdot f^2 \cdot L \cdot C}{f - 4\pi^2 \cdot f \cdot f_0 \cdot L \cdot C} \tag{7.6}$$

7.1.3.2 Measurements of voltage- and frequency-dependent behaviour of loads

In addition to simple parallel circuits of basic elements, it is also possible to measure the voltage- and frequency-dependent behaviour of real EL. This has, for example, been done in [8] for several low-voltage networks. The resulting (7.7) and (7.8) depict the measured behaviour and the load model is called L_LM in the following.

$$P_{LM}(U, f) = P_0 \cdot \left(\frac{U}{U_0}\right)^{1.46} \cdot \left(1 + 0.1 \cdot \left(\frac{f - f_0}{f_0}\right)\right) \tag{7.7}$$

$$Q_{LM}(U, f) = Q_0 + \underbrace{P_0 \cdot \left(0.91 \cdot \left(\frac{U - U_0}{U_0}\right) - 1.35 \cdot \left(\frac{f - f_0}{f_0}\right)\right)}_{\Delta Q(U, f)} \tag{7.8}$$

From (7.7), it can be seen that, as expected, few directly connected machines were present in the networks considered, as there is only a very small frequency dependence of the active power. The reactive power, on the other hand, shows an indirect proportionality to the frequency. This behaviour is due to the frequency dependence of inductances and capacities in the network. The investigated grids were therefore dominated by inductances. With the reactive power according to (7.8), a subgrid can change from capacitive to inductive behaviour due to voltage or frequency changes and vice versa. This behaviour cannot be reproduced with other standard load models (exponential or ZIP model, for example).

7.1.4 Approach for simple islanding scenarios

In the electrical grid, the process of unintentional islanding is very complex as a large number of DG units and EL influence each other. In order to be able to carry out basic investigations and to understand the fundamentals of unintentional islands, a simple one-busbar-arrangement is used. For this, the DG unit and EL are connected to a common busbar, as can be seen in Figure 7.2. As the detection methods are the focus in this chapter, the DG unit is connected to the busbar with its decoupling unit and additional islanding detection methods, if applicable.

7.1.5 Definition of the non-detection zone

For different combinations of DG units and EL, simulations or measurements with more or less balanced active and reactive power can be carried out. The powers ΔP and ΔQ exchanged with the upstream network before islanding (see Figure 7.2) are obtained with (7.9) and (7.10). These equations are also valid for larger networks with more DG units and EL. However, the grid losses have to be considered in those

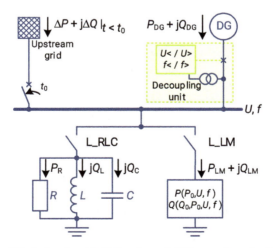

Figure 7.2 Simplified circuit with DG unit and different electrical load models on a common busbar

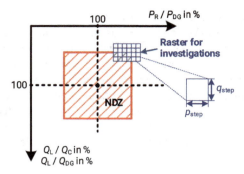

Figure 7.3 Schematic of a non-detection zone (NDZ) and a possible raster for simulations or measurements

cases as well. Not only the perfectly balanced cases with $\Delta P \approx 0$ and $\Delta Q \approx 0$, but also larger active and reactive power differences should be considered.

$$\Delta P = \sum_i P_{\text{DG }i} - \sum_j P_{\text{EL }j} \tag{7.9}$$

$$\Delta Q = \sum_i Q_{\text{DG }i} - \sum_j P_{\text{EL }j} \tag{7.10}$$

Cases in which an electrical island could not be terminated within 5 s are assigned to the so-called **non-detection zone (NDZ)**, as these are referred to as stable or quasi-stable electrical islands, as explained in Section 7.1.1. To illustrate the NDZ, the ratios $P_{\text{EL}}/P_{\text{DG}}$ and $Q_{\text{EL}}/Q_{\text{DG}}$ (or sometimes Q_L/Q_C) are more commonly used instead of ΔP and ΔQ, as the ratio in per cent is easier to interpret and compare. The NDZ is shown schematically in Figure 7.3 and is used in many studies on unintentional islanding [9–16].

7.1.6 NDZ calculation of a simple arrangement

For the simple arrangement of DG units and EL in Figure 7.2, it is possible to carry out analytical studies on unintentional islanding. In the following, this is exemplarily done for a DG unit with a full-sized converter connection to the grid and the load L_RLC. The power feed-in of the DG unit is assumed to be constant because of a fast current control. The decoupling unit of the DG unit is taken into account with the voltage and frequency limits that are presented later in Table 7.1. In this analytical study, however, only the steady-state end values of voltage and frequency are calculated; therefore, it cannot be assessed which protection criterion would respond first. Within the scope of the evaluation, it is therefore assumed that the frequency protection reacts faster than the voltage protection. Since only the steady-state end values of U and f are calculated, and thus no short-term violations of the voltage and frequency limits are recorded, the NDZs determined analytically are slightly larger than the NDZs resulting from electromagnetic transients (EMT) simulations or measurements.

Generally, the power balances from (7.9) and (7.10) must be fulfilled in an islanded grid. Since there is no longer a connection to the upstream grid, the ΔP and ΔQ shares are omitted and the power balances in (7.11) and (7.12) result for this specific arrangement.

$$0 = P_{\text{DG}} - P_{\text{R}} \tag{7.11}$$

$$0 = Q_{\text{DG}} + Q_{\text{C}} - Q_{\text{L}} \tag{7.12}$$

7.1.6.1 Active power

For the load L_RLC, the active power balance can be set up in (7.13) and converted to the voltage, (7.14) results. This relationship shows that the voltage resulting in the islanded grid only depends on the active power fed in by the DG units and absorbed by the load. It also shows that due to the lack of inertia of large electrical machines in many electrical islands, the correlation between frequency and active power observed in the interconnected grid does not occur or only occurs to a minimal extent.

A restriction of this relationship results from the $P(f)$-specification according to Section 7.1.2. This means that P_{DG} has to be reduced if 50.2 Hz is exceeded. To show the influence of the $P(f)$-specification, the calculation is carried out with both activated and deactivated active power reduction.

$$0 = P_{\text{DG}} - \frac{U^2}{R} \tag{7.13}$$

$$U = \sqrt{P_{\text{DG}} \cdot R} \tag{7.14}$$

Due to the additional frequency dependency of P_{DG} following the $P(f)$-specification, the calculation of f and U in the islanded network can no longer be done in a single calculation step. Instead, an iterative calculation with a termination criterion is done (both U and f change by less than 0.1% per calculation step).

7.1.6.2 Reactive power

With the load L_RLC, the reactive power balance in (7.15) results. By solving the quadratic equation, it can be converted to the frequency in (7.16). With the assumption $Q_{\text{DG}} = 0$, i.e. no reactive power component of the DG units, the calculation simplifies to (7.17). As expected, this corresponds to the resonant frequency of the parallel connection of inductance and capacitance. The frequency in an islanded electrical network without reactive power contribution of the DG units is thus exclusively dependent on the elements L and C. A voltage change in the islanded grid changes inductive and capacitive reactive power similarly due to the parallel circuit of L and C. With a constant reactive power by the DG units (zero or otherwise), the reactive power balance can only be obtained by changing the frequency in the islanded grid because the voltage is already defined by the active power balance as shown in (7.14).

$$0 = Q_{\text{DG}} + U^2 \omega C - \frac{U^2}{\omega L} \tag{7.15}$$

198 *Intended and unintended islanding of distribution grids*

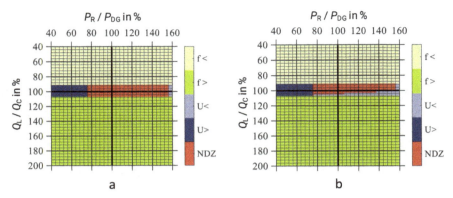

Figure 7.4 *Analytical determined NDZ with L_RLC: (a) without P(f)-specification; (b) with P(f)-specification*

$$f = \frac{Q_{DG}}{4\pi \cdot U^2 \cdot C} + \frac{1}{2\pi}\sqrt{\frac{Q_{DG}^2}{4 \cdot U^4 \cdot C^2} + \frac{1}{L \cdot C}} \tag{7.16}$$

$$f = \frac{1}{2\pi\sqrt{L \cdot C}} \tag{7.17}$$

7.1.6.3 Analytical calculation

The resulting NDZ in the case of a DG unit with a full-sized converter connection and the load L_RLC is illustrated in Figure 7.4. Figure 7.4(a) clearly shows, that for L_RLC, the voltage depends exclusively on the active power balance while the frequency depends exclusively on the reactive power balance. The disconnections in the case of unbalanced reactive power are therefore mainly carried out by the frequency criterion and in the case of very unbalanced active power by the voltage criterion.

If the $P(f)$-specification is also taken into account, the reduced NDZ in Figure 7.4(b) results. The additional disconnections result from the fact that in case of over frequency ($Q_L > Q_C$), the fed-in active power P_{DG} is additionally reduced. According to (7.14), this leads to a drop in the voltage in the islanded grid and thus to the $U<$ threshold violated in some situations. When modelling loads with the approach L_RLC, the $P(f)$-specification therefore causes a reduction of the NDZ. Modelling the load through L_RLC thus leads to lower requirements for additional islanding detection methods. Therefore, L_RLC does not seem to be the worst case when the specific behaviour of DG units, resulting from the connection requirements, is taken into account.

7.1.7 Influence of P(f) and real load model

In addition to the analytical analysis with L_RLC, a detailed EMT simulation has been done with a DG unit that is connected via a full-sized converter and the realistic

Unintentional islanding detection   199

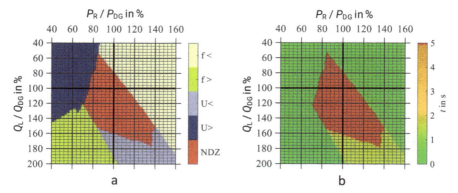

Figure 7.5 NDZ with feed-in via a full-sized converter and L_LM: (a) disconnection criterion; (b) detection time

load L_LM from Section 7.1.3.2. The $P(f)$-specification is active in this investigation. The results of this simulation series with only basic voltage and frequency protection of the decoupling unit are shown in Figure 7.5. Around 10,000 simulations are carried out with different load and generation combinations. In the diagram, each of the coloured squares represents the result of a simulation, whereby the step sizes p_{step} and q_{step} are 1%, as shown in the raster in Figure 7.3.

If the electrical island could be detected and switched off within 5 s after it was disconnected from the upstream grid, the disconnection criterion is shown in Figure 7.5(a) and the detection time in Figure 7.5(b). Non-detected electrical islands are marked as red boxes. It can be seen that there are also very unbalanced cases (e.g. 50% deviation for P and Q) in which detection is not successful. A major reason for this is the $P(f)$-specification which can lead to the stabilisation of unintentional electrical islands.

The influence of the $P(f)$-specification can be explained with the curves in Figure 7.6 for the case $P_R/P_{DG} = 80\%$ and $Q_L/Q_{DG} = 110\%$. At $t = 0$ s, the disconnection from the upstream grid takes place. Since $P_R < P_{DG}$ applies, the voltage in the islanded grid in Figure 7.6(a) initially rises very quickly. To achieve a balanced reactive power, the frequency also increases in Figure 7.6(b). Without $P(f)$-specification, frequency values above 51.5 Hz occur and thus a disconnection of the DG unit from the grid takes place. With $P(f)$-specification, on the other hand, the power fed by the P_{DG} is reduced when 50.2 Hz is exceeded at time t_1. As a result, the voltage in the islanded grid drops, whereby a balanced reactive power balance is achieved without violating the frequency criterion. Thus, no disconnection can occur and the case must be assigned to the NDZ.

7.2 Purpose and principle of detection methods

The occurrence of electrical islands is undesirable because they can result in severe problems and even dangers:

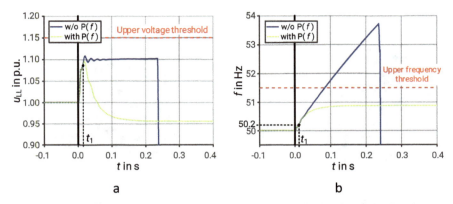

Figure 7.6 Processes during an unintentional electrical island with and without P(f)-specification for L_LM with $P_R/P_{DG} = 80\%$ and $Q_L/Q_{DG} = 110\%$: (a) voltage; (b) frequency

- Voltage and frequency cannot be controlled or influenced by the network operator
- Absence of voltage in the islanded grid is not ensured
- The probability of a successful automatic reclosing is reduced because the arc extinction is hindered by the continuous feeding of the fault
- The liability for damage as a result of the unintentional islanding (e.g. because of asynchronous re-connection) has to be addressed
- The step and touch voltage after single-phase faults could exceed maximum limits in the unintentional island
- The clearing of a fault could be impossible or at least non-selective

In Chapter 13, these challenges are explained in more detail using a real-world example. Because of these problems and dangers, numerous island detection methods (IDMs) have been developed [17–19]. These methods usually pursue three goals:

- Detection of all possible islands under real network conditions (different network topologies, types of DG unit, combination of loads, etc.)
- Little or no impact on voltage quality
- Low costs

7.2.1 Islanding protection with voltage and frequency thresholds

Basic islanding protection can be achieved by voltage and frequency thresholds that are applied to DG units, as explained in Section 6.6.

In Germany, in accordance with the technical guidelines [6,7], each system is equipped with a decoupling unit. This unit has the task of disconnecting the DG units from the mains in the event of violations of voltage or frequency thresholds. The

Table 7.1   *Exemplary voltage and frequency thresholds and associated delay times*

Criterion	Setting	Delay time in ms
$f >$	51.5 Hz	100
$f <$	47.5 Hz	100
$U >$	1.15 U_n	100
$U <$	0.80 U_n	1,500
$U <<$	0.45 U_n	300

settings of this decoupling unit are specified by the connection guidelines. Exemplary values are given in Table 7.1.

It should be noted that some standards are currently being revised. To ensure the stability of the overall system, DG units are becoming increasingly important and a disconnection of the units in the event of a disturbance is being increasingly delayed so that a participation in system stability or recovery can be achieved. It is therefore to be expected that both the limits of decoupling protection and the delay times will be extended in the future.

7.2.2   Additional detection methods

IDMs can be divided into active and passive methods, according to Figure 7.7. Basic decoupling protection, which includes voltage and frequency monitoring, must be present in all DG units.

Passive islanding detection methods (PIDMs) are often based on the evaluation of measured variables and neither influence the operation of the DG units nor do they lead to changes in the electrical network. IDMs based on communication technology also belong to the passive methods since necessary shutdowns of plants are carried out without first actively changing voltage, frequency or other network parameters. Basic islanding protection with voltage and frequency monitoring, must be provided in all DG units as explained in Section 7.2.1.

Active islanding detection methods (AIDMs), on the other hand, require either an adaptation of the DG control system or additional equipment in the electrical network. IDMs that can be implemented in the DG units are, for example, targeted changes to the phase angle, frequency or amplitude of the current fed by the DG units. The aim of this is to change the voltage or frequency during unintentional islanding in such a way that the thresholds of the voltage and frequency protection are violated, while in interconnected operation there should only be a slight influence. Other active methods also evaluate the reaction of the grid to a changed feed-in of the DG units, for example, additionally fed harmonics in the current.

Other possibilities for active detection of islanding using an intervention in the electrical grid include impedance switching, which is intended to disturb the power

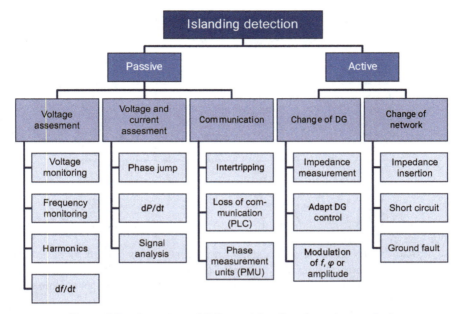

Figure 7.7   Overview of different islanding detection methods

balance, and the insertion of short circuits or ground faults in order to achieve the disconnection of the DG units via protection relays.

7.3   Description of selected islanding detection methods

There are countless types of IDMs in the literature, but many methods are based on the same principles. In the following, selected IDMs are presented in more detail with their mode of operation. At this point, it is impossible to give a complete presentation of all IDMs since countless nuances of the methods exist, and there are also many possible combinations. Further IDMs can be found in [17–20], for example.

7.3.1   Voltage and frequency thresholds – PIDM

The basic IDM is provided by the decoupling unit required via connection requirements according to Section 7.2.1. In addition, many active IDM require voltage and frequency thresholds for a systematic disconnection, as changes in U and f are forced. This procedure is therefore seen as a basic IDM that is part of each DG unit.

7.3.2   Detection of voltage harmonics – PIDM

Converters in the network generate harmonics that are superimposed on the normal current waveform. Normally, the upstream grid has a small network impedance so that the resulting voltage harmonics and, therefore, the total harmonic distortion

(THD) is very low. However, in islanded operation, the current harmonics cause significantly larger voltage harmonics due to a larger grid impedance, so a large THD can be used as a criterion for detecting an electrical island. In addition to the harmonics of the DG unit, there is also an influence from the no-load operation of the upstream transformer if it is located within the island. The harmonics in the current can cause a further increase in the voltage THD due to the nonlinearities and hysteresis of the transformer.

In practice, the problem of this method is the determination of a suitable threshold for island detection. Often, the harmonic content in normal operation is far below the threshold, and it cannot be guaranteed that there are values above the threshold in islanded operation. The low-pass behaviour of some loads or deliberate filtering of harmonics often further limits the applicability of this method. The fact that the introduction of harmonics into the grid is to be reduced further by improved converters and control strategies is also contrary to the mode of operation on which this method is based.

7.3.3 Rate of change of frequency – PIDM

This IDM monitors the gradient of the frequency of the voltage to decide whether an unintentional island occurred or not. Techniques such as the continuous Fourier transform or a phase-locked loop (PLL) are used to estimate the frequency and the rate at which it changes. As already described, the frequency of the voltage normally changes when an electrical island occurs. Since this change occurs relatively quickly due to the small amount of inertia in the electrical island, it can be used as a criterion for detection, which can even respond before the over- and under-frequency thresholds are violated. Figure 7.8 illustrates an exemplary application of this criterion.

To prevent inaccuracies caused by minor phase jumps during switching events in the grid, a measurement window of 80 ms has proven to be optimal [18]. A threshold is set so that slower fluctuations in the network frequency, which are not an indication

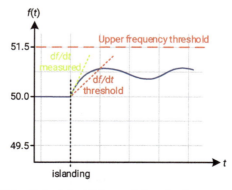

Figure 7.8 Principle of determining the frequency change

204   *Intended and unintended islanding of distribution grids*

of islanding, do not lead to false tripping. Practical experience has shown 0.3 Hz/s to be the optimum threshold [21]. As the system inertia decreases with the shutdown of large power plants, it is to be expected that larger frequency gradients will also occur in normal grid operation in the future. To avoid unintentional disconnection, larger rate of change of frequency (RoCoF) limits will therefore be required.

7.3.4   Phase jump detection – PIDM

With phase jump detection, a sudden change in the voltage phase angle is interpreted as an indication of an island occurrence [22–24]. To detect a phase jump, a tracked angular reference or reference half-wave duration must be continuously calculated, which:

- is present long enough to detect an occurring phase jump
- is tracked quickly enough so that slow changes of the phase angle are not cumulatively considered as an island occurrence during normal grid operation

One simple approach is to calculate the phase jump via the voltage zero crossings of the 50-Hz fundamental. Equation (7.18) is used to calculate the duration of a half-wave. From the deviation to a reference half-wave duration $T_{H\,ref}$, an angular difference can be calculated with (7.19).

$$T_{H\,actual} = \left| t_{last\ positive\ zero\ crossing} - t_{last\ negative\ zero\ crossing} \right| \tag{7.18}$$

$$\Delta\varphi = 180° \cdot \frac{T_{H\,actual} - T_{H\,ref}}{T_{H\,ref}} \tag{7.19}$$

An overreaction may occur due to the presence of noise in the measurement signal, DC voltage components or due to oscillations during load changes. On the other hand, an underreaction is possible in case of processes that develop over several periods. To avoid over- and underreactions, a summed angle difference with several half-wave phase angle deviations can be calculated using (7.20). With this equation, the last k-values and the actual value of $\Delta\varphi$ are summed up.

$$\Delta\varphi_{sum} = \sum_{v=-k}^{0} \Delta\varphi_v \tag{7.20}$$

7.3.5   Communication – PIDM

7.3.5.1   Intertripping

With this IDM, every relevant switching operation in the upstream grid is transmitted to all large DG units via communication links. By receiving information about a switch opening taking place and the local voltage measurement, each DG unit can reliably determine whether it is in normal or islanded operation. Depending on regional conditions, communication links could include fibre optic cables (FOC), normal copper lines or radio links. In the simplest case, each switch opening could lead to the uncoupling of all relevant DG units.

This cost-intensive measure allows the DG units to be controlled by the grid operator, which can be a great grid operational advantage not only to prevent islanding, but additionally during and after system disturbances.

7.3.5.2 Use of power line carrier communications

By using power line carrier communications (PLCC), a high-frequency signal generated by a transmitter can be introduced into the subgrid. All DG units can detect and evaluate this continuous signal via a receiver. As long as the signal can be received, normal operation is assumed. In the event of a grid separation, on the other hand, the signal is interrupted so that the DG units can disconnect from an electrical island that may have occurred without any further necessary measurements.

In extreme cases, it may happen that certain non-linear loads, such as motors or even transformers, in an islanding network generate similar harmonics or subharmonics as the PLCC. Therefore, the PLCC signal should contain a small amount of information or variation to prevent malfunction of this IDM. Thus, the possibility of random imitation of the signal would be eliminated.

7.3.6 Impedance measurement – AIDM

In this IDM, the amplitude of the current fed into the grid by the DG unit is deliberately reduced by di_{DG} at certain intervals. As long as the upstream grid is connected, this reduction in the feed-in will cause no or only an insignificant reduction in the terminal voltage, so that (7.21) results in a minimal impedance. In islanded operation, however, the voltage is no longer supported by the upstream grid and accordingly follows the current waveform fed from the DG unit. The measured impedance from the ratio in (7.21) therefore becomes larger than in the normal state and can be an indication that islanding has occurred. To prevent an island from being detected in normal operation, a threshold must be set below which no detection may take place.

$$Z = \frac{du_{DG}}{di_{DG}} \tag{7.21}$$

Although this method works very well for a single DG unit in an electrical island, it loses its effectiveness with a larger number of DG units in the grid. Since, in this case, all DG units would cyclically reduce their feed-in, the current dips overlap. The determinable value of du_{DG} becomes smaller with each additional DG unit and is finally no longer sufficient to detect the difference between normal operation and islanding. This is shown in [17], for example.

By synchronising the DG units, the problem can be reduced but not eliminated. Another problem is the not insignificant amount of flicker and, in some cases, instability that is caused by a larger number of DG units with this IDM.

7.3.7 Frequency shift – AIDM

With the frequency shift, an active attempt to change the frequency of the network is made. The aim is to violate the thresholds of frequency protection if an unintentional island occurs [17,25–27]. For this purpose, the quantity chopping fraction cf

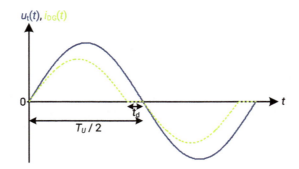

Figure 7.9 Injected current with frequency modified by t_d

is introduced, which results from the ratio of a dead time t_d to half the period of the voltage ($T_U/2$) at the terminals according to (7.22). With t_d, the period duration of the current to be fed in is adjusted, as shown in the example in Figure 7.9. If the DG is connected to a weak grid (e.g. an electrical island), the grid voltage follows this change in the period duration of the current, and the grid frequency changes. This will eventually lead to a disconnection due to an over- or under-frequency trip.

Electrical islands in which a resonant circuit with a large quality factor (Q factor) and a resonant frequency close to 50 Hz occurs are problematic. This can lead to the frequency of the voltage being stabilised and the method thus not being effective.

$$cf = \frac{2 \cdot t_d}{T_U} \tag{7.22}$$

In another form, especially for three-phase inverters, this is achieved by a phase angle transformation of the reference currents $i_{d\,ref}$ und $i_{q\,ref}$ in dq-components, since a specific dead time per phase is not easily realisable for three-phase inverters. Here, instead of a dead time according to (7.22), a phase offset φ_f is determined depending on the frequency deviation as shown in (7.23) [25,27,28].

$$\varphi_f = \frac{\pi}{2}\left(cf + K_f \cdot \frac{f - f_t}{\text{Hz}}\right) \tag{7.23}$$

In the converter, this is realised with the angle transformation in (7.24). As the equation shows, the angle change influences both active and reactive power injection to the grid. In particular, an increase in active power is not always possible, which partially limits the applicability of this method.

$$\begin{aligned} i^*_{d\,ref} &= i_{d\,ref} \cdot \cos\varphi_f - i_{q\,ref} \cdot \sin\varphi_f \\ i^*_{q\,ref} &= i_{d\,ref} \cdot \sin\varphi_f + i_{q\,ref} \cdot \cos\varphi_f \end{aligned} \tag{7.24}$$

Unintentional islanding detection   207

7.3.8   Phase shift – AIDM

With the phase shift, in contrast to the frequency shift, the phase angle of the current to be fed in is changed directly as a function of the frequency [17, 29–31]. In this case, $i_{d\,ref}$ is not varied, but only $i_{q\,ref}$, and thus the reactive power is adjusted. The change of the current angle to be injected φ^*_{DG} is not linear with the frequency, but is mostly given by a sine function, as shown in (7.25). The angle $\Delta\varphi_{max}$ represents the maximum permissible angle change to the operating point $\varphi_{DG\,0}$, which should occur at the frequency deviation Δf_{max}.

$$\varphi^*_{DG} = \varphi_{DG\,0} + \Delta\varphi_{max} \cdot \sin\left(\frac{\pi}{2} \cdot \frac{f - f_r}{\Delta f_{max}}\right) \tag{7.25}$$

7.3.9   Q(f) control – AIDM

This IDM, which is very similar to the phase shift, should lead to a disturbance of the reactive power balance in the island network case as soon as smaller frequency deviations occur. For this purpose, the reactive power provided is adjusted by ΔQ in the case of frequency deviations according to (7.26). The strength of the response to frequency deviations is determined by the parameter K_q with the unit kvar/Hz [32]. Inductive reactive power (under excited operation) is generated by the DG when frequency increases, and capacitive reactive power (overexcited operation) is generated when frequency drops. In contrast to the phase shift, with this method, the response of the inverter is independent of the active power currently being fed in, since the amount of reactive power is changed rather than the power angle. Exceptions are situations where the rated power is already fed in. Since inverters cannot be overloaded, the active power would have to be reduced if a frequency deviation occurs. This can be achieved, for example, by limiting only i_d and not i_q in the current limiter if the rated current would be is exceeded.

$$\Delta Q = K_q \cdot (f - f_r) \tag{7.26}$$

7.3.10   Modulation of $\cos\varphi/\sin\varphi$ – AIDM

In this IDM, the reference angle φ_{ref} of the terminal voltage that is determined by the PLL of the DG is modified by adding an angle according to (7.27) [33]. The signal $\sin\varphi^*_{ref}$ thus receives an additional share of the second harmonic, as can be read from (7.28) after transformation with the addition theorem. This changes neither the amplitude nor the zero crossing of the angle.

$$\sin\varphi^*_{ref} = \sin(\varphi_{ref} + k \cdot \sin\varphi_{ref}) \tag{7.27}$$

$$\sin\varphi^*_{ref} \cong \sin\varphi_{ref} + \frac{k}{2}\sin(2 \cdot \varphi_{ref}) \tag{7.28}$$

The PWM utilises the modified reference angle, which leads to an influence on the voltage at the point of connection. After the park transformation of the measured voltage at the connection point has been carried out, this leads to a 50-Hz component of the q-component of the voltage. This component can be filtered out and evaluated with a bandpass filter. The difference between this component and a fixed or floating reference value can be used as a decision-making criterion for unintentional islanding.

7.3.11 Impedance insertion – AIDM

To disturb the power balance in an electrical island, this IDM additionally switches an impedance on or off after each switch opening where an electrical island could potentially occur [34]. The switching is performed with a delay to avoid unintentionally balancing an already unbalanced power state.

Existing capacitor banks or compensation coils can be used for this operation if integrated into an appropriate logic.

7.3.12 Fault throwers – AIDM

7.3.12.1 Earth fault

If this IDM is used by the grid operator to avoid islanding, a subsequent earthing switch in a single phase is closed each time a breaker is opened as shown in Figure 7.10. As a result, an earth fault is introduced into the possibly occurring electrical island, which causes the neutral voltage displacement (NVD) in the network to increase. The fault can therefore be detected by a voltage transformer that can measure the displacement voltage.

A major drawback with this approach is that all single-phase earth faults in the grid during non-islanded operation also lead to at least a temporary shutdown of DG in the affected grid. This can only be avoided in solidly or low-impedance earthed networks as the NVD is low and the fault is quickly switched off. In these networks, this IDM would only react if the grounding transformer is not part of the electrical island.

7.3.12.2 Short circuit

A straightforward method is inserting a three-phase short circuit after disconnection from the upstream grid. The short-circuit current (in the case of directly coupled generators) or the significant voltage dip in the entire subgrid (in the case of DG with full-sized converters) will ensure the decoupling of all DG in the potential electrical island.

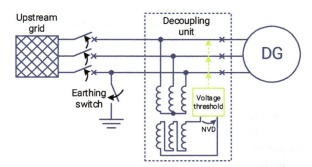

Figure 7.10 Measurement of the neutral voltage displacement

However, this places a large stress on the DG and other equipment. All equipment must be able to briefly feed to a short circuit whenever disconnected from the upstream grid. Although all facilities and equipment should, in principle, be able to do this, the increasing frequency of short circuits due to short-circuiting after grid disconnection can lead to faster wear of equipment and facilities.

7.4 Evaluation of detection methods

7.4.1 Evaluation criteria

To be able to compare different IDMs with each other, reliable and generally applicable evaluation criteria are required. For this purpose, two new parameters that were developed in [1] can be calculated for each method. The NDZ, described in Section 7.1.5, is used as a base for those parameters.

7.4.1.1 Effectiveness of the detection

The first parameter A_{NDZ} is a measure of the NDZ size in the examined cases. The parameter is obtained by (7.29). For each DG or EL combination i, the number of undetected cases with $N_{NDZ\,i}$ is summed. The total area of the NDZ, $A_{NDZ\,IDM}$, is then obtained by multiplying the sum of N_{NDZi} by the step sizes for active and reactive power (p_{step} and q_{step}).

To be able to directly estimate the improvement achieved by an IDM, the parameter size of NDZ ($S\,I_{NDZ}$ for short), according to (7.30), can be used. With this approach, $A_{NDZ\,IDM}$ is related to a suitable reference value. The basic voltage and frequency protection of the decoupling unit (see Figure 7.2) represents a useful reference method here. Accordingly, the voltage and frequency protection alone results in $S\,I_{NDZ} = 1.0$ and additional IDM must reduce this value as much as possible.

$$A_{NDZ\,IDM} = \frac{1}{n} \sum_{i=1}^{n} N_{NDZi} \cdot p_{step} \cdot q_{step} \tag{7.29}$$

$$S\,I_{NDZ} = \frac{A_{NDZ\,IDM}}{A_{NDZ\,ref}} \tag{7.30}$$

7.4.1.2 Speed of the detection

The second evaluation parameter is the average detection time t_D, which measures of how fast successful detections have been. In order to calculate this time, the cumulative frequency function of the detection times has to be estimated. The cumulative frequency function represents the times at which an electrical island could be detected in different combinations of load and generation. This is shown as an example in Figure 7.11. A value of $F(2s) = 0.6$ means that 60% of all electrical islands have been detected after 2 s. A maximum of 5 s is allowed as the time until successful detection after disconnection from the upstream grid. Therefore, detections after this time are not considered. The threshold of 5 s is used as it is assumed as the limit of an unstable island (see Section 7.1.1), but other values can be applied as well. In the case of several scenarios to be investigated (e.g. different kinds of loads), several

210 *Intended and unintended islanding of distribution grids*

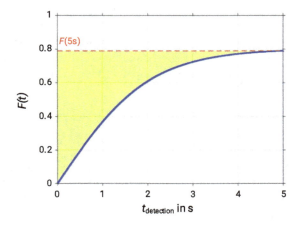

Figure 7.11 Generic cumulative frequency function (adapted from [35])

cumulative frequency functions $F_i(t)$ can be combined into a single function $F_{tot}(t)$ using (7.31).

$$F_{tot}(t) = \frac{\sum_{i=1}^{n} F_i(t)}{n} \tag{7.31}$$

With $F_{tot}(t)$, the second evaluation parameter $t_{D\,IDM}$ can be calculated via (7.32) as the average time of successful island detections. The magnitude of $t_{D\,IDM}$ is proportional to the time area marked with orange in Figure 7.11. It can thus already be estimated by a visual comparison of the cumulative frequency functions of different IDMs. For this parameter, a normalisation to the reference time $t_{D\,ref}$ can be performed according to (7.33), too. The parameters $t_{D\,IDM}$ and thus also TI_{NDZ} may increase for some IDM since some IDM reduce the NDZ, but the additionally detected cases can sometimes have larger detection times. Additional IDM must always reduce the parameter SI_{NDZ} and, at best, also TI_{NDZ} to values below one.

$$t_{D\,IDM} = \frac{\int_0^{5\,s} (F_{tot}(5\,s) - F_{tot}(t))\,dt}{F_{tot}(5\,s)} \tag{7.32}$$

$$TI_{NDZ} = \frac{t_{D\,IDM}}{t_{D\,ref}} \tag{7.33}$$

7.4.2 Comparison of detection methods

As an example of the presented evaluation criteria, the graph in Figure 7.12 displays the cumulative distributions of detection times for three different IDMs. The detection times were obtained through simulations with the arrangement in Figure 7.2 using the load L_LM and a generation unit that is connected by an inverter. As shown in Figure 7.3 for the determination of the NDZ,

Unintentional islanding detection 211

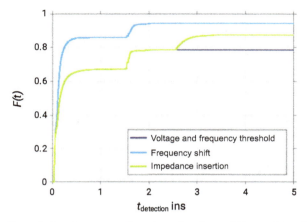

Figure 7.12 Cumulative frequency functions for three different islanding detection methods (adapted from [35])

Table 7.2 Evaluation criteria for three different IDM (adapted from [35])

IDM	A_{NDZ} in sq%	SI_{NDZ}	t_D in ms	TI_{NDZ}
Reference method: voltage and frequency thresholds	4,208	1.00	345	1.00
Frequency shift	1,126	0.27	250	0.72
Impedance insertion	2,496	0.59	586	1.70

different active and reactive power imbalances between 40 and 160% and a step size of 1% were applied, resulting in about 14.000 simulations per curve.

With this, the effectiveness of the IDM after 5 s and the detection times can be compared. The evaluation parameters for the investigated load model and IDM, as determined by these cumulative distributions, can be found in Table 7.2.

7.4.2.1 Detection with voltage and frequency thresholds

The reference parameters for detection with only voltage and frequency thresholds without additional IDM are $A_{NDZ\,ref} = 4,208$ sq% and $t_{D\,ref} = 345$ ms.

7.4.2.2 Comparison of the IDM

Using the normalised comparison parameters SI_{NDZ} and TI_{NDZ} presented in Table 7.2, it is possible to conduct a quantitative evaluation of the different IDM.

Using the frequency shift reduces the size of the NDZ and detection time significantly. While the Impedance Insertion does lead to a reduction of SI_{NDZ}, it also causes an increase in the average detection time t_D. This is because the impedance was only switched on after 2.5 s in this case, as can be seen in Figure 7.12. However, a faster switching process could potentially stabilise already unbalanced power

212 *Intended and unintended islanding of distribution grids*

balances, which would normally lead to a tripping of the under-voltage criterion and therefore should be avoided. As seen from the SI_{NDZ} values, the switched impedance was insufficient to prevent all unintentional electrical islands.

While both additional IDMs can significantly reduce the NDZ, there are still many generation and load combinations that cannot be detected. This shows that the real load behaviour depicted with L_LM is a more critical case than the parallel circuit L_RLC.

As can be seen from this small example, more information can be deducted by using the cumulative frequency function and the evaluation criteria compared to a simple NDZ. In [1], the application of these evaluation criteria to additional IDM and with different scenarios is carried out.

References

[1]   Palm S. *Untersuchung und Bewertung von Verfahren zur Inselnetzerkennung, -prognose und -stabilisierung in Verteilnetzen, 1.* Norderstedt: BoD – Books on Demand, 2019.

[2]   Li C, Savulak J, and Reinmuller R. Unintentional islanding of distributed generation – operating experiences from naturally occurred events. *IEEE Trans. Power Delivery*, 2014;29(1):269–274, doi: 10.1109/TPWRD. 2013.2282264.

[3]   Pazos FJ. Operational experience and field tests on islanding events caused by large photovoltaic plants. *International Conference on Electricity Distribution*, 2011, pp. 1–4.

[4]   Cullen N, Thornycroft J, and Colinson A. Risk analysis of islanding of photovoltaic power systems within low voltage distribution networks. Report IEA PVPS T5-08: 2002, 2002.

[5]   Yazdani A. and Iravani R. *Voltage-Sourced Converters in Power Systems.* Hoboken, NJ: Wiley, 2010.

[6]   VDE, VDE-AR-N 4105: Generators connected to the low-voltage distribution network. 2018.

[7]   VDE, "VDE-AR-N 4110: Technical requirements for the connection and operation of customer installations to the medium voltage network (TAR medium voltage)." 2018.

[8]   Palm S, Schegner P, and Schnelle T. Measurement and modeling of voltage and frequency dependences of low-voltage loads. *IEEE Power & Energy Society General Meeting*, IEEE, Jul. 2017, pp. 1–5, doi: 10.1109/ PESGM.2017.8273781.

[9]   Ropp M, Begovic M, and Rohatgi A. Analysis and performance assessment of the active frequency drift method of islanding prevention. *IEEE Trans. Energy Convers.*, 1999;14(3):810–816, doi: 10.1109/60.790956.

[10]   Begovic M, Ropp M, Rohatgi A, and Pregelj A. Determining the sufficiency of standard protective prevention in grid-connected PV systems. University Center of Excellence for Photovoltaics Conference Papers, Nov. 1998, pp. 2519–2524.

REFERENCES 213

[11] Zeineldin HH and Kirtley JL. A simple technique for islanding detection with negligible nondetection zone. *IEEE Trans. Power Delivery*, 2009;24(2): 779–786, doi: 10.1109/TPWRD.2009.2013382.

[12] Vieira JCM, Freitas W, and Morelato A. An investigation on the non-detection zones of synchronous distributed generation anti-islanding protection. *IEEE Trans. Power Delivery*, 2008;23(2):593–600, doi: 10.1109/TPWRD.2007.915831.

[13] Fan Y, and Li C. Analysis on non-detection zone of the islanding detection in photovoltaic grid-connected power system. *APAP, IEEE*, Oct. 2011, pp. 275–279, doi: 10.1109/APAP.2011.6180527.

[14] Raipala O, Maäkinen A, Repo S, and Jaärventausta P. The effect of different control modes and mixed types of DG on the non-detection zones of islanding detection. *CIRED 2012 Workshop: Integration of Renewables into the Distribution Grid*, IET, 2012, pp. 237–237, doi: 10.1049/cp.2012.0823.

[15] Bruschi J, Raison B, Cadoux Besanger F, and Grenard S. Impact of new European grid codes requirements on anti-islanding protections: a case study. *CIRED*, Lyon, 2015.

[16] Arguence O, Cadoux F, and Raison B. Influence of electronic-based loads on unwanted islanding. 2017.

[17] Bower W, and Ropp M. Evaluation of islanding detection methods for photovoltaic utility – interactive power systems. 2002.

[18] Econnect. Assessment of islanded operation of distribution networks and measures for protection. 2001.

[19] Li C, Cao C, Cao Y, Kuang Y, Zeng L, and Fang B. A review of islanding detection methods for microgrid. *Renew. Sustain. Energy Rev.*, 2014:35:211–220, doi: 10.1016/j.rser.2014.04.026.

[20] Cebollero JA, Cañete D, Martín-Arroyo S, García-Gracia M, and Leite H. A survey of islanding detection methods for microgrids and assessment of non-detection zones in comparison with grid codes. *Energies*, 2022;15(2):460. doi: 10.3390/en15020460.

[21] Chowdhury SP and Crossley PA. Islanding protection of distribution systems with distributed generators – a comprehensive survey report. *2008 IEEE Power and Energy Society General Meeting – Conversion and Delivery of Electrical Energy in the 21st Century*, IEEE, Jul. 2008, pp. 1–8, doi: 10.1109/PES.2008.4596787.

[22] Pena P, Etxegarai A, Valverde L, Zamora I, and Cimadevilla R. Synchrophasor-based anti-islanding detection. *2013 IEEE Grenoble Conference*, Grenoble: IEEE, Jun. 2013, pp. 1–6, doi: 10.1109/PTC.2013. 6652280.

[23] Xu M, Melnik RVN, and Borup U. Modeling anti-islanding protection devices for photovoltaic systems. *Renew Energy*, 2004:29(15):2195–2216, doi: 10.1016/j.renene.2004.04.005.

[24] Ropp M, Begovic M, Rohatgi A, Kern G, Bonn R, and Gonzalez S. Determining the relative effectiveness of islanding detection methods using phase criteria and nondetection zones. *IEEE Trans. Energy Convers.*, 2000;15(3):290–296, doi: 10.1109/60.875495.

214 *Intended and unintended islanding of distribution grids*

[25] Wang X, Freitas W, Xu W, and Dinavahi V. Impact of DG interface controls on the Sandia frequency shift antiislanding method. *IEEE Trans. Energy Convers.*, 2007;22(3):792–794, doi: 10.1109/TEC.2007.902668.

[26] Tran-Quoc T, Le TMC, Kieny C, and Bacha S. Behaviour of grid-connected photovoltaic inverters in islanding operation. *PowerTech*, Trondheim, 2011, pp. 1–8, doi: 10.1109/PTC.2011.6019267.

[27] Zeineldin HH and Kennedy S. Sandia frequency-shift parameter selection to eliminate nondetection zones. *IEEE Trans. Power Delivery*, 2009;24(1): 486–487, doi: 10.1109/TPWRD.2008.2005362.

[28] Pazos FJ, Romero-Cadaval E, González E, Delgado I, and Monreal J. Failure analysis of inverter based anti-islanding systems in photovoltaic islanding events. *CIRED*, Stockholm, 2013, pp. 10–13.

[29] Lloyd G, Hosseini S, An C, Chamberlain M, Dysko A, and Malone F. Experience with accumulated phase angle drift measurement for islanding detection. *CIRED*, Stockholm, 2013, pp. 10–13.

[30] Robitaille M, Agbossou K, and Doumbi ML. Modeling of an islanding protection method for a hybrid renewable distributed generator. *Canadian Conference on Electrical and Computer Engineering*, IEEE, 2005, pp. 1485–1489, doi: 10.1109/CCECE.2005.1557259.

[31] Hung G-K, Chang C-C, and Chen C-L. Automatic phase-shift method for islanding detection of grid-connected photovoltaic inverters. *IEEE Trans. Energy Convers.*, 2003;18(1):169–173, doi: 10.1109/TEC.2002.808412.

[32] De Mango F, Liserre M, and Aquila AD. Overview of anti-islanding algorithms for PV systems. Part II: Active methods. *International Power Electronics and Motion Control Conference*, IEEE, Aug. 2006, pp. 1884–1889, doi: 10.1109/EPEPEMC.2006.4778680.

[33] Ciobotaru M, Agelidis VG, Teodorescu R, and Blaabjerg F. Accurate and less-disturbing active antiislanding method based on PLL for grid-connected converters. *IEEE Trans. Power Electron.*, 2010;25(6):1576–1584, doi: 10.1109/TPEL.2010.2040088.

[34] Bejmert D and Sidhu TS. Investigation into islanding detection with capacitor insertion-based method. *IEEE Trans. Power Delivery*, 2014;29(6):2485–2492, doi: 10.1109/TPWRD.2014.2347032.

[35] Palm S and Schegner P. Evaluation and comparison of islanding detection methods by extended analysis of the non detection zone. *CIRED International Conference on Electricity Distribution*, Madrid, 2019, pp. 1–5.

Chapter 8

Planning methods for secure islanding

Agnes M. Nakiganda[1] and Petros Aristidou[2]

Traditionally, distribution grids have been modelled as either passive elements or aggregated loads, due to their lack of participation in power, frequency, and voltage control. With major transformations in the design and operation of the low-voltage (LV) and medium-voltage (MV) distribution networks (DNs) including the proliferation of distributed energy resources (DERs) and the requirement for them to provide active support to the power grid, DNs are now active. Microgrids (MGs) present a revolutionary step in the electric power system infrastructure designed as independent power distribution networks with the ability to split from the bulk network and form self-sufficient islands.

An MG, as defined in [1], refers to *'a group of interconnected loads and distributed energy resources with clearly defined electrical boundaries that acts as a single controllable entity with respect to the grid and can connect and disconnect from the grid to enable it to operate in both grid-connected or island modes'*.

Diverse operating conditions or unexpected outages, either in the main grid or in the MG, can sometimes trigger unscheduled islanding of the latter (meaning the MG is disconnected from the main grid without prior warning or preparation). To guarantee the islanded MG's continued supply of demand and security, operators must ensure that the islanded MG can supply its loads for a certain amount of time within the secure operation limits (voltage, frequency, and loading). These are usually called adequacy and steady-state (or static) security constraints. Moreover, the unscheduled islanding of MGs can also violate the transient security constraints related to voltage and frequency, that is, right after the islanding and before the system stabilises to a new steady state. These violations can trigger a cascade of disconnections and eventually the MG collapse. Thus, operators must ensure not only the *static security and adequacy* of the MG after the disconnection but also the *transiently secure transition* of the MG from grid-connected to islanded mode. Figure 8.1 shows the frequency inside an MG after islanding with a transiently secure and an insecure behaviour, respectively.

---

[1]Department of Wind and Energy Systems, Technical University of Denmark (DTU), Denmark
[2]Department of Electrical Engineering, Computer Engineering, and Informatics, Cyprus University of Technology, Cyprus

216 *Intended and unintended islanding of distribution grids*

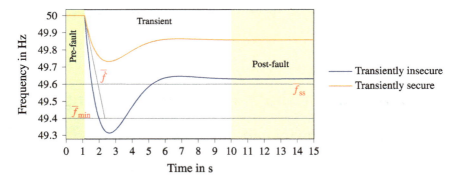

Figure 8.1 *MG centre of inertia frequency response after an unscheduled islanding event with and without transient constraints*

The behaviour of the MG (both dynamic and static) after abrupt islanding is dependent on the pre-fault operating conditions, the available units in the MG after islanding, and their control capabilities. Thus, system operators need to embed appropriate constraints in their MG planning problems (operational and investment) to guide the system to solutions that are economical and secure (both transiently and in the steady state) after islanding. In this chapter, we present some of the mathematical models used for secure operational and investment planning of MGs.

In Section 8.2, we will first discuss the operational planning formulation that allows dispatching the *existing* units internal to the MG in the most economical way while ensuring pre-fault and post-islanding *steady-state* security and adequacy. The reformulations required to facilitate the computationally efficient solution of the operational planning model are also discussed in the same section. When the existing MG units cannot ensure secure steady-state conditions post-islanding, investment planning methods are necessary to select the most economical investment candidates that will help secure the system's operation. The formulation required for this is shown in Section 8.3. Finally, Section 8.4.1 introduces the dynamic security constraints required to ensure the secure transition of the MG from grid-connected to islanded mode.

8.1 Notations used in this chapter

Let $\mathbf{j} = \sqrt{-1}$, $|\bullet|$ denote the magnitude, $\bullet^*$ complex conjugate while $\underline{\bullet}/\overline{\bullet}$ represent lower/upper bounds of the quantity $\bullet$. Let us assume a balanced radial network represented by a connected graph $G(N, L)$, with $N := \{1, \ldots, N\}$ denoting the set of network nodes with index 1 defined as the point of common coupling (PCC) to the transmission network. Set $L \subseteq N \times N$ designates the set of network branches. The MG hosts several generators and loads, where $G \subseteq N$ indicates the subset of nodes with distributed energy resources (DERs), and $D \subseteq N$ the subset of nodes with load demand. $D^v \subseteq D$ and $D^f \subseteq D$ being the set of loads with variable and fixed demand, respectively. Set

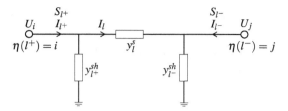

Figure 8.2 The Π model of the line and notation used in OPF formulation

C includes all CIGs, C^i being the set of generators connected to node i, where $C \subseteq G$. $C^{d/v}$ is a set of CIGs with droop or Virtual Synchronous Machine (VSM) control schemes. Set S consists of SGs, where $S \subseteq G$, S^i being the set of generators connected to node i. O is a set of representative days and T is a set of hours in a representative day or planning horizon. Θ represents the feasible solution space for all the optimisation variables.

The active and reactive power injections at each bus i are defined by $s_i = p_i + \mathbf{j}q_i$. The power injections are of the bulk grid import (export), DERs, and loads defined by: s^{imp} (s^{exp}), $s_g : g \in G$, and $s_i^d \in D$, respectively. Set G^i represents all generators connected to node i. The voltage at each bus is defined by $U_i = |U_i|\angle\theta_i$ with the square of voltage magnitude denoted by $u_i = |U_i|^2$. Bus voltage at the PCC node in grid-connected mode is considered fixed at $V_1 = 1\angle 0°$ pu.

Each branch $l \in L$ is represented by a Π model (Figure 8.2) with the sending and receiving ends denoted by l^+ and l^-, respectively, connected by two adjacent nodes $\eta(l^+) = i$ and $\eta(l^-) = j$. Set $L^{\eta(l^+)}$ includes all lines connecting from downstream of a node/line in the *from* direction indexed by m, while $L^{\eta(l^-)}$ includes all lines connecting from upstream in the *to* direction indexed by n. y_l^s is the series admittance given by $y_l^s = g_l + \mathbf{j}b_l = 1/(r_l + \mathbf{j}x_l) = 1/z_l$ while $y_{l^+}^{sh} = \mathbf{j}b_{l^+}^{sh}$ and $y_{l^-}^{sh} = \mathbf{j}b_l^{sh}$ are the shunt admittances at the sending and receiving ends, respectively ($y_{l^+}^{sh} = y_{l^-}^{sh} = \mathbf{j}b_l^{sh}/2$).

The active and reactive power flows into the line at the sending (receiving) end are denoted by $S_{l^+} = P_{l^+} + \mathbf{j}Q_{l^+}$ ($S_{l^-} = P_{l^-} + \mathbf{j}Q_{l^-}$). $I_{l^+} = |I_{l^+}|\angle\vartheta_{l^+}$ ($I_{l^-} = |I_{l^-}|\angle\vartheta_{l^-}$) is the current flowing into the line from sending (receiving) nodes while I_l is the current in the longitudinal section. The square of current flow is denoted by $f_{l^+} = |I_{l^+}|^2$, $f_{l^-} = |I_{l^-}|^2$, and $f_l = |I_l|^2$ in each case. Each time step in the planning horizon T is indexed by $t \in T$.

8.2 Operational planning with static secure islanding constraints

In this section, we describe the mathematical optimisation problem that allows to perform operational planning, while considering the pre- and post-islanding steady-state security and adequacy constraints. This is a sub-problem of the security-constrained AC optimal power flow family of problems [2], extended to consider the specific topology of MGs (and distribution networks in general) and the post-islanding security and adequacy constraints.

218   *Intended and unintended islanding of distribution grids*

8.2.1   *Day-ahead optimal planning model with static islanding constraints*

The compact formulation of this problem is defined as:

$$\underset{\chi}{\text{minimise}} : \Theta(\chi) \tag{8.1a}$$

$$\text{subject to} : h^{gc}(\chi) = 0 \tag{8.1b}$$

$$g^{gc}(\chi) \leq 0 \tag{8.1c}$$

$$h^{isl}(\chi) = 0 \tag{8.1d}$$

$$g^{isl}(\chi) \leq 0 \tag{8.1e}$$

where χ represents both the steady-state operational variables (nodal voltages, power flows in the distribution lines, and power consumption by the load in the network) and the control variables, which include the active and reactive power injections by the different generation units and the power exchange with the grid.

To ensure the static security of the system both pre- and post-islanded, constraints related to the *grid-connected* and *islanded* states of the MG should be included (denoted by superscripts $\{\bullet^{gc}\}$ and $\{\bullet^{isl}\}$, respectively). The objective and constraints are detailed below, with only the differences between (8.1b)/(8.1d) and (8.1c)/(8.1e) highlighted.

8.2.1.1   Objective

The objective function (8.1a) is a cost minimisation function that includes costs related to the *grid-connected* and *islanded* state, as:

$$\underset{\chi \in \Omega^{\text{MG}}}{\min} \; \Theta^{gc}(\chi^{gc}) + \Theta^{isl}(\chi^{gc}, \chi^{isl}) \tag{8.2}$$

where $\Omega^{\text{MG}} = \left\{ \chi = [\chi^{gc}, \chi^{isl}] \; \middle| \; \chi^{gc} \in \Omega^{gc} \; ; \; \chi^{isl} \in \Omega^{isl} \right\}$

During grid-connected operation, the objective can take the form:

$$\begin{aligned}
\Theta^{gc}(\chi^{gc}) = \sum_{t \in T} & \left(\left(C^{\text{imp}^{\text{P}}} \cdot p_t^{\text{imp}} - C^{\text{exp}^{\text{P}}} \cdot p_t^{\text{exp}} \right) \right. \\
& \left. + \left(C^{\text{imp}^{\text{Q}}} \cdot q_t^{\text{imp}} - C^{\text{exp}^{\text{Q}}} \cdot q_t^{\text{exp}} \right) \right) \\
& + \sum_{t \in T} \sum_{g \in \{S,C\}} \left(C_g^{\text{P}} \cdot p_{gt} + C_g^{\text{Q}} \cdot q_{gt} \right) + \sum_{t \in T} \sum_{d \in D^{\text{v}}} \left(C_d^{\text{v}} \cdot p_{dt}^{\text{v}} \right)
\end{aligned} \tag{8.3}$$

In this mode, the active/reactive power imported/exported from/to the grid costs $C^{\text{imp}^{\text{P/Q}}} / C^{\text{exp}^{\text{P/Q}}}$, respectively, while the generation costs of the different generators are defined by $C_g^{\text{P/Q}}$. Additionally, penalty costs C_d^{v} are applied to the use of variable loads due to the inconvenience of shifting demand away from the consumer's preferred time.

On the other hand, the islanded operation costs, defined in (8.4), consider the penalty cost C_d incurred when a load is curtailed during the emergency operation

Planning methods for secure islanding   219

at time t, when the MG operates as an island. Of course, when the system is not islanded, this cost is zero.

$$\Theta_t^{\mathrm{isl}}(\chi^{\mathrm{isl}}) = \sum_{d \in D} \left(C_d \cdot \left(1 - z_{dt}\right) \cdot \left(p_{dt}^{\mathrm{f}} + p_{dt}^{\mathrm{v}}\right)\right) \tag{8.4}$$

Binary z_{dt} indicates the connection status of a load, i.e. $z_{dt} = 0$ when the load is disconnected or curtailed and $z_{dt} = 1$ when the load is connected. Superscripts $\{\bullet^{\mathrm{f}}\}$ and $\{\bullet^{\mathrm{v}}\}$ represent the fixed and variable loads, respectively.

8.2.1.2 AC power flow model

Equations (8.1b) and (8.1d) relate to the network steady-state model, normally represented by the AC power flow equations. The constraints include both the power flow and power balance constraints and are defined as:

$$s_{it}^{\mathrm{d}} - s_{t|i=1}^{\mathrm{imp}} + s_{t|i=1}^{\mathrm{exp}} - \sum_{g \in G^i} s_{gt} = \sum_{\eta(l^+)=i} S_{l^+} + \sum_{\eta(l^-)=i} S_{l^-} \qquad \forall i, t \tag{8.5a}$$

$$S_{l^+} = U_{\eta(l^+)t} \left(I_{l^+}\right)^*, \qquad S_{l^-} = U_{\eta(l^-)t} \left(I_{l^-}\right)^*, \qquad \forall l, t \tag{8.5b}$$

$$I_{l^+} = y_l^s \left(U_{\eta(l^+)} - U_{\eta(l^-)}\right) + y_l^{sh} U_{\eta(l^+)}, \qquad \forall l, t \tag{8.5c}$$

$$I_{l^-} = y_l^s \left(U_{\eta(l^-)} - U_{\eta(l^+)}\right) + y_l^{sh} U_{\eta(l^-)}, \qquad \forall l, t \tag{8.5d}$$

In (8.5), the complex form of the non-linear AC OPF formulation is presented taking into account the line shunts. Equation (8.5a) enforces the power balance at each node in the network while (8.5b)–(8.5d) define the active and reactive power flows at both ends of each line.

When the microgrid is operating in the post-islanded mode, the constraints in (8.5) are applied as stated corresponding to (8.1d); however, the power import from and export to the grid in (8.5b) is set to zero, i.e. $s_{t|i=1}^{\mathrm{imp}} = s_{t|i=1}^{\mathrm{exp}} = 0$.

8.2.1.3 Static operational/technical constraints

Equation (8.1c) includes the different techno-economic operational constraints applied to the network such as line loading constraints, voltage constraints, and generation limits on the different units. These are explicitly defined in the following. It should be noted that these constraints need to be satisfied both in the pre- and post-islanded system.

The constraints for the islanded operation are indicated by (8.1e). Note that where modifications are required in the islanding mode for a constraint, the changes have been clarified in the text.

Constraints on grid power exchange: Defined only for the grid-connected mode, the bounds on power imports and exports to the grid take the form:

$$\begin{aligned} 0 \le p_t^{\mathrm{imp}} \le \overline{p}_t^{\mathrm{imp}} \cdot z_t^{\mathrm{P}}, \quad 0 \le p_t^{\mathrm{exp}} \le \overline{p}_t^{\mathrm{exp}} \cdot (1 - z_t^{\mathrm{P}}), \\ 0 \le q_t^{\mathrm{imp}} \le \overline{q}_t^{\mathrm{imp}} \cdot z_t^{\mathrm{Q}}, \quad 0 \le q_t^{\mathrm{exp}} \le \overline{q}_t^{\mathrm{exp}} \cdot (1 - z_t^{\mathrm{Q}}), \end{aligned} \qquad \forall t \in T \tag{8.6}$$

220 *Intended and unintended islanding of distribution grids*

where the auxiliary binary decision variables z_t^P/z_t^Q prevent the simultaneous import and export of active/reactive power.

Constraints on loads: Power consumption of load at each node can be fixed (s_{dt}^f) or variable (s_{dt}^v), i.e. flexible load, the limitations applied to the load variations over the planning horizon include:

$$[p/q]_{dt} = ([p/q]_{dt}^f + [p/q]_{dt}^v) \cdot z_{dt}, \qquad \forall d \in D, \, t \in T \tag{8.7a}$$

$$0 \le p_{dt}^f \le \overline{P}_d^f, \quad 0 \le q_{dt}^f \le \overline{q}_d^f, \qquad \forall d \in D, \, t \in T \tag{8.7b}$$

$$\sum_{t \in T} p_{dt}^d = E_d, \qquad \forall d \in D \tag{8.7c}$$

$$p_{dt} \le E_d - \sum_{\tau=1}^{t-1} p_{d\tau}, \qquad \forall d \in D, \, t \in T \tag{8.7d}$$

where the binary variable z_{dt} indicates the connection status of the load and E_d the total energy demand of load d in the defined planning horizon typically equal to 24 hours. In grid-connected mode, this value is forced to one, i.e. all load should be served in its entirety. On the contrary, in islanded mode, z_{dt} can take on a value of one $\Longrightarrow$ load-connected or zero $\Longrightarrow$ load-curtailed. Constraint (8.7b) ensures the total power demand does not exceed the peak load defined at a given node, this is defined based on the maximum consumption at a node, i.e. based on the capacity of the distribution transformer at the node. Note that flexible loads relate to large consumers with the ability to upwardly/downwardly adjust consumption when required within their maximum/minimum consumption limits.

In grid-connected mode, constraint (8.7c) ensures that the defined energy consumption E_d for the day is met. During islanded operation, (8.7d) ensures that only unserved load consumption is met. The fraction of unserved load at any hour given the daily requirement is defined by the right-hand side of (8.7d).

Constraints on power production from synchronous generators:

$$0 \le p_{st} \le \overline{P}_s, \qquad \forall s \in S, t \in T \tag{8.8a}$$

$$\underline{q}_s \le q_{st} \le \overline{q}_s, \qquad \forall s \in S, t \in T \tag{8.8b}$$

$$- \mathrm{rd}_s \le p_{st} - p_{s(t-1)} \le \mathrm{ru}_s, \qquad \forall s \in S, t \in T \tag{8.8c}$$

The limits to the power output of the synchronous generators (SGs) are defined in (8.8a)–(8.8b). The inter-hour ramp-up ru_s and ramp-down rd_s limits of the SG units are ensured in (8.8c). These constraints are applied in both pre- and post-islanding operation modes of the network.

Constraints on power production from converter interfaced generators: In line with the current grid code, the operation of renewable energy units at a power factor less than one is allowed, thus the limits on power production from Converter Interfaced Generators (CIGs) are defined as:

$$0 \le p_{ct} \le \overline{P}_{ct}, \qquad \forall c \in C, t \in T, \tag{8.9a}$$

$$0 \le q_{ct} \le \tan(\overline{\phi}) \cdot \overline{P}_{ct}, \qquad \forall c \in C, t \in T, \tag{8.9b}$$

where ϕ is the phase angle and $\cos(\overline{\phi})$ is the minimum acceptable power factor of the CIG defined by the grid code. The active and reactive power limits are defined in (8.9a) and (8.9b), respectively. The limitation adopted in (8.9b) is flexible as it allows for reactive power generation at the maximum level, $\overline{q}_{ct}$, even at instances with low active power injection [3]. Note that the maximum available power of each CIG unit, $\overline{p}_{ct}$, is dependent on the weather conditions and nominal power of the unit. These constraints are applied in both the grid-connected and islanded modes.

Constraints on steady-stage voltage: Voltage levels at each node in steady-state operation should be maintained within the normal operation region defined by grid standards. In both grid-connected and islanded modes, this is ensured by:

$$\underline{U} \leq U_{it} \leq \overline{U}, \quad U_{t|i=1} = 1 \text{ pu}, \qquad \forall i \in N, \ t \in T \tag{8.10}$$

In grid-connected mode, the voltage at the PCC is maintained by the stiff grid. However, during the islanded mode, the voltage is controlled by the DERs present in the MG. This may be realised using a single-master or with a multi-master operating mode [4]. Voltage control can also be ensured by the SG units in the network, which can provide the voltage reference when the power supply to/from the grid is lost. Moreover, the CIG units provide further voltage and reactive power support based on (8.9b).

Constraints on line loading: The transmission capacity of each line is maintained within secure bounds in both pre- and post-islanding modes as:

$$0 \leq f_{tl^+} \leq \overline{f}_l, \quad 0 \leq f_{tl^-} \leq \overline{f}_l, \qquad \forall l \in L, \ t \in T \tag{8.11}$$

8.2.2 Network modelling for power flow constraints

The network model presented by the AC power flow equations in (8.5) is non-linear and non-convex due to (8.5b). Therefore, the model can only be solved through the adoption of non-linear programming (NLP) techniques, which introduces challenges in computational performance, scalability, and global optimality. On the other hand, convex relaxations have been proposed for this type of problem that can yield global optimal bounds to the original non-convex AC OPF [5]. The exactness of the relaxations depends on the tightness of the envelope and defined sufficient conditions, thus providing a lower bound on the objective at the least.

Some commonly used convex formulations that can be applied to radial distribution networks are presented in this subsection. These adopt the branch flow model (BFM) due to its desirable numerical characteristics in relation to radial networks [6]. This model formulates the power flow equations in terms of active and reactive power flows, squared current magnitude flows, and squared voltage magnitude at each node as indicated in [6].

A comparison regarding the accuracy and performance of these methods is detailed in [5].

8.2.2.1 Adapted DistFlow relaxation

In the DistFlow (DF) model [7,8], the AC power flow equations are redefined and represented using the constraints (8.12a)–(8.12d). Constraint (8.12c) relates to the SOCP relaxation applied to the equality constraint (8.5a)–(8.5b) in the non-linear model resulting in its convexification. The relaxed model in (8.12) can be used with either zero line shunts (hereinafter denoted as DF), i.e. line parameter $b_l^{sh} = 0$, or with non-zero line shunts (hereinafter denoted as $DF_{w/s}$), i.e. line parameter $b_l^{sh} > 0$. The sufficient conditions for the exactness of this model are defined in [9–11].

$$S_{tl^+} = s_{t\eta(l^-)}^{d} - s_{t|\eta(l^-)=1}^{imp} + s_{t|\eta(l^-)=1}^{exp} - \sum_{g \in G^{\eta(l^-)}} S_{gt}$$

$$+ \sum_{m \in L^{\eta(l^+)}} S_{tl_m^+} + z_l f_{lt} - \mathbf{j}(u_{\eta(l^+)t} + u_{\eta(l^-)t})\frac{b_l^{sh}}{2}, \qquad \forall l,t \qquad (8.12a)$$

$$u_{\eta(l^-)t} = u_{\eta(l^+)t} + |z_l|^2 f_{lt} - 2\operatorname{Re}\left(z_l^*\left(S_{tl^+} + \mathbf{j}u_{\eta(l^+)t}\frac{b_l^{sh}}{2}\right)\right), \qquad \forall l,t \qquad (8.12b)$$

$$f_{lt} \geq \frac{|S_{tl^+} + \mathbf{j}u_{\eta(l^+)t}\frac{b_l^{sh}}{2}|^2}{u_{\eta(l^+)t}} \text{ or } \frac{|S_{tl^-} - \mathbf{j}u_{\eta(l^-)t}\frac{b_l^{sh}}{2}|^2}{u_{\eta(l^-)t}}, \qquad \forall l,t \qquad (8.12c)$$

$$\overline{f}_l u_{\eta(l^+)t} \geq |S_{tl^+}|^2, \qquad \overline{f}_l u_{\eta(l^-)t} \geq |S_{tl^-}|^2, \qquad \forall l,t \qquad (8.12d)$$

8.2.2.2 Modified Lin-DistFlow relaxation

A modified linear approximation of the DistFlow formulation in (8.12), defined as LinDistFlow, is presented in (8.13). Here, the power flow equations are defined as in (8.13a)–(8.13b) with the assumption that line losses indicated by the square of current flow are negligible in comparison with the active and reactive power flows (i.e. $f_{lt} << S_{tl} \therefore f_{lt} \simeq 0$) [12]. A modification to the initial model in [12] to include line flow limits using constraints (8.13c)–(8.13f) is provided. Moreover, to maintain the linear nature of the problem, the quadratic line flow constraint (8.12c) is linearised using a convex polygon defined by inner approximations of the thermal loading circle [13,14]. Thus, constraint (8.12c) is replaced by its piece-wise approximation as defined in (8.13c)–(8.13f) where parameter $a_d = \sqrt{2} - 1$ is the derivative of the lines constructing the eight segments of the convex approximation. This model provides an upper bound on voltage and a lower bound on power flows in the network [12].

$$S_{tl^+} + s_{t\eta(l^-)} = \sum_{m \in L^{\eta(l^+)}} S_{tl_m^+}, \qquad \forall l,t \qquad (8.13a)$$

$$u_{\eta(l^-)t} = u_{\eta(l^+)t} - 2\left(r_l P_{tl^+} + x_l Q_{tl^+}\right), \qquad \forall l,t \qquad (8.13b)$$

$$-\overline{S}_l \leq P_{tl^+} + a_d Q_{tl^+} \leq \overline{S}_l, \qquad \forall l,t \qquad (8.13c)$$

$$-\overline{S}_l \leq P_{tl^+} - a_d Q_{tl^+} \leq \overline{S}_l, \qquad \forall l,t \qquad (8.13d)$$

$$-\overline{S}_l \leq a_d P_{tl^+} + Q_{tl^+} \leq \overline{S}_l, \qquad \forall l,t \qquad (8.13e)$$

$$-\overline{S}_l \leq a_d P_{tl^+} - Q_{tl^+} \leq \overline{S}_l, \qquad \forall l,t \qquad (8.13f)$$

8.2.2.3 Extended DistFlow relaxation with line shunts

Equation (8.14) presents a variant of the BFM relaxation for the AC power flows considering non-zero line shunts [7]. Unlike the model presented in (8.12) where current flow is only defined in the longitudinal section of the Π model in Figure 8.2, the current and power flows here are defined at both ends of the line. This enhances the non-violation of the line ampacity limits in the physical network [15]. Here, (8.14a)–(8.14e) define the power flow equations while the line flows are constrained by (8.14f). Parameter α_{l^+} is defined as $\alpha_{l^+} = 1 + z_l y_{l^+}^{sh}$. Similar to the model in (8.12), constraint (8.14b) has been relaxed from an equality (8.5a)–(8.5b) to inequality thus obtaining an SOCP relaxation of the non-convex power flow. Sufficient conditions for the exact SOCP relation of this model are detailed in [7].

$$s_{it} = \sum_{\eta(l^+)=i} S_{l^+} + \sum_{\eta(l^-)=i} S_{l^-} \qquad\qquad \forall i, t \qquad (8.14\text{a})$$

$$u_{t\eta(l^+)} \cdot f_{tl^+} \geq |S_{tl^+}|^2 \quad \text{or} \quad u_{t\eta(l^-)} \cdot f_{tl^-} \geq |S_{tl^-}|^2, \qquad \forall l, t \qquad (8.14\text{b})$$

$$|\alpha_{l^+}|^2 u_{t\eta(l^+)} - u_{t\eta(l^-)} = 2\,\mathrm{Re}(\alpha_{l^+} z_l^* S_{tl^+}) - |z_l|^2 f_{tl^+}, \qquad \forall l, t \qquad (8.14\text{c})$$

$$|\alpha_{l^-}|^2 u_{t\eta(l^-)} - u_{t\eta(l^+)} = 2\,\mathrm{Re}(\alpha_{l^-} z_l^* S_{tl^-}) - |z_l|^2 f_{tl^-}, \qquad \forall l, t \qquad (8.14\text{d})$$

$$\alpha_{l^+}^* u_{t\eta(l^+)} - z_l^* S_{tl^+} = \left(\alpha_{l^-}^* u_{t\eta(l^-)} - z_l^* S_{tl^-}\right)^*, \qquad \forall l, t \qquad (8.14\text{e})$$

$$0 \leq f_{l^+} \leq (\overline{f}_l), \qquad 0 \leq f_{l^-} \leq (\overline{f}_l), \qquad \forall l, t \qquad (8.14\text{f})$$

8.2.2.4 Augmented DistFlow with line shunts

The formulation in (8.15) enhances that in (8.12) by adding a new set of constraints, i.e. "augmentations", defined by (8.15a)–(8.15j) with conditions on its sufficient operation defined in [8]. The augmentations create inner approximations (restrictions) for the feasible space of the problem that ensures a tighter envelope for the original relaxation in (8.12). The tighter envelopes are achieved by introducing security constraints based on auxiliary variables for both the line power flows and node voltages. In (8.15), auxiliary variables defined by superscripts $\hat{\bullet}/\check{\bullet}$ indicate the lower/upper bound on the associated variable.

However, while the set of security constraints improves the feasibility of the model, it creates a larger set of optimisation variables that widen the solution space. Sufficient conditions for exactness i.e. AC feasibility of this model include the requirement for strictly increasing cost function for active power in the objective and various conditions on the network parameters defined in [8].

$$(8.12\text{a}) - (8.12\text{d})$$

$$\hat{S}_{tl^+} + s_{t\eta(l^-)} = \sum_{m \in L^{\eta(l^+)}} \hat{S}_{tl_m^+} - \mathbf{j}\left(\check{u}_{\eta(l^+)t} + \check{u}_{\eta(l^-)t}\right)\frac{b_l^{sh}}{2}, \qquad \forall l, t \qquad (8.15\text{a})$$

$$\hat{S}_{tl^-} + s_{t\eta(l^-)} = \sum_{m \in L^{\eta(l^+)}} \hat{S}_{tl_m^+}, \qquad \forall l, t \qquad (8.15\text{b})$$

$$\check{u}_{\eta(l^-)t} = \check{u}_{\eta(l^+)t} - 2\,\mathrm{Re}\left(z_l^*\left(S_{tl^+} + \mathbf{j}\check{u}_{\eta(l^+)t}\frac{b_l^{sh}}{2}\right)\right), \qquad \forall l, t \qquad (8.15\text{c})$$

$$\check{S}_{tl^+} + s_{t\eta(l^-)} = \sum_{m \in L^{\eta(l^+)}} \check{S}_{tl_m^+} + z_l \check{f}_{lt} - \mathbf{j}(u_{\eta(l^+)t} + u_{\eta(l^-)t})\frac{b_l^{sh}}{2}, \qquad \forall l, t \qquad (8.15\mathrm{d})$$

$$\check{S}_{tl^-} + s_{t\eta(l^-)} = \sum_{m \in L^{\eta(l^+)}} \check{S}_{tl_m^+}, \qquad \forall l, t \qquad (8.15\mathrm{e})$$

$$\check{f}_l u_{\eta(l^+)t} \geq \max\left(\left|\hat{P}_{tl^+}\right|^2, \left|\check{P}_{tl^+}\right|^2\right) + $$
$$\max\left(\left|\hat{Q}_{tl^+} + \check{u}_{\eta(l^+)t}\frac{b_l^{sh}}{2}\right|^2, \left|\check{Q}_{tl^+} + u_{\eta(l^+)t}\frac{b_l^{sh}}{2}\right|^2\right), \qquad \forall l, t \qquad (8.15\mathrm{f})$$

$$\check{f}_l u_{\eta(l^-)t} \geq \max\left(\left|\hat{P}_{tl^-}\right|^2, \left|\check{P}_{tl^-}\right|^2\right) + $$
$$\max\left(\left|\hat{Q}_{tl^-} - \check{u}_{\eta(l^-)t}\frac{b_l^{sh}}{2}\right|^2, \left|\check{Q}_{tl^-} - u_{\eta(l^-)t}\frac{b_l^{sh}}{2}\right|^2\right), \qquad \forall l, t \qquad (8.15\mathrm{g})$$

$$\overline{f}_l u_{\eta(l^+)t} \geq \max\left(\left|\hat{P}_{tl^+}\right|^2, \left|\check{P}_{tl^+}\right|^2\right) + \max\left(\left|\hat{Q}_{tl^+}\right|^2, \left|\check{Q}_{tl^+}\right|^2\right), \qquad \forall l, t \qquad (8.15\mathrm{h})$$

$$\overline{f}_l u_{\eta(l^-)t} \geq \max\left(\left|\hat{P}_{tl^-}\right|^2, \left|\check{P}_{tl^-}\right|^2\right) + \max\left(\left|\hat{Q}_{tl^-}\right|^2, \left|\check{Q}_{tl^-}\right|^2\right), \qquad \forall l, t \qquad (8.15\mathrm{i})$$

$$\hat{P}_{tl^+} \leq \check{P}_{tl^+} \leq \overline{P}_{l^+}, \qquad \hat{Q}_{tl^+} \leq \check{Q}_{tl^+} \leq \overline{Q}_{l^+}, \qquad \forall l, t \qquad (8.15\mathrm{j})$$

8.3 Investment planning with static secure islanding constraints

In this section, we expand on the operational planning problem presented in the previous sections to investigate investment planning methods. In this case, the units available in the MG are not known, and the most cost-effective combination of investment units is selected to ensure static and dynamic security. Moreover, while the day-ahead operational planning problem was solved over 24 hours using forecasted generation and consumption values, the investment planning problem is solved over multiple *representative days* trying to capture all the potential variations that the MG might meet.

8.3.1 Changes in the objective

The investment planning model of an MG under static operational constraints in grid-connected and islanded mode can be presented as a min–max–min optimisation problem with the compact form defined as:

$$\min_{\chi \in \Omega^{MG}} \Theta^{\mathrm{inv}}(\chi^{\mathrm{inv}}) + \Theta^{\mathrm{gc,opr}}(\chi^{\mathrm{inv}}, \chi^{\mathrm{gc,opr}}) + \|\check{\Theta}^{\mathrm{isl,opr}}(\chi^{\mathrm{inv}}, \chi^{\mathrm{gc,opr}}, \chi^{\mathrm{isl,opr}})\|_\infty, \quad (8.16)$$

where $\Omega^{MG} = \{\chi = [\chi^{\mathrm{inv}}, \chi^{\mathrm{gc,opr}}, \chi^{\mathrm{isl,opr}}] \mid \chi^{\mathrm{inv}} \in \Omega^{\mathrm{inv}} ; \chi^{\mathrm{gc,opr}} \in \Omega^{\mathrm{gc,opr}} ; \chi^{\mathrm{isl,opr}} \in \Omega^{\mathrm{isl,opr}}\}$, Θ^{inv} represents the vector of investment variables as a function of uncertain parameters (i.e. representative days), $\chi^{\mathrm{gc,opr}} \in \Theta^{\mathrm{gc,opr}}(\chi^{\mathrm{inv}})$ represents the vector of operation variables in grid-connected mode as a function of investment variables and uncertain

Planning methods for secure islanding 225

parameters, and $\chi^{\text{isl,opr}} \in \Theta^{\text{isl,opr}}(\chi^{\text{inv}}, \chi^{\text{gc,opr}})$ represents the vector of operation variables in islanded mode as a function of investment decisions, operation variables in grid-connected mode, and uncertain parameters.

The term $\Theta^{\text{inv}}(\chi^{\text{inv}})$ in the objective function is defined as:

$$\Theta^{\text{inv}} = \sum_{g \in \{S,C\}} \left(C_g^{\text{inv}} \cdot z_g \right) + \sum_{l \in L} \left(C_l^{\text{inv}} \cdot z_l \right), \tag{8.17}$$

and includes the total investment/reinforcement costs of generators/lines throughout the planning horizon. The optimisation variables $\chi^{\text{inv}} = \{z_g, z_l\}, \forall g \in \{S, C\} \wedge \forall l \in L$

Moreover, $\breve{\Theta}^{\text{isl,opr}} = [\min \Theta_{11}^{\text{isl,opr}}, \dots, \min \Theta_{n^T,n^O}^{\text{isl,opr}}]$, $n^T = |T|$, $n^O = |O|$, and $\|\breve{\Theta}^{\text{isl,opr}}\|_\infty = \max(\min_{\forall t, \forall o} \Theta_{to}^{\text{isl,opr}})$. Hence, the objective function (8.16) minimises the total investment costs Θ^{inv}, the *"expected"* total operation costs in the grid-connected mode $\Theta^{\text{gc,opr}}$ for all hours ($t \in T$) of all representative days ($o \in O$), and the *"worst-case"* total penalty costs of disconnecting loads from MG in islanded mode $\Theta^{\text{isl,opr}}$ for all hours in all representative days.

The min–max–min objective function (8.16) can be rewritten as a single minimisation problem by using the auxiliary variable γ:

$$\min_{\chi \in \Omega^{\text{MG}}} \Theta^{\text{inv}}(\chi^{\text{inv}}) + \Theta^{\text{gc,opr}}(\chi^{\text{inv}}, \chi^{\text{gc,opr}}) + \gamma \tag{8.18a}$$

$$\text{s.t.} \quad \gamma \geq \Theta_{to}^{\text{isl,opr}}(\chi^{\text{inv}}, \chi^{\text{gc,opr}}, \chi^{\text{isl,opr}}), \ \forall t \in T, o \in O, \tag{8.18b}$$

The optimisation problem (8.18) is a mixed-integer linear programming (MILP) problem, and as such can be solved by available software packages to obtain optimal investment and operation decisions in grid-connected and islanded mode.

8.3.2 Changes in the constraints

The vector of investment variables (χ^{inv}) in (8.16) consists of a set of binary variables (0/1) assigned to each of the investment candidates. The candidates considered in this case are SGs, CIGs, and power transmission lines. Therefore, constraints (8.8), (8.9), and (8.11) in both pre- and post-islanding modes are modified to include binary variables z_s, z_c, and z_l for each candidate SG, CIG, and power line, respectively, as follows:

Modification of (8.8):

$$0 \leq p_{st} \leq \overline{p}_s \cdot z_s, \qquad\qquad \forall s \in S, t \in T \tag{8.19a}$$

$$\underline{q}_s \leq q_{st} \leq \overline{q}_s \cdot z_s, \qquad\qquad \forall s \in S, t \in T \tag{8.19b}$$

$$-rd_s \leq p_{st} - p_{s(t-1)} \leq ru_s, \qquad\qquad \forall s \in S, t \in T \tag{8.19c}$$

For all candidate SGs, decision z_s can take on a value of zero if not installed and one if the unit is installed. For existing SG units this value is always set to one.

Modification of (8.9):

$$0 \leq p_{ct} \leq \overline{p}_{ct} \cdot z_c, \qquad\qquad \forall c \in C, t \in T, \tag{8.20a}$$

$$0 \leq q_{ct} \leq \tan(\overline{\phi}) \cdot \overline{p}_{ct} \cdot z_c, \qquad\qquad \forall c \in C, t \in T, \tag{8.20b}$$

226 *Intended and unintended islanding of distribution grids*

The decision to install a CIG unit is determined by the binary variable z_c. If installed $z_c = 1$ and if not installed $z_c = 0$. Similarly, for all existing units z_c is always set to one.

Modification of (8.11):

$$0 \le f_{tl^+} \le \overline{f}_l^0 \cdot z_l^0 + \overline{f}_l \cdot z_l, \qquad \forall l \in L,\ t \in T \tag{8.21}$$

$$0 \le f_{tl^-} \le \overline{f}_l^0 \cdot z_l^0 + \overline{f}_l \cdot z_l, \qquad \forall l \in L,\ t \in T \tag{8.22}$$

$$z_l^0 + z_l = 1, \qquad \forall l \in L \tag{8.23}$$

The initial status of the line is represented by the binary variable $z_l^0 = 1$ and $z_l = 0$ if there is no requirement for reinforcement at loading capacity $\overline{f}_l^0$. If, however, the line loading limits are violated, a new line with a higher loading capacity $\overline{f}_l$ can be utilised by setting binary variable $z_l = 1$ and $z_l^0 = 0$.

All other constraints remain as previously detailed, however, each constraint is now enforced not only for each period but for every representative day in set O.

8.3.3 Representative days

As mentioned above, the investment planning problem is usually solved using data (generation, consumption, etc.) from one or more representative days. These days are extracted from historical data and sometimes enhanced with predicted operating data (e.g. future energy scenarios). The most secure approach would be to use all the historical data (e.g. one year) and ensure the objective is minimised over all the data and the constraints hold for all the data. However, due to tractability issues, it may not be feasible to consider the entire yearly patterns. Thus, a sufficient number of representative daily patterns (i.e. representative days) is used instead of the yearly patterns for correlated load demands and RES power productions through appropriate data clustering techniques to characterise their increasing/decreasing variations during the planning horizon proficiently.

In general, data clustering techniques offer a means for grouping patterns exhibiting consistent behaviour, thus, allowing for a reduced number of scenarios to be utilised instead of the entire input data set. Different clustering techniques fall into two common groups, i.e. hierarchical and partitional algorithms [16]. An example of a clustering technique is the agglomerative hierarchical clustering (AHC) technique [16,17] that can be used to obtain scenarios (i.e. representative days) characterising the uncertainty of correlated load demands and RES power productions during the entire planning horizon. Using a bottom-up approach, this technique merges data based on a proximity (similarity) measure, i.e. aggregating from individual points to the most high-level cluster [17].

Taking an example of 24-hour patterns for a single year of data, the data clustering procedure groups daily patterns consisting of 72 distinct entries corresponding to the 24 hourly values each of normalised load demand, solar power production, and wind power production for all 365 days. In other words, the input data set is a (365×72)-dimensional matrix wherein each row of the matrix pertains to daily profiles of load demands, solar power productions, and wind power productions during

Planning methods for secure islanding   227

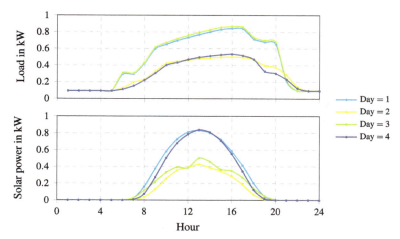

Figure 8.3 Demand and solar power generation patterns (four representative days)

a specific day of the planning horizon. All data can be normalised using the system base power, thus taking on values between '0' and '1'.

Some sample profiles of four representative days for load and PV power generation are depicted in Figure 8.3. These were extracted from the yearly historical data of PV generation in Texas for the year 2016 [18].

8.4 Incorporating transient islanding constraints in planning problems

The previous sections presented the MG operation and investment planning problems considering only static security constraints. However, as highlighted in the introduction of this chapter, it is important to ensure that the MG will survive the islanding transients. The islanding capability of MGs is critical in enhancing resilience by ensuring continuity and mitigating interruptions of energy supply to consumers in the event of extreme weather conditions or significant faults in the bulk transmission grid [19,20].

The successful island creation, especially following disastrous events, is subject to the secure transient performance of DERs, thus ensuring the survivability of the MG. However, unlike traditional bulk grids, MGs are inherently faced with a lack of rotational inertia and damping capability affecting their security in the event of significant power imbalance, and, more importantly, islanding from the main grid [21,22]. An MG is considered secure if all equipment (e.g. lines and generators) operate within their technical limits and tolerances, avoiding subsequent network disconnections and associated risk of cascading failures [23].

228   *Intended and unintended islanding of distribution grids*

The transient behaviour of MGs (and electric power systems in general) is described by the time-domain solution of an *Initial-Value Problem of Differential-Algebraic Equations* (IVP DAE) over the period of interest T:

$$F(\chi^{\text{inv}}, \chi^{\text{gc,opr}}, x, \dot{x}, \tau) = 0, \qquad \tau \in [0, T]$$

where x are the *differential-algebraic states* of the IVP DAE problem. The structure of the DAE depends on the investment and operation variables ($\chi^{\text{inv}}, \chi^{\text{gc,opr}}$) variables, while the initial values of the DAE model (x_0) depend only on the pre-fault operational decision variables ($\chi^{\text{gc,opr}}$). The period of interest (T) when considering the transient behaviour of the system is usually in the range of 15–20 s. It should be noted that the DAEs F are non-linear equations due to the nature of the electric power system models.

Thus, to incorporate the transient constraints in either the operation or investment planning problems presented previously, the compact formulation of this problem is restructured as a dynamic optimization problem:

$$\underset{\chi}{\text{minimise}} : \Theta(\chi) \tag{8.24a}$$

$$\text{subject to} : h^{gc}(\chi) = 0 \tag{8.24b}$$

$$g^{gc}(\chi) \leq 0 \tag{8.24c}$$

$$h^{isl}(\chi) = 0 \tag{8.24d}$$

$$g^{isl}(\chi) \leq 0 \tag{8.24e}$$

$$F(\chi, x, \dot{x}, \tau) = 0, \ \tau \in [0, T] \tag{8.24f}$$

$$\rho(\chi, x, \dot{x}, \tau) \leq 0, \ \tau \in [0, T] \tag{8.24g}$$

where ρ denotes the transient security limits imposed on the system variables over the time-period T. Examples of such constraints are the fault ride through (FRT) curves shown in Figure 8.4.

Such problems cannot be directly solved by existing optimisation solvers. It is necessary to pre-process the problem and convert it to a more standard format or develop a decomposition algorithm to split the DAEs from the remaining optimization problem. There are three main approaches to solving this kind of problem:

1. Discretise the DAEs with a preselected time step h and embed the resulting algebraic equations in place of (8.24f). Then, solve the resulting mixed-integer non-linear programming (MINLP) problem. While this is a straightforward approach that maintains a lot of the details of the system, it is very computationally intensive and impossible for real MG applications. Examples of software that can automatically achieve this can be found in [24].

    For instance, let's assume we analyse a 15-second transient response after the unscheduled MG islanding with a time step of 10 ms (a usual time step for transient analysis simulations). This results in 1,500 discretised points. Even a small 3-bus MG with four generators is modelled with more than 130 DAEs. Which

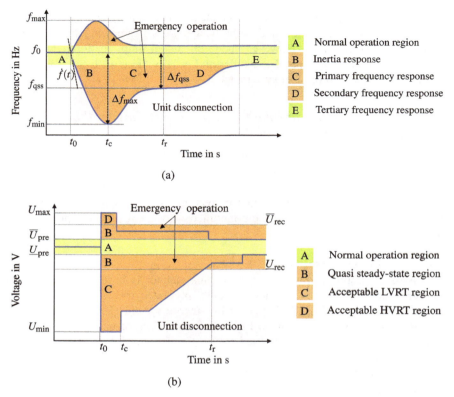

Figure 8.4 Fault ride through diagrams to ensure the transient security of a system after a fault. (a) frequency FRT diagram and (b) voltage FRT diagram.

leads to more than 200,000 new variables introduced in the optimisation problem. Moreover, due to the discretisation schemes used (e.g. trapezoidal method, and BDF), the variables are coupled between each time step.

2. Use a decomposition strategy to extract (8.24f) into a sub-problem and use an external dynamic simulator to simulate the transients. Then, extract some security constraints and embed them into the planning problem in an iterative solution strategy. An example of this approach can be found in [25].

    This approach allows incorporating higher accuracy in the transient response (protections, wide-area controls, etc.) and offers high computational performance since it exploits specialised simulation software. However, due to the use of external complicated simulation software, any mathematical proofs related to the convergence and optimality of the solution are extremely challenging.

3. First, replace (8.24f) with a *simplified/reduced model* capturing only part of the dynamic response that we are interested in. Then, use a decomposition strategy to solve the simplified/reduced model externally and extract some security constraints to be embedded into the planning problem in an iterative solution strategy. An example of this approach can be found in [26].

230 *Intended and unintended islanding of distribution grids*

This approach makes it easier to provide proof of convergence and optimality due to the use of simplified models for the dynamics. Moreover, it has high computational performance. However, due to the model simplification/reduction, it focuses only on one type of dynamics and usually the models are not able to capture protections or other system details.

8.4.1 *Incorporating transient frequency constraints in the investment planning problem*

In this subsection, we present one example of incorporating the transient frequency constraints in an MG investment planning problem using the third methodology listed in the previous section. First, the simplified model capturing the frequency dynamics after the islanding is detailed in Section 8.4.1.1. Then, the formulations of the decomposition strategy and the modified investment planning problem are presented in Section 8.4.1.2.

8.4.1.1 Frequency response model

The frequency response model adopted in this chapter is based on the uniform representation of frequency transients initially presented in [27] for a system of only SGs and modified in [28,29] for a low-inertia system. The dynamic model illustrated by the block diagram in Figure 8.5 comprises of both traditional SGs (indexed by $i \in S$) and CIGs (indexed by $c \in C$). The impact of *grid-supporting* CIGs providing frequency support via droop ($d \in C^d \subseteq C$) and VSM ($v \in C^v \subseteq C$) control is also included, as these are the two most common control approaches in the literature [30,31].

By analysis of the block diagram in Figure 8.5, the s-domain transfer function $G(s)$ between the active power change $\Delta P_e(s)$, where positive values corresponding to a net load decrease, and the CoI frequency deviation $\Delta f(s)$ can be derived as:

$$
G(s) = \frac{\Delta f(s)}{\Delta P(s)} = \left(\underbrace{(sM_s + D_s)}_{\text{SGs swing dynamics}} + \underbrace{\sum_{i \in S} \frac{K_i(1 + sF_iT_i)}{R_i(1 + sT_i)}}_{\text{SGs turbine \& governor response}} \right.
$$
$$
\left. + \underbrace{\sum_{d \in C^d} \frac{K_d}{R_d(1 + sT_d)}}_{\text{Droop-based CIGs}} + \underbrace{\sum_{v \in C^v} \frac{sM_v + D_v}{1 + sT_v}}_{\text{VSM-based CIGs}} \right)^{-1}.
$$
(8.25)

where M_s and D_s denote the aggregated normalised inertia and damping of SGs, respectively, while K_i, F_i, T_i, and R_i refer to the mechanical gain factor, the fraction of total power generated by the SG turbine, turbine time constant, and droop of SG i, respectively. Parameters K_d, R_d, and T_d define the power gain factor, droop, and time constant of the droop-based CIG indexed by d, respectively, and M_v, D_v, and T_v denote the virtual inertia constant, virtual damping constant, and time constant of the VSM-based CIGs denoted by index v.

It is noteworthy to mention that while different generators can have slightly distinct transient frequency responses, the dynamics described by the CoI swing

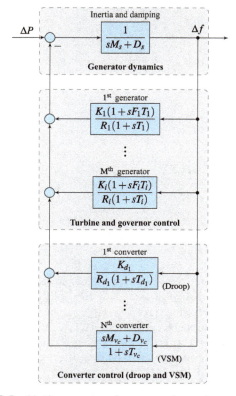

Figure 8.5 Uniform system frequency dynamics model [29]

equation with aggregate inertia M_s and damping D_s has been shown to adequately capture the system behaviour [27,28]. Also, droop-based CIGs consider only the damping capability of the converter (i.e. $D = 1/R_d$) while the VSM-based CIGs consider both the damping and the 'inertia' capability of the converter (i.e. D_v and M_v, respectively) [28].

Assuming that the time constants ($T_i \approx T$) of all SGs are orders of magnitude higher than the ones of converters [32], one can approximate $T \gg T_{d,v} \approx 0$, which transforms the transfer function (8.25) into:

$$G(s) = \frac{1}{MT} \frac{1+sT}{s^2 + 2\zeta\omega_n s + \omega_n^2}, \tag{8.26}$$

where $\omega_n = \sqrt{\frac{D+R_s}{MT}}$ and $\zeta = \frac{M+T(D+F_s)}{2\sqrt{MT(D+R_s)}}$. The parameters are calculated as follows:

$$M_s = \sum_{i \in S} M_i \frac{P_i}{P_s^{\text{base}}}, \quad D_s = \sum_{i \in S} D_i \frac{P_i}{P_s^{\text{base}}}, \tag{8.27a}$$

232 *Intended and unintended islanding of distribution grids*

$$R_s = \sum_{i \in S} \frac{K_i}{R_i} \frac{P_i}{P_s^{\text{base}}}, \qquad F_s = \sum_{i \in S} \frac{K_i F_i}{R_i} \frac{P_i}{P_s^{\text{base}}}, \tag{8.27b}$$

$$M_c = \sum_{v \in C^v} M_v \frac{P_{c_v}}{P_c^{\text{base}}}, \qquad D_c = \sum_{v \in C^v} D_v \frac{P_{c_v}}{P_c^{\text{base}}}, \tag{8.27c}$$

$$R_c = \sum_{d \in C^d} R_d \frac{P_{c_d}}{P_c^{\text{base}}}, \tag{8.27d}$$

$$M = \frac{M_s P_s^{\text{base}} + M_c P_c^{\text{base}}}{P_s^{\text{base}} + P_c^{\text{base}}}, \tag{8.27e}$$

$$D = \frac{D_s P_s^{\text{base}} + D_c P_c^{\text{base}} + R_c P_c^{\text{base}}}{P_s^{\text{base}} + P_c^{\text{base}}}, \tag{8.27f}$$

where P_i and P_c denote the active power capacity of the SG unit i and the CIG unit c, respectively. Furthermore, P_s^{base} and P_c^{base} represent the base power of all SG and CIG units connected to the system, that is, $P_s^{\text{base}} = \sum_{i \in S} P_i$ and $P_c^{\text{base}} = \sum_{c \in C} P_c$, respectively.

Following a disturbance, the dynamic frequency response is characterised by the instantaneous rate of change of frequency (RoCoF) ($\dot{f}_{\max}$) and frequency nadir ($\Delta f_{\max}$), whereas the steady-state response is governed by the constant frequency deviation from a pre-disturbance equilibrium (Δf_{ss}). Assuming a step-wise disturbance in the active power $\Delta P_e(s) = -\Delta P/s$, where ΔP is the net power change, the time-domain expression for frequency deviation ($\omega(t) \equiv \Delta f(t)$) can be derived as follows:

$$\omega(t) = -\frac{\Delta P}{M}\left(\frac{1}{T\omega_n^2} + \frac{1}{\omega_d}e^{-\zeta\omega_n t}\left(\sin \omega_d t - \frac{1}{\omega_n t}\sin \omega_d t + \phi\right)\right), \tag{8.28}$$

where $\omega_d = \omega_n\sqrt{1 - \zeta^2}$ and $\phi = \sin^{-1}\left(\sqrt{1 - \zeta^2}\right)$.

The maximum RoCoF occurs at $t_r = 0^+$, the instance of the disturbance, i.e. $\dot{\omega}_{max} = \dot{\omega}(t_r)$. The time-domain expression for RoCoF can be, therefore, obtained by solving $\dot{\omega}(t_r)$ and is derived as:

$$\dot{f}_{\max} = \dot{f}(t_0^+) = -\frac{\Delta P}{M}, \tag{8.29}$$

The frequency nadir described in (8.30) occurs at the time instance t_m when $\dot{\omega}(t_m) = 0$, this is derived as:

$$\Delta f_{\max} = -\frac{\Delta P}{D + R_s}\left(1 + \sqrt{\frac{T(R_s - F_s)}{M}}e^{-\zeta\omega_n t_m}\right), \tag{8.30}$$

with $t_m = (1/\omega_d)\tan^{-1}\left(\omega_d/\left(\omega_n\zeta - T^{-1}\right)\right)$.

Finally, the quasi-steady-state frequency given in (8.31) is derived from (8.28) for $t \to \infty$ as:

$$\Delta f_{\text{ss}} = -\frac{\Delta P}{D + R_s}, \tag{8.31}$$

It is clear that the aggregate system parameters such as M, D, R_g, and F_g have a direct impact on frequency performance. In particular, RoCoF and steady-state deviation are explicitly affected by M and (D, R_g), respectively, while frequency nadir has a non-linear dependency on all four system factors. With the increasing penetration of CIGs and subsequent decommissioning of conventional SGs, these parameters are drastically reduced and can compromise the overall frequency performance. To prevent the accidental activation of load-shedding, under/over frequency, and RoCoF protection relays, the proposed three-stage solution algorithm, described in the following, imposes limits on the aforementioned frequency metrics to account for low levels of inertia and damping and their impact on the frequency response after an MG islanding. The analytical model and closed-form expressions defining the different metrics have been verified and studied in [27,28,33].

8.4.1.2 A decomposition algorithm for solving an MG planning problem with dynamic constraints

A single-year mathematical formulation is considered to solve the inertia-aware investment and operational planning problem for MGs. In the proposed model, uncertain variations of load demands and renewable generations during the planning horizon are characterised by a sufficient number of representative days, obtained by utilising the k-means clustering technique [34]. Also, it is assumed islanding from the main grid may occur at each hour (indexed by $t \in T$) of every representative day (indexed by $o \in O$) to ensure the robustness of the optimal investment and operational plan against the worst-case unscheduled event in MG. Therefore, the compact formulation of the proposed planning model can be presented as follows:

$$\min_{\chi} \Theta^{\mathrm{inv}}(\chi^{\mathrm{inv}}) + \Theta^{\mathrm{gc,opr}}(\chi^{\mathrm{inv}}, \chi^{\mathrm{gc,opr}}) + \gamma \tag{8.32a}$$

subject to:

$$\gamma \geq \Theta_{to}^{\mathrm{isl,opr}}(\chi^{\mathrm{inv}}, \chi^{\mathrm{gc,opr}}, \chi^{\mathrm{isl,opr}}), \forall t \in T, o \in O, \tag{8.32b}$$

$$\Phi^{\mathrm{gc,opr}}(\chi^{\mathrm{inv}}, \chi^{\mathrm{gc,opr}}) \leq 0, \tag{8.32c}$$

$$\Phi^{\mathrm{isl,opr}}(\chi^{\mathrm{inv}}, \chi^{\mathrm{gc,opr}}, \chi^{\mathrm{isl,opr}}) \leq 0, \tag{8.32d}$$

$$\underline{\Delta f}^{\max} \leq \Delta f_{to}^{\max} \leq \overline{\Delta f}^{\max}, \qquad \forall t \in T, o \in O, \tag{8.32e}$$

$$\underline{\dot{f}}^{\max} \leq \dot{f}_{to}^{\max} \leq \overline{\dot{f}}^{\max}, \qquad \forall t \in T, o \in O, \tag{8.32f}$$

$$\underline{\Delta f}^{\mathrm{qss}} \leq \Delta f_{to}^{\mathrm{qss}} \leq \overline{\Delta f}^{\mathrm{qss}}, \qquad \forall t \in T, o \in O, \tag{8.32g}$$

where scalar variables are indicated by non-bold symbols while vectors/matrices are indicated by bold symbols. Also, χ denotes the vector of all decision variables relating to the investment χ^{inv}, grid-connected operation $\chi^{\mathrm{gc,opr}}$, and islanded operation $\chi^{\mathrm{isl,opr}}$ of the MG, i.e. $\chi = [\chi^{\mathrm{inv}}, \chi^{\mathrm{gc,opr}}, \chi^{\mathrm{isl,opr}}]$.

The objective function (8.32a) minimises the total investment costs (i.e. Θ^{inv}), the *"expected"* total operational costs in the grid-connected mode (i.e. $\Theta^{\mathrm{gc,opr}}$) for all hours of all representative days, and the *"worst-case"* total penalty costs of disconnecting load demands from MG in the islanded mode for the worst operational

234   *Intended and unintended islanding of distribution grids*

hour among all hours of all representative days (i.e. $\Theta^{\text{isl,opr}} \equiv \gamma$), as detailed in Section 8.3. The auxiliary variable γ is used to minimize the penalty costs for the worst operational hour among all hours of all representative days in the islanded mode (i.e. $\Theta^{\text{isl,opr}} \equiv \gamma \geq \Theta^{\text{isl,opr}}_{t,o}$ $\forall t \in T, \forall o \in O$) rather than the aggregated penalty costs for all hours of all representative days.

Constraints (8.32c) and (8.32d) correspond to the static linearised operational limitations in the grid-connected and islanded modes, respectively, where the power flow constraints are modeled using the convex relaxation in (8.14). Note that equality constraints (i.e. $a = b$) can be included in (8.32c) and (8.32d) by opposite inequality constraints (i.e. $a \geq b$ and $a \leq b$). Also, constraints (8.32e), (8.32f), and (8.32g) represent the transient frequency security limitations to ensure the adequacy of the operational reserve for frequency support in the event of islanding from the main grid.

In this work, it is assumed that the transient frequency response, as described in (8.30), (8.29), and (8.31) and utilised in (8.32e), (8.32f), and (8.32g), respectively, depends on the amount of power exchange with the main grid at the time of islanding of the MG, i.e. $-\Delta P = p_{to}^{\text{grid}}$. Therefore, the frequency support provided by a generator g, given its investment status z_g, is a function of its control parameters, i.e. $M(z_g)$, $D(z_g)$, $F_g(z_g)$, and $R_g(z_g)$ in addition to the amount of power exchange with the main grid at the time of islanding. The extended formulation of the investment planning model utilised in this chapter can be found in [26].

Although the static operational constraints in the grid-connected and islanded modes are linear, the investment planning problem described in (8.32) is a MINLP problem due to the inclusion of discrete/continuous investment/operational variables as well as non-linear and non-convex transient security constraints. To solve this intractable MINLP optimisation problem, a computationally tractable decomposition strategy based on the notion of Benders decomposition is described in detail in the following.

This section outlines a computationally tractable decomposition strategy for solving the inertia-aware MG planning problem in (8.32) using dual cutting planes. The solution approach is developed based on the notion of Benders decomposition to tackle the MINLP problem in (8.32) through a four-step iterative procedure, as illustrated in Figure 8.6. The master problem is related to the MG planning problem under linear static security constraints, while the sub-problem relates to the transient frequency security feasibility. The dual cutting planes utilised in this algorithm capture the sensitivity of the solution of the sub-problem to the variables/solution of the master problem. These sensitivities are then used in the computation of the master problem at the next iteration, therefore, capturing the impact of the optimal decisions of the master problem on the optimal solution of the sub-problem. Both decomposition algorithms are presented in the sequel.

In this algorithm, the complicating variables* separating the master problem and the sub-problem are the power exchange of the MG with the main grid (i.e. p_{to}^{grid}) and frequency control parameters for different generators (i.e. $\check{M}_s$, $\check{D}_s$, $\check{F}_s$, and $\check{R}_s$ for SGs,

---

*Complicating variables are specific variables in a complex optimisation problem preventing a tractable distributed solution if appropriate decomposition strategies are not adopted [35].

Planning methods for secure islanding 235

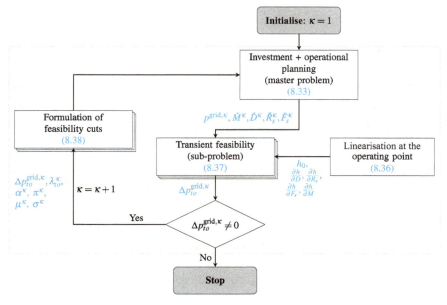

Figure 8.6 The proposed decomposition algorithm for an MG planning problem with dynamic constraints

$\check{M}_c$, $\check{D}_c$, and $\check{R}_c$ for CIGs, and $\check{M}$ and $\check{D}$ for both SGs and CIGs). The accent ˇ is used to differentiate a non-normalized parameter • from its normalized counterpart ǒ. The tasks at each iteration are detailed as follows:

Step 1: Initial formulation of the master problem

Initially, at iteration $\kappa = 1$, the master problem, which is a relaxation of (8.32), is solved to obtain feasible values of the complicating variables. It can be formulated as given below:

$$\min_{\chi} \Theta^{\text{inv}}(\chi^{\text{inv}}) + \Theta^{\text{gc,opr}}(\chi^{\text{inv}}, \chi^{\text{gc,opr}}) + \gamma \tag{8.33a}$$

subject to:

$$\gamma \geq \Theta_{to}^{\text{isl,opr}}(\chi^{\text{inv}}, \chi^{\text{gc,opr}}, \chi^{\text{isl,opr}}), \quad \forall t \in T, o \in O, \tag{8.33b}$$

$$\Phi^{\text{gc,opr}}(\chi^{\text{inv}}, \chi^{\text{gc,opr}}) \leq 0, \tag{8.33c}$$

$$\Phi^{\text{isl,opr}}(\chi^{\text{inv}}, \chi^{\text{gc,opr}}, \chi^{\text{isl,opr}}) \leq 0, \tag{8.33d}$$

$$\check{M}_s = \sum_{i \in S} M_i P_i z_i, \qquad \check{D}_s = \sum_{i \in S} D_i P_i z_i, \tag{8.33e}$$

$$\check{R}_s = \sum_{i \in S} \frac{K_i}{R_i} P_i z_i, \qquad \check{F}_s = \sum_{i \in S} \frac{K_i F_i}{R_i} P_i z_i, \tag{8.33f}$$

236 *Intended and unintended islanding of distribution grids*

$$\check{M}_c = \sum_{v \in C^v} M_v P_{c_v} z_v, \quad \check{D}_c = \sum_{v \in C^v} D_v P_{c_v} z_v, \tag{8.33g}$$

$$\check{R}_c = \sum_{d \in C^d} R_d P_{c_d} z_d, \tag{8.33h}$$

$$\check{M} = \check{M}_s + \check{M}_c, \tag{8.33i}$$

$$\check{D} = \check{D}_s + \check{D}_c + \check{R}_c, \tag{8.33j}$$

$$P^{\text{base}} = P_s^{\text{base}} + P_c^{\text{base}} = \sum_{i \in S} P_i z_i + \sum_{c \in C} P_c z_c. \tag{8.33k}$$

where the vector χ includes all investment variables (i.e. $z_s, \forall i \in S, z_v, \forall v \in C^v$, $z_d, \forall d \in C^d$, and $z_c = \{z_v, z_d\}$). The proposed master problem (8.33) is a mixed-integer linear programming (MILP) problem and can be solved by available optimisation packages straightforwardly.

Step 2: Linearisation at each operating point
The Nadir constraint (8.32e) defined in (8.30) is highly non-linear and non-convex. To remedy this issue, before its application to the sub-problem, it is linearised around the operating point at each hour of every representative day (i.e. $\forall t \in T$ and $\forall o \in O$). For this purpose, $\Delta f_{to}^{\max}$ in (8.32e) can be rewritten as follows:

$$\Delta f^{\max} = \frac{P_{to}^{\text{grid}}}{h(D, R_s, F_s, M)}, \tag{8.34}$$

where:

$$h(D, R_s, F_s, M) = \frac{1}{\frac{1}{D+R_s}\left(1 + \sqrt{\frac{T(R_s-F_s)}{M}}e^{-\zeta\omega_n t_m}\right)}. \tag{8.35}$$

Following, a Taylor expansion is utilised to linearise the Nadir constraint (8.32e) at each iteration as given below:

$$\tilde{h}\Delta f^{\max} \le p_{to}^{\text{grid}} \le \tilde{h}\overline{\Delta f}^{\max}, \tag{8.36a}$$

$$
\begin{aligned}
h \approx \tilde{h} = h_0^\kappa &+ \frac{\partial h}{\partial D}(D - D^\kappa) + \frac{\partial h}{\partial R_s}(R_s - R_s^\kappa) \\
&+ \frac{\partial h}{\partial F_s}(F_s - F_s^\kappa) + \frac{\partial h}{\partial M}(M - M^\kappa),
\end{aligned} \tag{8.36b}
$$

where $\tilde{h}$ is an auxiliary variable, and hereafter, the superscript κ is used to differ fixed variables (e.g. D^κ) from non-fixed variables (e.g. D) at iteration κ of the decomposition algorithm. The Taylor expansion introduces an approximation error that lowers the accuracy of the expression. The proximity between the true and approximate expressions (i.e. $\Delta f_{\text{exact}}^{\max,\kappa}$ and $\Delta f_{\text{approx}}^{\max,\kappa}$, respectively) is computed using the absolute error $\varepsilon_{\text{abs}} = |\Delta f_{\text{exact}}^{\max,\kappa} - \Delta f_{\text{approx}}^{\max,\kappa}|$ and the relative error $\varepsilon_{\text{rel}} = \frac{|\Delta f_{\text{exact}}^{\max,\kappa} - \Delta f_{\text{approx}}^{\max,\kappa}|}{\Delta f_{\text{exact}}^{\max,\kappa}}$. Application of these metrics to the above first-order approximation for 1,000 scenarios indicated $\varepsilon_{\text{abs}} = 4.7878 \times 10^{-4}$ and $\varepsilon_{\text{rel}} = 0.2734\%$ on average. Although higher

Planning methods for secure islanding   237

order approximations can be used to obtain an even more accurate estimation of the model, it can result in the non-linearity and non-convexity of the optimization problem and further complexity. Also, the dynamic simulations indicated in the preceding sections further provide guarantees on the efficacy of the first-order approximation.

Step 3: Formulation of the sub-problem

To check whether the transient security constraints (8.32e)–(8.32g) are satisfied based on the solution of the master problem at each iteration or not, feasibility sub-problems are formulated. Given optimal values of complicating variables obtained from solving the master problem at iteration k (e.g. $p_{to}^{grid,k}$), the sub-problem for each hour t of every representative day o can be formulated as a linear programming (LP) problem using the slack variable Δp_{to}^{grid} as given below:

$$\min_{\Delta p_{to}^{grid}} \; |\Delta p_{to}^{grid}| \tag{8.37a}$$

subject to:

$$\tilde{h}\underline{\Delta f}^{max} \leq p_{to}^{grid} + \Delta p_{to}^{grid} \leq \tilde{h}\overline{\Delta f}^{max}, \tag{8.37b}$$

$$\tilde{h} = h_0^k + \frac{\partial h}{\partial D}\frac{(D - \check{D}^k)}{P^{base,\kappa}} + \frac{\partial h}{\partial R_s}\frac{(R_s - \check{R}_s^k)}{P_s^{base,\kappa}}$$
$$\qquad + \frac{\partial h}{\partial F_s}\frac{(F_s - \check{F}_s^k)}{P_s^{base,\kappa}} + \frac{\partial h}{\partial M}\frac{(M - \check{M}^k)}{P^{base,\kappa}}, \tag{8.37c}$$

$$\tilde{a}\underline{f}^{:max} \leq p_{to}^{grid} + \Delta p_{to}^{grid} \leq \tilde{a}\overline{f}^{:max}, \tag{8.37d}$$

$$\tilde{a} = \frac{M}{P^{base,\kappa}}, \tag{8.37e}$$

$$\tilde{b}\underline{\Delta f}^{qss} \leq p_{to}^{grid} + \Delta p_{to}^{grid} \leq \tilde{b}\overline{\Delta f}^{qss}, \tag{8.37f}$$

$$\tilde{b} = \frac{D}{P^{base,\kappa}} + \frac{R_s}{P_s^{base,\kappa}}, \tag{8.37g}$$

$$p_{to}^{grid} = p_{to}^{grid,\kappa} \qquad\qquad (\text{dual } \lambda_{to}), \tag{8.37h}$$

$$M = \check{M}^\kappa \qquad\qquad (\text{dual } \alpha), \tag{8.37i}$$

$$D = \check{D}^\kappa \qquad\qquad (\text{dual } \pi), \tag{8.37j}$$

$$R_s = \check{R}_s^\kappa \qquad\qquad (\text{dual } \mu), \tag{8.37k}$$

$$F_s = \check{F}_s^\kappa \qquad\qquad (\text{dual } \sigma), \tag{8.37l}$$

where the auxiliary variables $\tilde{h}$ in (8.37b)–(8.37c), $\tilde{a}$ in (8.37d)–(8.37e), and $\tilde{b}$ in (8.37f)–(8.37g) are used to include the constraints (8.32e), (8.32f), and (8.32g), respectively, in the sub-problem without any non-linear term. Also, $\lambda_{to}, \alpha, \pi, \mu,$ and σ are dual variables for the constraints fixing the complicating variables (i.e. the power exchange with the main grid, aggregated inertia, damping, droop, and turbine power

238 *Intended and unintended islanding of distribution grids*

fraction, respectively) in the sub-problem. These dual variables provide the sensitivity of the optimal solution obtained from solving the sub-problem to the optimal values of the complicating variables obtained from solving the master problem.

If the optimal solution in (8.37) is equal to zero for all hours of all representative days, this implies the feasibility of the master problem. In this case, the algorithm is terminated, and the solution of the master problem at the final iteration is the optimal solution. On the other hand, if the optimal solution in (8.37) is non-zero for even one hour of a specific representative day, i.e. $|\Delta p_{to}^{\text{grid}}| > 0$, this implies the infeasibility of the sub-problem given the values of the complicating variables. Physically, this is associated with violations of transient security constraints. To eliminate these violations, feasibility cuts are added to the master problem.

Step 4: Formulation of resilient feasibility cut

Given the dual variables obtained from solving each sub-problem at iteration κ (i.e. λ_{to}^{κ}, α^{κ}, π^{κ}, μ^{κ}, and σ^{κ}), the master problem in (8.33) is updated with the dual cutting planes if the sub-problem in (8.37) is infeasible, i.e. $|\Delta p_{to}^{\text{grid},\kappa}| > 0$. The cutting planes added to the master problem are defined as follows:

$$\Delta p_{to}^{\text{grid},\kappa} + \lambda_{to}^{\kappa}(p_{to}^{\text{grid}} - p_{to}^{\text{grid},\kappa}) + \alpha^{\kappa}(\check{M} - \check{M}^{\kappa})$$
$$+ \pi^{\kappa}(\check{D} - \check{D}^{\kappa}) + \mu^{\kappa}(\check{R}_s - \check{R}_s^{\kappa}) + \sigma^{\kappa}(\check{F}_s - \check{F}_s^{\kappa}) \leq 0, \, \forall t, \forall o, \forall \kappa \tag{8.38}$$

The dual cutting planes in the decomposition algorithm are associated with the grid power exchange and the unit control parameters. This implies that the sufficiency of frequency support is examined based on the level of the power exchange with the main grid (p_{to}^{grid}) in addition to the aggregated levels of inertia (M) and damping (D) of CIG and SG units plus the droop support (R_s) and the turbine power fraction (F_s) of SG units.

8.4.2 Case study results

In this subsection, the performance of the proposed decomposition algorithm detailed in Figure 8.6 is demonstrated on an 18-bus LV network [36] shown in Figure 8.7 and a 30-bus MV network [37,38] shown in Figure 8.8 and validated under various operating scenarios (see Figure 8.3). The 18-bus network consists of one SG unit already installed, and the investment candidates consist of one SG unit (SG_2), two *grid-supporting* PV units (PV_1, PV_2), and one *grid-feeding* PV unit (PV_3) with fixed power output. Candidates PV_1 and PV_2 provide VSM and droop control, respectively. The generator parameters are as described in Table 8.1. The transient frequency security constraints are enforced through thresholds imposed on frequency nadir ($\overline{\Delta f}^{\text{max}} = -\underline{\Delta f}^{\text{max}} = 0.6\,\text{Hz}$), RoCoF ($\overline{\dot{f}}^{\text{max}} = -\underline{\dot{f}}^{\text{max}} = 2\,\text{Hz/s}$), and quasi-steady-state frequency deviation ($\overline{\Delta f}^{\text{qss}} = -\underline{\Delta f}^{\text{qss}} = 0.2\,\text{Hz}$). Further network parameters can be found in [26].

We compare the performance of the proposed model (Figure 8.6) hereon denoted as the decomposed algorithm (A1) that accounts for both static and dynamic frequency constraints, to a benchmark algorithm (A0) with only static security constraints. Four 24-hour representative days (see 8.3) are used in this case study. The

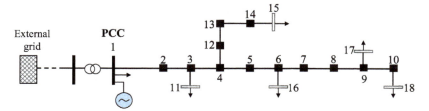

Figure 8.7 Modified CIGRE low-voltage 18-bus test network

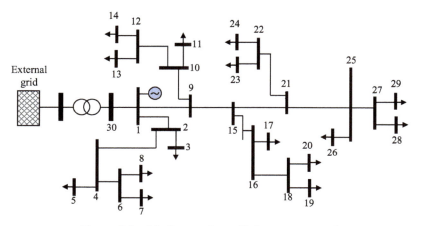

Figure 8.8 Medium-voltage 30-bus test network

Table 8.1 *Generator control parameters and investment costs*

	SG$_1$	SG$_2$	PV$_1$	PV$_2$	PV$_3$
Annualised investment cost ($)	–	40,000	70,000	65,000	60,000
Capacity (kW)	280	350	350	350	350
M (s)	14	14	14	–	–
D (p.u.)	0.9	0.9	0.9	–	–
K (p.u.)	1	1	1	1	–
R (p.u.)	0.03	0.03	–	0.05	–
F (p.u.)	0.35	0.35	–	–	–

Existing generator.
Candidate generators.

implementation is done in MATLAB®, with the optimisation model formulated in YALMIP [39] and solved by GUROBI [40].

8.4.2.1 Planning costs

The costs and planning decisions are compared with each of the algorithms in Table 8.2. It can be seen that A1 has 8.8% higher costs compared to the benchmark algorithm (A0). This is expected since the addition of more constraints always increases

240   *Intended and unintended islanding of distribution grids*

Table 8.2   Comparison of optimal costs and decisions, inertia support, and computational performance for each algorithm using four representative days

	Only static: A0	Static and transient: A1
Costs and decisions		
Total cost ($)	223,390	242,740
Investment cost ($)	61,000	131,000
Investment decisions	PV_3	PV_1, PV_3
Operational cost ($)	162,390	111,740
Demand shift penalty ($)	3,675	7,787
Demand disconnection penalty ($)	14,536	5,337
Frequency support		
M (s)	7.84	17.64
D (p.u)	0.50	1.13
Computational performance		
Number of iterations	–	3
Computation time (s)	612	3,438

the optimisation costs. The difference between the two costs can be considered as the incremental cost needed to ensure the transient frequency security of the system.

When transient frequency security constraints are applied to the problem (A1), the algorithm must minimize costs while ensuring that the level of frequency support in the network is adequate to eliminate violations. Recall that the transient frequency response depends on the aggregated levels of parameters M, D, R_s, and F_s provided by the installed units (see (8.27e) and (8.27f)).

Compared to A0, A1 selects an additional VSM-based CIG unit (PV_1) contributing to both the aggregated damping and inertia levels (see (8.27e) and (8.27f)). Droop-based CIG units contribute to frequency support only in the region of primary frequency response, and not during inertia response. As more frequency support is available from the units selected by A1, the preventive operational actions are kept to a minimum.

8.4.2.2   Dynamic performance

Based on the units installed by each algorithm, the total aggregated level of M and D are 7.84 s and 0.5 p.u. for A0 as compared to 17.64 s and 1.13 p.u. for A1. Figure 8.9 presents a box plot that indicates the variations in the measured values for each of the frequency security metrics for the 96 hours in four representative days. The security limit in each case is indicated by the dotted red line.

In the case of the Nadir values, averages of 49.61 Hz and 49.78 Hz are recorded in A0 and A1, respectively (see Figure 8.9(a)). In the case of the RoCoF values, depicted in Figure 8.9(b), an average of 1.59 Hz/s is achieved for A0 as compared to 0.79 Hz/s for A1. RoCoF is mainly dependent on the total inertia level (M) present in the network (see (8.29)). The solution provided by A0 provides inertia levels of

Planning methods for secure islanding 241

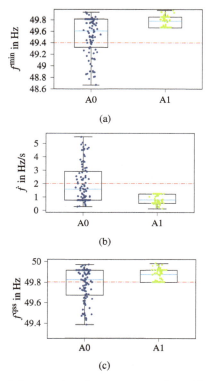

Figure 8.9 Variation of the transient frequency metrics with and without including frequency constraints for all the operating scenarios in the four representative days. (a) Nadir frequency, (b) rate of change of frequency (RoCoF), and (c) quasi-steady-state frequency.

7.84 s, provided mainly by the pre-installed SG. The results based on A1 include the additional installation of VSM-based unit PV_1 resulting in an inertia level of 17.64 s, and therefore, a better performance level compared to A0. On the other hand, the quasi-steady-state frequency is dependent on the aggregated D and R_s parameters, as indicated in (8.31). In Figure 8.9(c), averages of 49.82 Hz and 49.87 Hz are observed for A0 and A1, respectively. From Table 8.2, it can be seen that the aggregated damping levels are higher with the units installed by the A1 solution, hence providing better performance.

The impact of the inclusion of transient constraints on the active power exchange with the main grid is shown in Figure 8.10. For the MG, the power exchange with the main grid is usually the largest power injection, the loss of this power exchange may result in large frequency excursions. A0 only considers static constraints, while A1 vary the dispatch based on the sensitivities to both the power loss and available inertia. It is noteworthy to mention that during instances of power import, the disconnection from the main grid can result in potential under-frequency, while for power export, over-frequencies can be recorded.

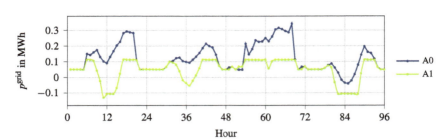

Figure 8.10 Variations in active power exchange with the main grid without inclusion (A0) and with inclusion (A1) of the transient frequency constraints during MG planning (−/+ indicate power export/import)

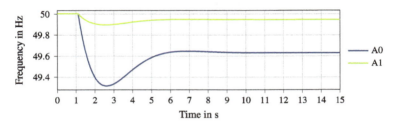

Figure 8.11 Variation in the CoI frequency response 14 s after MG islanding at hour 68 (see Figure 8.10) with and without including the transient frequency constraints during planning

The analytical performance, shown in Figure 8.9, is further validated using time-domain simulations. Given the optimal solution provided by the A0 and A1, Figure 8.11 indicates the frequency trajectories for the operational scenario at hour 68 in the analysis period. Note that based on Figure 8.10, this hour represents the highest power exchange from the grid, and thus, the worst-case mismatch in power if the MG is disconnected. The superiority of A1 over A0 is further validated in Figure 8.11 which compares the frequency trajectories of each technique using a time-domain simulation. The dynamic simulation is performed with PyRAMSES [41] with the grid disconnection occurring at time $t = 1$ s.

8.5 Concluding remarks

The most economical operation of the MG is not always secure against unscheduled islanding events. Both the transients right after the islanding event and the post-islanding steady-state can lead to instability or insecure operating results. In this chapter, we showcase different methods to enhance the operational and investment planning problems to ensure the steady-state and transient security of the MG after an unscheduled islanding event.

While many improvements have been made recently, the commercial software still does not incorporate transient security constraints. Moreover, many issues still need to be addressed, such as incorporating multiple dynamic security constraints in a computationally efficient manner, providing convergence and optimality certificates for the proposed decomposition methods, and bringing the research-oriented algorithms closer to industrial grade.

References

[1]   IEEE. IEEE Standard for the Specification of Microgrid Controllers. IEEE Std 20307-2017. 2018; pp. 1–43.

[2]   Capitanescu F, Glavic M, Ernst D, *et al.* Applications of security-constrained optimal power flows. In: *Modern Electric Power Systems Symposium*, MEPS06; 2006.

[3]   Kotsampopoulos P, Hatziargyriou N, Bletterie B, *et al.* Review, analysis and recommendations on recent guidelines for the provision of ancillary services by distributed generation. In: *2013 IEEE International Workshop on Intelligent Energy Systems* (IWIES); 2013. pp. 185–190.

[4]   Lopes JAP, Moreira CL, and Madureira AG. Defining control strategies for MicroGrids islanded operation. *IEEE Transactions on Power Systems*. 2006;21(2):916–924.

[5]   Nakiganda A, Dehghan S, and Aristidou P. Comparison of AC optimal power flow methods in low-voltage distribution networks. In: *Proceedings of the 2021 ISGT conference*; 2021.

[6]   Farivar M and Low SH. Branch flow model: relaxations and convexification – Part I. *IEEE Transactions on Power Systems*. 2013;28(3):2554–2564.

[7]   Zhou F and Low SH. A note on branch flow models with line shunts. *IEEE Transactions on Power Systems*. 2021;36(1):537–540.

[8]   Nick M, Cherkaoui R, Boudec JL, *et al.* An exact convex formulation of the optimal power flow in radial distribution networks including transverse components. *IEEE Transactions on Automatic Control*. 2018;63(3):682–697.

[9]   Low SH. Convex relaxation of optimal power flow – Part II: Exactness. *IEEE Transactions on Control of Network Systems*. 2014;1(2):177–189.

[10] Huang S, Wu Q, Wang J, *et al.* A sufficient condition on convex relaxation of AC optimal power flow in distribution networks. *IEEE Transactions on Power Systems*. 2017;32(2):1359–1368.

[11] Gan L, Li N, Topcu U, *et al.* Exact convex relaxation of optimal power flow in radial networks. *IEEE Transactions on Automatic Control*. 2015;60(1): 72–87.

[12] Baran ME, and Wu FF. Network reconfiguration in distribution systems for loss reduction and load balancing. *IEEE Transactions on Power Delivery*. 1989;4(2):1401–1407.

[13] Fortenbacher P and Demiray T. Linear/quadratic programming-based optimal power flow using linear power flow and absolute loss approximations. *International Journal of Electrical Power & Energy Systems*. 2019;107: 680–689.

244 *Intended and unintended islanding of distribution grids*

[14] Yang Z, Zhong H, Bose A, *et al.* A linearized OPF model with reactive power and voltage magnitude: a pathway to improve the MW-Only DC OPF. *IEEE Transactions on Power Systems.* 2018;33(2):1734–1745.

[15] Christakou K, Tomozei DC, Le Boudec JY, *et al.* AC OPF in radial distribution networks – Part I: On the limits of the branch flow convexification and the alternating direction method of multipliers. *Electric Power Systems Research.* 2017;143:438–450.

[16] Jain AK, Murty MN, and Flynn PJ. Data clustering: a review. *ACM Computing Surveys.* 1999;31(3):264–323.

[17] Bouguettaya A, Yu Q, Liu X, *et al.* Efficient agglomerative hierarchical clustering. *Expert Systems with Applications.* 2015;42(5):2785–2797.

[18] Sengupta M, Xie Y, Lopez A, *et al.* The national solar radiation data base (NSRDB). *Renewable and Sustainable Energy Reviews.* 2018;89:51–60.

[19] Liu X, Shahidehpour M, Li Z, *et al.* Microgrids for enhancing the power grid resilience in extreme conditions. *IEEE Transactions Smart Grid.* 2017;8(2):589–597.

[20] Zhou Y, Panteli M, Moreno R, *et al.* System-level assessment of reliability and resilience provision from microgrids. *Applied Energy.* 2018;230:374–392.

[21] Milano F, Dörfler F, Hug G, *et al.* Foundations and challenges of low-inertia systems (Invited Paper). In: *2018 Power Systems Computation Conference (PSCC)*; 2018.

[22] Zheng W, Crossley P, Xu B, *et al.* Transient stability of a distribution subsystem during fault-initiated switching to islanded operation. *International Journal of Electrical Power & Energy Systems.* 2018;97:418–427.

[23] Kundur P, Paserba J, Ajjarapu V, *et al.* Definition and classification of power system stability IEEE/CIGRE joint task force on stability terms and definitions. *IEEE Transactions on Power Systems.* 2004;19(3):1387–1401.

[24] Nicholson B, Siirola JD, Watson JP, *et al.* pyomo.dae: a modeling and automatic discretization framework for optimization with differential and algebraic equations. *Mathematical Programming Computation.* 2017 12;10(2). Available from: https://www.osti.gov/biblio/1421609.

[25] Nakiganda AM and Aristidou P. Resilient microgrid scheduling with secure frequency and voltage transient response. *IEEE Transactions on Power Systems.* 2022 Dec.

[26] Nakiganda AM, Dehghan S, Markovic U, *et al.* A stochastic-robust approach for resilient microgrid investment planning under static and transient islanding security constraints. *IEEE Transactions on Smart Grid.* 2022 Jan.

[27] Anderson PM and Mirheydar M. A low-order system frequency response model. *IEEE Transactions on Power Systems.* 1990;5(3):720–729.

[28] Markovic U, Chu Z, Aristidou P, *et al.* LQR-based adaptive virtual synchronous machine for power systems with high inverter penetration. *IEEE Transactions on Sustainable Energy.* 2019;10(3):1501–1512.

[29] Paturet M, Markovic U, Delikaraoglou S, *et al.* Stochastic unit commitment in low-inertia grids. *IEEE Transactions on Power Systems.* 2020.

Planning methods for secure islanding   245

[30]   Rocabert J, Luna A, Blaabjerg F, *et al*. Control of power converters in AC microgrids. *IEEE Transactions on Power Electronics*. 2012;27(11): 4734–4749.

[31]   Markovic U, Stanojev O, Aristidou P, *et al*. Partial grid forming concept for 100% inverter-based transmission systems. In: *2018 IEEE Power & Energy Society General Meeting (PESGM)*; 2018.

[32]   Markovic U, Stanojev O, Aristidou P, *et al*. Understanding small-signal stability of low-inertia systems. *IEEE Transactions on Power Systems*. 2021;36(5):3997–4017.

[33]   Aik DLH. A general-order system frequency response model incorporating load shedding: analytic modeling and applications. *IEEE Transactions on Power Systems*. 2006;21(2):709–717.

[34]   Dehghan S, Amjady N, and Aristidou P. A robust coordinated expansion planning model for wind farm-integrated power systems with flexibility sources using affine policies. *IEEE Systems Journal*. 2019.

[35]   Conejo AJ, Castillo E, Minguez R, *et al*. *Decomposition Techniques in Mathematical Programming: Engineering and Science Applications*. Springer Science & Business Media; 2006.

[36]   Nakiganda AM, Dehghan S, Markovic U, *et al*. A stochastic-robust approach for resilient microgrid investment planning under static and transient islanding security constraints. *IEEE Transactions on Smart Grid*. 2022.

[37]   Kägi-Kolisnyc E. *Distribution Management System Including Dispersed Generation and Storage in a Liberalized Market Environment*. EPFL; 2009.

[38]   Nakiganda AM and Aristidou P. Resilient microgrid scheduling with secure frequency and voltage transient response. *IEEE Transactions on Power Systems*. 2023;38(4):3580–3592.

[39]   Löfberg J. YALMIP: a toolbox for modeling and optimization in MATLAB. In: *2004 IEEE International Conference on Robotics and Automation* (IEEE Cat. No.04CH37508); 2004. pp. 284–289.

[40]   Gurobi Optimization, LLC. Gurobi Optimizer Reference Manual; 2023. Available from: https://www.gurobi.com.

[41]   Aristidou P, Lebeau S, and Van Cutsem T. Power system dynamic simulations using a parallel two-level Schur-complement decomposition. *IEEE Transactions on Power Systems*. 2016;31(5):3984–3995.

Chapter 9

Dynamic modelling for distribution grid analysis in the time domain

Christoph Brosinsky[1], Nayeemuddin Ahmed[2] and Harald Weber[2]

9.1 Introduction

During the last couple of years, there has been a rapid increase in the share of storages, flexible loads and inverter-based resources (IBRs), i.e. inverter-connected renewable generation. Consequently, this transition is transforming the distribution grids from solely passive consumers into active distribution grids. However, the unpredictable nature of IBRs poses challenges for transmission and distribution system operators in maintaining system security. With IBRs gradually replacing fossil-fired conventional power plants, the power generation from conventional power plants decreases relative to decentralized resources. The absence of rotating masses in the power grid leads to lower system inertia and a lack of frequency support. This decrease in system inertia increases dynamic phenomena and higher gradients in the rate of change of frequency (RoCoF) (see also Chapters 2 and 3). As a result, the time available to effectively mitigate the effects of disturbances diminishes from an operational perspective, causing the electric power system to be more susceptible to severe disturbances [1,2]. The effect of decreasing inertia on the system frequency response after a system disturbance is illustrated in Figure 9.1. Such disturbances in power systems with low-inertia conditions result in higher RoCoF values. This can lead to the system's frequency rapidly entering its contingency band or even reaching emergency conditions [3]. This can further deteriorate the mismatch between load and generation, which in turn impacts the system to maintain frequency stability. Consequently, system protection schemes such as load shedding, which can potentially lead to islanded operation of parts of the power system, may be activated much more frequently to mitigate potential power system stability problems.

Technical challenges associated with the progressive increase in IBRs also include voltage stability problems. In cases where the connection between the

---

[1]Distribution Management System Department, Operations Management Division, TEN Thüringer Energienetze GmbH & Co. KG, Erfurt, Germany
[2]Institute for Electrical Power Engineering, Faculty of Computer Science and Electrical Engineering, University of Rostock, Germany

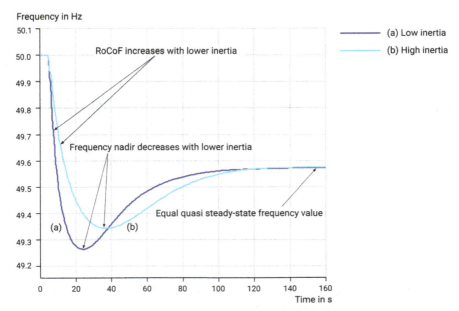

Figure 9.1   Qualitative comparison of power system frequency response for low- and high-inertia conditions

distributed (renewable) energy resource (DER) and the grid has a low short circuit ratio (SCR), i.e. a high impedance and a low short circuit capacity, a weak grid is created which is susceptible to voltage distortions and a lack of selective protection. In such cases, small disturbances can lead to voltage instability and oscillations, instability in inverter control, and even unit disconnection. As a result, IBRs may not be able to provide dynamic grid support under such conditions [4]. To deduce potential solutions for such issues, an in-depth understanding of the dynamic interactions between various IBRs and the associated factors, e.g. low system strength, inverter capacity and type of expected contingencies, is required.

The identification or investigation of such phenomena can be conducted with 'well-designed' and validated dynamic simulation models. This raises questions regarding the level of modelling detail required for such in-depth simulation studies. In general, the complexity and dimension of the power system models increase due to the inclusion of additional state variables* from the necessary IBR models. Furthermore, the corresponding model parameters of the IBR units are rarely known, which leads to uncertainty regarding the accuracy and reliability of the simulation results. Such factors have a significant impact on the investigation results involving weak systems, as they are highly sensitive to incorrect parameterization [5].

---

*State variables are quantities that govern the behaviour of a system. They can represent various aspects (such as angular deviation, angular velocity and state of charge) depending on the nature of the simulation.

Dynamic modelling for distribution grid analysis in the time domain 249

Suitable models of IBRs and other grid elements are the key to investigating the quasi-stationary and dynamic behaviour of the transitioning power system. Ideally, for comparison between the field data and simulation results, system operators need simple open-source models to run simulations. However, to protect their intellectual property, manufacturers and suppliers of IBR and grid-connected generation systems usually provide detailed but encrypted black-box models. These models should comply with the grid codes and emulate the physical system behaviour as far as possible.

In Germany, DERs connected to the medium, high and extra-high voltage grid with nominal power exceeding 1 MVA are subject to certification per **VDE-AR-N 4110, VDE-AR-N 4120** and **VDE-AR-N 4130**, respectively (see also Chapter 4). The current grid code requirements demand that IBRs are equipped with grid support capabilities such as voltage and frequency support as well as fault ride-through (FRT) in case of grid perturbations. In practice, the simulation models are validated by measurements obtained through specialized field tests, which provide information about the dynamic behavior of IBRs. These tests encompass a variety of scenarios, including voltage and frequency step changes, as well as voltage ride-through tests, to assess the IBR's performance under different operational conditions. Modelling dynamic IBR behaviour requires knowledge of IBR internal control loops, which are usually not disclosed by manufacturers to protect intellectual property. Alternatively, simulations can be carried out with generic models that are used by researchers and can generally fulfil the grid codes.

The requirements for DER grid connection are defined by national and international standards. These include **EN 50438, IEC 61727, IEC 62116, DIN EN 62116, VDE 0126-2, VDE AR-N 4105**, as well as the applicable national guidelines, and **IEEE 1547**. Some of the primary requirements of these systems concerning the provision of regulation services include:

- Frequency-dependent active power output, according to the requested frequency droop
- Feeding of capacitive reactive current in the event of a fault, i.e. low voltage ride-through (LVRT), according to specified droop for dynamic voltage support
- Static voltage maintenance and reactive power output through suitable characteristic curves, i.e. $Q(U)$, $Q(P)$, $\cos \varphi$ or setpoints (Q_{set} and $\cos \varphi_{set}$)

The validation of these required DER characteristics is usually conducted using a model-based certification process. The models provided by the plant manufacturer contain the controllers and system parameters to reflect the real (physical) unit characteristics. The fulfilment of the requirements is assured with a certificate of compliance. This demonstrates the ability of the generation unit to provide ancillary services and ensures the necessary controllability anticipating future power system operation requirements.

As most DERs are connected to the distribution grid, modelling such networks requires further exploration. Since standard DER models from the literature can only reproduce the dynamic behaviour to a limited extent, standard models must be adapted or new models must be developed in order to reproduce the behaviour of dynamic distribution networks.

250 *Intended and unintended islanding of distribution grids*

As part of the efforts focused on understanding the effect of high IBR penetration on the grid, a significant proportion of the research is dedicated to analysing the so-called Islanded System Operation [6]. In contrast to unintentional islanding incidents arising from abrupt faults, changes in load distribution, or circuit breaker activation due to a primary utility power failure, intentional islanding involves a premeditated procedure utilized mainly for system maintenance and operational concerns.

Intentionally islanded power systems, often comprise a single or a combination of generation units, such as hydroelectric power plants (HPP), photovoltaic (PV) or wind power plants (WPP) along with energy storages (e.g. supercapacitor, battery or hydrogen). Depending on its capacity, the energy storage unit is used for peak shaving[†] or as a long-term energy storage unit. Today, such storage systems often also offer frequency control on demand. For island operation, storage and generation units within the island grid require a load/frequency controller, a voltage regulator, and information and communications technology (ICT) systems for coordination.

9.1.1 Modelling of power system dynamics

Dynamic simulations have been an essential tool in transmission system analysis to investigate system stability in the time domain and ensure its stability and reliability. The increasing share of IBRs has emphasized the need for such investigations also in distribution grids. Depending on the type of dynamic phenomenon being investigated, the complexity of the utilized mathematical model for the simulations varies. The highest level of complexity and detail are present in accurate models which require exact knowledge of all the associated system components. Comparatively, average or approximate models only highlight part of the system (associated with the investigated dynamic phenomenon) in higher detail. As expected, accurate models are much more computationally intensive than their counterparts. Hence, the decision of selecting the level of model detail is always based on a trade-off between computation time and modelling accuracy.

In general, there are two approaches to simulating power system dynamics: root mean square (RMS) and electromagnetic transient (EMT). While EMT simulations represent instantaneous values of electric quantities, an RMS simulation utilises phasors or effective values (magnitude and angle, where necessary). RMS and EMT modelling approaches focus on different power system phenomena (see Table 9.1) occurring on different time scales (microseconds up to minutes or hours). While detailed investigations of system phenomena such as resonance or inverter-driven stability require EMT simulation models, RMS models may be appropriate for investigations of the rotor angle, voltage and frequency stability [7,8].

When it comes to modelling IBRs, RMS simulation models achieve faster execution times. This is a consequence of multiple approximations in the model representation which include simplified PLL dynamics and power electronic device switching. However, this simplification compromises accuracy and additionally

---

[†]Peak shaving, in this case, refers to the process of meeting maximum load demands using energy storages such as batteries. The battery is charged and stores excess energy during periods of low electricity demand or when connected to the public grid, depending on market signals and correspondingly low electricity prices. Next, during peak load hours or when prices are high, the battery is discharged to reduce the overall demand from the grid.

Dynamic modelling for distribution grid analysis in the time domain 251

Table 9.1 Selected dynamic power system phenomenon from IBR and their time frames

Time frames (modelling approach)	Dynamic phenomena in power systems
Microseconds to milliseconds (EMT)	• Wave propagation phenomena in high-voltage transmission lines caused by lightning strikes or switching operation, initial response to faults and the behaviour of protective relays. • Modelling of control systems in detail, especially when it is of interest to accurately represent the fast control responses of inverters.
Milliseconds to seconds (EMT)	• Response of the power system to transient events, including the recovery phase, i.e., electromagnetic changes caused by protection systems. • Interactions between electrical machines and power electronic devices and the interconnected power system.
Seconds to minutes (RMS)	• Voltage and frequency control actions to maintain secure system operating conditions. • Electromechanical dynamics caused by post-disturbance oscillations of the generator's rotating masses and governing machines. • Interaction between inverter-based resources and the rest of the power system.
Seconds to hours (RMS)	• Long-term behaviour of power systems, including steady-state operation, load behaviour and the effects of seasonal variations on system performance. • Adaptation of power generation to load variations. • Studies on prime mover control and automatic power generation. • Thermodynamic changes in boiler control of steam power plants. • Studies on the cumulative effects of repeated dynamic events over long periods of time.

ignores other factors such as grid dynamics, system unbalances (i.e. negative and zero sequence components), and the presence of harmonics. On the other hand, EMT simulation models accurately represent the behaviour of power electronic devices and include precise PLL dynamics. They incorporate high-frequency switching, fast inverter controls, and adverse controller interactions. However, modelling all aspects in detail leads to longer execution times compared to RMS-type simulations. In summary, EMT simulations are best suited for investigating fast transient phenomena in the microsecond to millisecond range, while RMS simulations are more appropriate for longer timeframes, focusing on steady-state and quasi-steady-state behaviors. The choice between EMT and RMS simulations depends on the specific dynamic phenomena of interest and the timeframes involved.

9.1.2 Modelling of electromagnetic transient

To investigate phenomena, that involve instantaneous values or electromagnetic transients, detailed equipment and specific controller models need to be applied within

252   *Intended and unintended islanding of distribution grids*

an EMT analysis tool. These EMT models are used to analyse transient power system events, particular system conditions or equipment interactions such as [9]:

- Weak system conditions (low SCR or inertia)
- Detailed inverter and collector system design
- Detailed equipment and system interaction studies
- Unbalanced faults in three-phase electrical networks
- Switching transients, inrush phenomena and protection
- Synchronisation and controller interaction

To conduct EMT simulations, inductors (with inductance L) and capacitors (with capacitance C) are modelled using differential equations. These take the frequency-independent instantaneous values into account, as represented by the equations (9.1) and (9.2).

$$u(t) = L\frac{di(t)}{dt} \tag{9.1}$$

$$i(t) = C\frac{du(t)}{dt} \tag{9.2}$$

For EMT modelling, a system of differential-algebraic equations (DAE) needs to be solved. It can be described as follows:

$$\dot{x} = f(x, y, z, u) \tag{9.3}$$

$$\dot{z} = h(x, y, z) \tag{9.4}$$

$$0 = g(x, y, z, u). \tag{9.5}$$

where (9.3) is the set of the differential equations f describing the dynamics of all the components and controllers in the system except for the resistive, inductive and capacitive components of the network. The network dynamics of these three components are described by the differential equations h in (9.4). The algebraic equations g of the system are stated in (9.5). Here, $\dot{x}$ refers to the algebraic states of the power system network, y refers to the output variables of the DAE system from the preceding time step, $\dot{z}$ refers to the dynamic states, and u the control input variables.

To process the numerical integration routines for solving the DAE systems in EMT simulations time steps are in the range of a few microseconds up to 10 milliseconds. Hence, the computational effort and the required time to simulate a three-phase electrical network, considering the frequency dependence of impedance, the switching transients, harmonics, unbalanced network components and power electronic devices including the detailed controls and protection systems is relatively high. For this reason, simulations in real-time dependent environments prefer to apply RMS simulations to assess the dynamic power system behaviour and stability criteria. For this reason, instead of detailed EMT simulations, RMS simulations are preferably used for simulations in real-time dependent environments (e.g. Dynamic Security Assessment applications) to evaluate the dynamic behaviour of the energy system and the stability criteria.

Dynamic modelling for distribution grid analysis in the time domain   253

9.1.3   Modelling of electromechanical interactions (RMS)

Due to their advantages concerning simulation time and computational effort, power system dynamic stability studies for large interconnected power systems often apply RMS simulation tools. Dynamic phenomena, such as electro-mechanical oscillations, the rotor-angle stability of synchronous generators, or voltage and frequency stability can be sufficiently investigated by applying a phasor representation, which only considers the magnitude and phase of the complex quantities of the power system [10]. The phasor-based RMS simulations apply transformed sinusoidal signals, i.e. differential equations are substituted by algebraic equations, which are easier and faster to solve and analyze (see also Section 9.1.5). However, the phasor representation assumes constant phasor rotation frequency (e.g. $f_n = 50$ Hz), as the frequency dependency is neglected in the phasor representation. In phasor representation, the differential equations describing the grid dynamics transform into algebraic equations. Thus, the frequency-dependent components in the network equations are removed entirely resulting in the following simplifications:

- The rate of change of current through an inductor is considered to be infinite. Consequently, the current in the inductor is assumed to change instantaneously.
- The rate of change of voltage through a capacitor is assumed to be infinite, resulting in an instantaneous change of voltage in the capacitor.
- The frequency used for the calculation of the capacitive and inductive reactances is assumed to be constant, neglecting variations in the actual frequency.

These simplifications significantly reduce computational effort compared to EMT simulations, which is exemplarily shown in (9.6)–(9.9). Here, this simplification is done for the resistor element in the network equations but is analogous for inductors and the capacitors in the network equations. However, the simplification also leads to zero dynamics in the network equations. Hence, transients are removed and studies on resonances cannot be conducted using RMS simulation models.

$$U\cos(\omega t + \theta) = R \cdot I \cdot \cos(\omega t + \theta) \tag{9.6}$$

$$U e^{\omega t + j\theta} = R \cdot I \, e^{\omega t + J\theta} \tag{9.7}$$

$$U e^{\omega t} e^{j\theta} = R \cdot I \, e^{\omega t} e^{j\theta} \tag{9.8}$$

$$U e^{j\theta} = R \cdot I \, e^{j\theta} \tag{9.9}$$

Thus, voltages and currents are derived by solving the DAE describing the power system by impedances or admittances, instead of solving the detailed frequency and voltage-dependent DAE of all components in the system model. The inductors and capacitors are modelled algebraically by bulk phasor impedances. An inductance L is expressed as an (inductive) reactance as given in (9.10), where the voltage precedes the current and ω represents the angular frequency of the oscillation. A capacitance C is expressed as a negative (capacitive) reactance as given in (9.11) where the voltage lags behind the current.

$$X_L = \omega L \tag{9.10}$$

$$X_C = -\frac{1}{\omega C} \tag{9.11}$$

254 *Intended and unintended islanding of distribution grids*

In consequence, the differential equations h of the RMS models are transformed into algebraic equations by setting $\dot{z}$ equal to zero. Hence, the DAE system for RMS modelling is generally formulated as follows:

$$\dot{x} = f(x, y, z, u) \tag{9.12}$$

$$0 = h(x, y, z) \tag{9.13}$$

$$0 = g(x, y, z, u). \tag{9.14}$$

This simplification results in a grid model that can reflect the system dynamics close to the nominal frequency (e.g. $f_n = 50$ Hz) but gets inaccurate for frequencies far from the nominal or fundamental frequency f_n. In consequence, the results of the RMS simulation model can only represent system conditions around the nominal frequency f_n of typically 0.1 Hz to 3 Hz. The accuracy of the network model decreases rapidly for phenomena with frequencies significantly outside this range [9]. Hence, for RMS simulations with significant frequency changes, the results may not be reliable as machine equations and network equations can become inaccurate.

Besides neglecting the network dynamics, RMS-type studies often assume a symmetrical three-phase network and consider only the positive sequence components (see also symmetrical components). Furthermore, RMS models typically omit the inner current control loops of inverters, fast switching transients, phase-locked loop models, and various other controller and inverter details. These simplifications allow a significantly shorter simulation time, applying time-step sizes in the range of a few milliseconds up to 100 ms, in comparison to a three-phase or even asymmetrical three-phase EMT simulation. Thus, RMS models can only represent the slower power system dynamics, which exclude the transient RLC power system dynamics and fast equipment controllers. As RMS studies cannot investigate transients that are faster than the dynamics close to the fundamental frequency, they are mainly applied for time-constrained large-scale power systems simulations (e.g. dynamic security assessment) and positive sequence (balanced) systems.

9.1.4 *Model initialisation and numerical integration*

As outlined in Sections 9.1.2 and 9.1.3, dynamic power system simulation models are represented by sets of DAE that describe the dynamic behaviour of the system. These DAE used to describe continuous-time systems are converted into discrete-time systems and solved numerically using dedicated simulation software. A variety of numerical integration methods with variable step size or fixed step size are available for this purpose, and further information on these methods can be found in the literature (e.g. in [11–13]). For power system simulation studies, the explicit modified Euler method with fixed step size is often used due to its ease of implementation and simplicity [14].

Dynamic simulations typically start from a stable operational state (steady-state condition $[\dot{x}, \dot{z}] = 0$). To obtain valid results, it is essential to ensure the accuracy of the initial simulation conditions. If the initial conditions are not set correctly, the simulation may fail to converge or exhibit unstable behaviour. Correct initial conditions ensure that the simulation starts from a stable operating point and converges to

Dynamic modelling for distribution grid analysis in the time domain **255**

a valid solution. Initialisation errors are typically caused by exceeding the equipment limits defined in the system model. Other causes for invalid initialisation can be incorrect model parameters. For example, an incorrect reactance value for a generator can result in an unrealistic excitation voltage. The initial conditions are calculated for each DAE system representing a dynamic equipment model. These models include, e.g., generators, exciters, governors, inverters, and controllers. As the augmented bus admittance matrix is usually constructed based on a power flow solution, it is recommended to check the plausibility of the steady-state power flow solution before applying it as the basis for model initialisation. At each time step of the simulation, the set of differential equations, f, is integrated and the set of algebraic equations, g, is solved. If a topological change occurs, the augmented bus admittance matrix (see Section 9.2.1) is re-factorized and the network equations in g are recalculated. The variables of interest are stored and time is advanced using the optimal step size depending on the simulation mode (RMS and EMT) and the chosen numerical integration method (variable step size or fixed step size). The mathematical methods to solve the differential algebraic equation (DAE) systems by numerical integration are described in detail e.g. in [15].

9.1.5 Transformation methods

To analyze an alternating current power system or complex electrical machinery, mathematical transformations are typically employed to separate variables and solve equations involving time-varying quantities by relating all variables to a common reference frame. While in an EMT simulation, the instantaneous values are calculated by solving time-domain circuit equations, the RMS values can be derived from the EMT simulation solution by transformation or modelled directly in the RMS reference frame model. For electric power system simulations, transformations between the abc, stationary $\alpha\beta0$ reference frames, and rotating dq0 reference frames are applied to analyse or control three-phase technologies, including machines and inverters [16]. The three-phase reference frame, in which the vectors a, b, and c are co-planar quantities that are oriented at an angle of 120 degrees to each other, can be converted into direct signals to simplify the calculation. The Clarke transformation is a mathematical technique used in power systems to convert three-phase quantities (often represented in the *abc* coordinates) into two components in an orthogonal stationary coordinate system (commonly represented as $\alpha - \beta$ coordinates). In the symmetrical case, the zero component is omitted, which means that the remaining $\alpha - \beta$ coordinates are equivalent to the space vector. The components are available within the network equations and can be used directly in the RMS model. This transformation is particularly useful in analyzing and controlling three-phase systems, such as inverter-based systems, where it simplifies the analysis and control algorithms.

$$\begin{bmatrix} I_\alpha \\ I_\beta \\ I_0 \end{bmatrix} = \begin{bmatrix} \frac{2}{3} & -\frac{1}{3} & -\frac{1}{3} \\ 0 & \frac{1}{\sqrt{3}} & -\frac{1}{\sqrt{3}} \\ \frac{1}{3} & \frac{1}{3} & \frac{1}{3} \end{bmatrix} \cdot \begin{bmatrix} I_a \\ I_b \\ I_c \end{bmatrix} \tag{9.15}$$

In general, two transformation methods are commonly employed. The Clarke transformation is primarily used to simplify the representation of three-phase quantities by converting them into a two-phase stationary reference frame ($\alpha - \beta$), facilitating analysis and control. The Park transformation converts the two components in the $\alpha - \beta$-coordinate system into an orthogonal rotating reference coordinate system (dq) aligned with the system's angular frequency. This transformation is especially valuable for controlling rotating machines like motors and generators. Usually, the Clarke transformation is applied first to convert *abc* quantities into $\alpha - \beta$ coordinates, and the Park transformation is applied subsequently to further simplify control and analysis in the common dq reference frame. The implementation of these two transformations one after the other simplifies the calculations by converting alternating currents and voltages into direct current signals. The transformation requires a reference angle, often obtained from the synchronous angular frequency of the system. The direct (d) and quadrature (q) components represent the real and imaginary parts of the transformed signal in the rotating reference frame. The three reference frames are illustrated in Figure 9.2. In the frequently assumed symmetrical case, the dq components are calculated by ((9.16)).

$$\begin{bmatrix} I_d \\ I_q \end{bmatrix} = \begin{bmatrix} \cos\theta(t) & \sin\theta(t) \\ -\sin\theta(t) & \cos\theta(t) \end{bmatrix} \cdot \begin{bmatrix} I_\alpha \\ I_\beta \end{bmatrix} \tag{9.16}$$

For all rotating systems, the *dq* coordinate system is used as a common reference to relate the vectors in the individual coordinate systems applied for the *i*-th rotating machine or inverter model in the modelled power system.

$$\begin{bmatrix} i_{d,i} \\ i_{q,i} \end{bmatrix} = \begin{bmatrix} \cos(\delta(t) + \vartheta(t)) & \sin(\delta(t) + \vartheta(t)) \\ -\sin(\delta(t) + \vartheta(t)) & \cos(\delta(t) + \vartheta(t)) \end{bmatrix} \cdot \begin{bmatrix} I_{d,i} \\ I_{q,i} \end{bmatrix} \tag{9.17}$$

Here, the angle $\theta(t)$ represents the difference between the d-axis of the rotating coordinate system and the q-axis of the static reference system, whereas

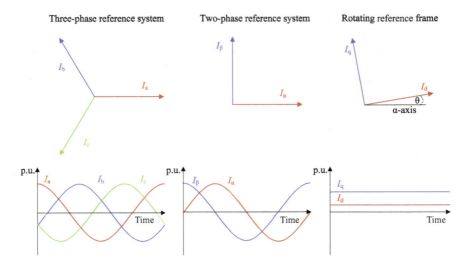

Figure 9.2 Illustration of the reference frames for space vector representation: (a) abc, (b) $\alpha - \beta$ and (c) dq-reference frame

Dynamic modelling for distribution grid analysis in the time domain 257

the quadrature-axis leads the direct-axis which conforms to the IEEE standard definition [17].

9.2 Power system modelling

9.2.1 The electric grid model

To create the model of the interconnected power system, the dynamic component models formulated as differential equations must be linked to the electric grid model. For RMS simulations, the grid model can be described by algebraic equations (see Section 9.1.3). Here, to solve the model equations efficiently all system variables are expressed in a common reference frame. To conduct EMT simulations, at least the model components relevant to the investigated scenario must be described in detail by differential equations as outlined in Section 9.1.2.

The electrical grid is represented in the form of the nodal admittance matrix $\underline{Y}_{bus}$. According to the system topology the nodes are interconnected by branch elements (i.e. transmission lines or transformers), which link the nodes i and j. Thus, all off-diagonal elements $\underline{Y}_{bus,ij}$ are zero if there is no linking branch element between nodes i and j. For the sake of numerical effort, $\underline{Y}_{bus}$ is expressed as a sparse matrix. Ohm's law reads as

$$\underline{\mathbf{i}}_i = \begin{bmatrix} \underline{i}_1 \\ \underline{i}_2 \\ \vdots \\ \underline{i}_i \end{bmatrix} = \begin{bmatrix} \underline{Y}_{11} & \underline{Y}_{12} & \cdots & \underline{Y}_{1i} \\ \underline{Y}_{21} & \underline{Y}_{22} & \cdots & \underline{Y}_{2i} \\ \vdots & \vdots & \ddots & \vdots \\ \underline{Y}_{i1} & \underline{Y}_{i2} & \cdots & \underline{Y}_{ii} \end{bmatrix} \begin{bmatrix} \underline{u}_1 \\ \underline{u}_2 \\ \vdots \\ \underline{u}_i \end{bmatrix} = \mathbf{Y}_{bus}\underline{\mathbf{u}}_i. \tag{9.18}$$

To resolve the algebraic equations (in RMS simulations) for the respective terminal voltage $\underline{u}_t$, the load admittance matrix $\underline{\mathbf{Y}}_l$ and the generator stator admittance $\underline{\mathbf{Y}}_g$ are added as shunt elements to the node admittance matrix $\underline{\mathbf{Y}}_{bus}$ by addition, to create the augmented admittance matrix $\underline{\mathbf{Y}}_{aug}$. In case the load changes (i.e. the amount of real and reactive power) during the time-domain simulation, the load admittance value is updated accordingly. As the simplified static load model, represented by shunt admittances are passive elements, they have zero current injection. Thus, the voltages can be calculated by solving a system of linear equations.

The injected currents by the generators are considered in the network matrix $\underline{\mathbf{Y}}$ by g generator terminal nodes within the given number m of network nodes. As the passive loads are modelled by shunt admittances, the load currents injected into the nodes are zero and the network equation can be represented by

$$\begin{pmatrix} \underline{i}_{m,1} \\ \vdots \\ \underline{i}_{m,i} \\ 0 \\ \vdots \\ 0 \end{pmatrix} = \begin{bmatrix} \underline{Y}_{gg} & \cdots & \underline{Y}_{gm} \\ \vdots & \ddots & \vdots \\ \underline{Y}_{mg} & \cdots & \underline{Y}_{mm} \end{bmatrix} \cdot \begin{pmatrix} \underline{u}_{t,1} \\ \vdots \\ \underline{u}_{t,m} \\ \underline{u}_1 \\ \vdots \\ \underline{u}_i \end{pmatrix}. \tag{9.19}$$

258   *Intended and unintended islanding of distribution grids*

The created augmented admittance matrix $\underline{\mathbf{Y}}_{aug}$ is applied to resolve the algebraic states. By solving (9.18) the bus voltages u_i at the i-th bus are subsequently determined.

To model dynamic states in the system, it is necessary to incorporate active components like dynamic loads and converters through the use of differential equations. The level of detail depends on the phenomena to be investigated (see Section 9.1.1). Assuming steady-state conditions, dynamic device models like generators, dynamic loads, and converters are transformed into Norton equivalents, which are then added to the grid matrix. This allows the composition of the augmented admittance matrix $\underline{\mathbf{Y}}_{aug}$.

$$\underline{\mathbf{Y}}_{aug} = \underline{\mathbf{Y}}_{bus} + \underline{\mathbf{Y}}_{load} + \underline{\mathbf{Y}}_{generator} + \underline{\mathbf{Y}}_{inverter} \tag{9.20}$$

Before utilizing an integration rule to address the composite DAE model of synchronous machines, exciters, governors, dynamic loads or inverter models in the time domain, it is essential to initialize these components. To expedite the initialization process and avoid long initialization times, time-domain simulations are frequently initiated from a steady-state operating condition, which can be obtained through state estimation or power flow calculation. The description of the network model can be realised by a simplified bus-branch approach, where only network buses, terminals, and branches are considered. A more complex full topology representation, i.e. a node-breaker model also includes switches and breakers, including their status (open/closed).

9.2.2   Transmission and distribution line models

Several approaches to model transmission and distribution lines exist [18]. The most common approaches are briefly introduced and delineated as follows.

1.   Π-section model: The lumped parameter Π-section model is widely used in power system studies due to its simplicity and suitability to model power flows and electro-mechanical transients. However, the precision of the Π-section model remains constrained within a delimited frequency spectrum around the nominal system frequency.
2.   Bergeron model: The Bergeron model is also a constant frequency model, but it is more accurate than the Π-model. It considers the travelling waves through the line by application of distributed inductances (L) and capacitances (C). Conceptually akin to an infinite series of Π-Sections, the Bergeron model distinguishes itself by the amalgamation of a lumped resistance component, strategically positioned at half the distance along the line, with quarter portions positioned at each terminal end [19]. The model is sensitive to line parameter variations and has drawbacks in the modelling of high-frequency effects.
3.   Frequency-dependent model: The frequency-dependent model accurately represents the transmission line or cable over the entire frequency range. The line resistances, inductances, capacitances and conductances are calculated and obtained from the physical line geometry and are frequency-dependent. The travelling waves are accurately represented in this model and the travelling wave

Dynamic modelling for distribution grid analysis in the time domain 259

speed is also frequency dependent, making it suitable to model electromagnetic transients over the full frequency spectrum.

Transmission line and cable parameters are frequency-dependent. Therefore, considering frequency dependency and the full frequency spectrum adds significant computational costs to the power system simulation, since the line and cable parameters require recalculation at every time step during the simulation. For EMT simulations, the utilization of the Bergeron model is generally recommended over a Π-section equivalent when the system length is sufficient to allow for the propagation of travelling waves to traverse the entire length of the line within a single time step (travelling speed is approximately the speed of light). Thus, overhead line models for EMT simulations should be represented by the Bergeron model or by a full spectrum frequency-dependent model to involve travelling wave and frequency effects.

9.2.3 Load modelling

It is important to have detailed information about the load composition to accurately model the load dynamics, as loads can significantly influence the system voltage and dynamic stability. A simplified approach is to represent loads by a constant admittance, and to integrate them into the nodal admittance matrix $\underline{\mathbf{Y}}_{\text{bus}}$. One widely accepted model to consider voltage or frequency dependency of loads is the exponential dynamic load model, also known as the composite *ZIP* model given in (9.21). The static ZIP model expresses the load in exponential form, whereby the exponent determines the voltage dependency of the load ($[n_p, n_q] = 0$: for constant power (P), $[n_p, n_q] = 1$: for constant current (I), $[n_p, n_q] = 2$: for impedance characteristics (Z)). To consider voltage or frequency dependency in the static ZIP model, a second term as given in (9.21) can be considered.

$$ P = P_{\text{r}}\left(\frac{U}{U_{\text{r}}}\right)^{n_p} \left[\frac{1+T_{\text{p1}}}{1+T_{\text{p2}}} + k_{\text{p}}\frac{f-f_{\text{n}}}{f_{\text{n}}} \right] \qquad Q = Q_{\text{r}}\left(\frac{U}{U_{\text{r}}}\right)^{n_q} \left[\frac{1+T_{\text{q1}}}{1+T_{\text{q2}}} + k_{\text{q}}\frac{f-f_{\text{n}}}{f_{\text{n}}} \right] \qquad (9.21) $$

With: f_{n} – nominal frequency, f – frequency, U_{r} – rated voltage of the load (at connection point), P_{r} – rated active power of the load, Q_{r} – rated reactive power of the load, n_{p} – active power coefficient (function of the voltage), n_{q} – reactive power coefficient (function of the voltage), k_{p} – active power coefficient (function of the frequency), k_{q} – reactive power coefficient (function of the frequency), $T_{\text{p1,2}}$ – active power time constants of the load model and $T_{\text{q1,2}}$ – reactive power time constants of the load model. The active and reactive power coefficients or the time constants can be set to zero, to model a load which is constant, frequency- or voltage-independent.

In today's power system, industrial loads containing rotating machinery and drives, are mostly interfaced by converters, thus the behaviour can be substituted by converter characteristics. Load models that need to represent dynamic behaviour, such as rotating machinery (motors) or converter-interfaced devices need to be modelled by differential equations. As there are usually more loads than generators within a modelled power system, a high dimensional model containing a high number of

260   *Intended and unintended islanding of distribution grids*

dynamic states is derived. To avoid unsolvable model complexity for large power systems, dynamic load models with a low number of differential equations are preferred. Thus, the simplest load model which approximates the load behaviour in a good manner should be applied. It should be noted that a coupling between the equation systems for active and reactive power should be considered for accurate load modelling.

Many composite loads react very fast and quickly reach a new steady state following changes in voltage and frequency. Here, the application of static load models can be accepted since the approximation of dynamics is acceptable. Simulation studies involving phenomena such as oscillations, voltage stability, or long-term stability require an in-depth modelling of the dynamic behaviour. The dynamic behaviour of loads, in response to voltage and frequency deviations after disturbances, needs to be modelled by differential equations [20,21]. The same applies to industrial power systems where the load includes a high share of rotating machinery. For RMS simulations, the dynamic composite load model developed by WECC can be adapted [22]. In EMT simulations, the load model is usually described by the three elements inductance, capacitance, and resistance, whereby the ratios between these must be specified depending on the load type. Understanding load modelling and load behaviour will become increasingly important in the future, as controllable loads enable flexibility in the form of operational degrees of freedom to operate power systems.

9.2.4   Modelling of conventional generation

In the bulk power system as well as in islanded distribution grids, the electric frequency needs to be maintained by frequency-regulating power plants. Traditionally, this task is performed by hydropower or turbine-driven synchronous generators within thermal power plants (TPPs). A turbine-driven synchronous generator can be regarded as a dynamic system in which electromagnetic and electromechanical processes take place and interact. These interactions can be modelled in different degrees of detail.

Depending on the model order, the number of rotor coils on the direct axis (d-axis) and the quadrature axis (q-axis) and corresponding state variables vary from one to six [23]. Generally, higher-order models consist of a higher number of state variables within the set of differential equations. The classical synchronous machine model (2nd order), e.g., neglects all state variables of the rotor coils, thus the field voltage and the corresponding rotor flux linkage is assumed to be constant [24,25]. The dynamics of the classical synchronous machine model are only dependent on the two states δ and ω in the swing equation (see Chapter 2), thus it is of second order. Therefore, it is applied for representing "remote" machines in very large interconnected power systems, and due to its simplicity and computational efficiency applied to screen for contingencies and ranking for transient stability limit search applications [25]. The most common model types described in the literature (e.g. in [10,24,25]) and their application to power system analysis are listed in Table 9.2. Odd numbered model orders are not considered here. These often represent salient pole generator types [25].

Investigations on power system stability by modelling power systems have shown, that stability mechanisms are only correctly exhibited if synchronous

Dynamic modelling for distribution grid analysis in the time domain 261

Table 9.2 Classification of dynamic generator models

Model order	Modelled details	Model application
2nd order	Classical model: voltage source behind transient reactance	Simplified machine dynamics, low-fidelity. Preferred for modelling power plants without significant impact on the investigated system
4th order	Field circuit with one equivalent damper on q-axis	Dynamic stability studies on electro-mechanical phenomena in large interconnected power systems
6th order	Two additional dampers on q-axis	Detailed stability studies considering flux and sub-transient reactance in rotor dynamics and electromagnetic phenomena
8th order	Field circuit with two equivalent damper circuits on d-axis and three dampers on q-axis	Detailed studies on electromagnetic transients (instantaneous values)

machine models of fourth or higher orders are applied [26]. For studies involving large perturbations (e.g. severe faults, such as three-phase short circuits) it is common practice to apply sixth-order models [23]. For studies of large power systems, "it can be adequate to use fourth-order models if the machine parameters are correctly determined" [27]. While higher-order models promise better results, the determination of their parameter sets requires more attention [28]. The traditional power system dynamics are mostly influenced by synchronous generators and their associated regulators and controllers, such as the turbine governor, the automatic voltage regulator (AVR) comprising the excitation system and the power system stabilizer (PSS). Dynamic models for turbine-governors and excitation system models for power system stability studies are described in [29,30].

9.2.5 Thermal power plant

The TPP model used in the network represented in Section 9.3.1 is described here. It emulates the behaviour of a conventional TPP operating in fixed pressure mode during RMS simulations and is adopted from [31]. The composite frame to model the TPP in DIgSILENT PowerFactory is shown in Figure 9.3. It comprises a subset of models describing the dynamics of its turbine, boiler, governor, voltage regulator, and synchronous generator.

Figure 9.4 represents the governor model dynamics in the TPP. Based on the difference between the measured and reference frequency (f_{ref}), the governor determines the necessary change of the valve position, which regulates the steam flow rate into the turbine. Here, the valve dynamics are represented via a PT1 controller with a time delay (T_V). The initial valve position (aT_0) is determined from the initialisation

262 Intended and unintended islanding of distribution grids

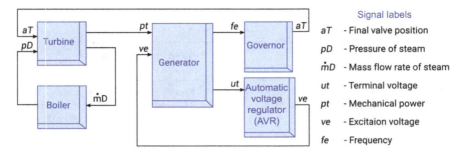

Figure 9.3 *Composite frame for the synchronous generator representing a thermal power plant*

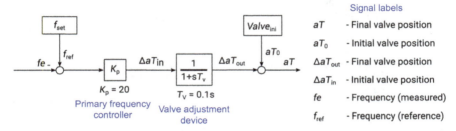

Figure 9.4 *Model definition of the thermal and hydroelectric power plant governor*

routine based on the active power setpoint derived from the initial load flow solution. This illustration only shows the presence of the proportional primary frequency controller. However, for the investigations performed in Section 9.3.1, a central AGC (secondary frequency controller) is also included to restore the frequency to the nominal value.

The valve position from the governor and the pressure of steam from the boiler are inputs to the turbine model, as shown in Figures 9.4–9.6. The product of these two signals determines the mass flow rate ($\dot{m}_D$) of steam, Figure 9.5. Part of this steam (α_{HD}) flows through the high-pressure turbine. The rest ($1 - \alpha_{HD}$) flows through a reheater (not modelled here for simplicity) and then the low-pressure turbine. Since the steam enters the high-pressure turbine first, its action is much faster and the associated time constant (T_{HD}) is much lower than that of the low-pressure turbine (T_{ND}). The sum of the power output from these two turbine stages determines the mechanical power input to the generator. The steam enthalpy remains constant since it can be assumed that the temperature control circuits maintain the steam temperature at its nominal value. The boiler/steam generator of the TPP is described in Figure 9.6. Its input is the mass flow rate of steam, which comes from the output of the turbine.

The AVR model used in the TPP is described in Figure 9.7. The measured terminal voltage of the generator serves as the input signal. Its difference from the

Dynamic modelling for distribution grid analysis in the time domain 263

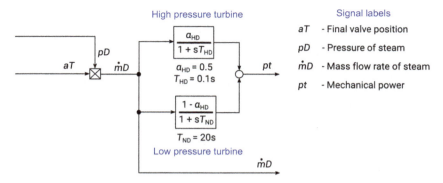

Figure 9.5 Model definition of the thermal power plant turbine

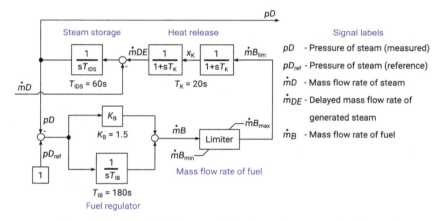

Figure 9.6 Model definition of the thermal power plant boiler

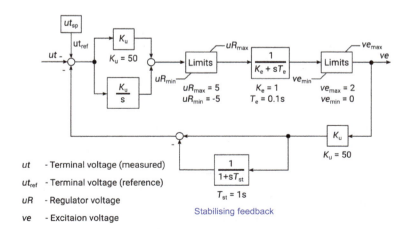

Figure 9.7 Model definition of the thermal and hydroelectric plant AVR

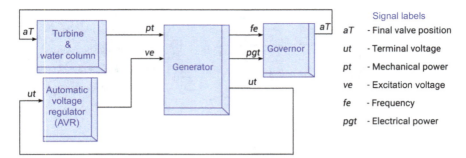

Figure 9.8   Composite frame for the synchronous generator representing a hydroelectric power plant

reference terminal voltage determines the required excitation (field) voltage, which is then fed back to the generator for added controller stability.

9.2.6   Hydroelectric power plant

The HPP model used in the RMS investigations represented in Section 9.3.1 is also adopted from [31]. The composite frame to model this plant in DIgSILENT PowerFactory is shown in Figure 9.8. The power plant comprises a subset of models describing the dynamics of the hydro turbine and water column, governor, AVR, and synchronous generator. The HPP utilizes the same governor and AVR model as the TPP (see Figures 9.4 and 9.7) but also has a secondary frequency controller as exhibited in the results presented in Section 9.3.2.

The two main parts of a hydroelectric power station are its water column and turbine. The modelling of the penstock can be quite complicated owing to the presence of hydraulic transients. This arises from the pressure fluctuations in the water column caused by a change in the mean flow conditions. In the analysis of such transients, the pressure in the pipelines is expressed in terms of the height of the water column (h). Additionally, the flow velocity of water is represented by the mass flow rate of water, (q). Both these quantities are functions of time (t) and displacement (x) of the water along the pipe.

Figure 9.9, which represents the model of the water column and turbine is based on (9.22)–(9.26). These equations, written in per unit, are derived from the fundamental Euler and Continuity equations. As exhibited in the figure, the block diagram takes the head loss (h_R) into account, which is present due to the friction between the water and the pipe walls. Head losses can result from both boundary losses, i.e. the effect of shear forces, and form losses which originate from recirculating eddies due to the geometry of the pipe. T_W is referred to as the water start-up time constant which represents the time taken by the water to reach steady state speed. T_L represents the time taken by the pressure wave to travel the length of the penstock. T_L^* represents a modified value of the wave propagation time constant which

Dynamic modelling for distribution grid analysis in the time domain 265

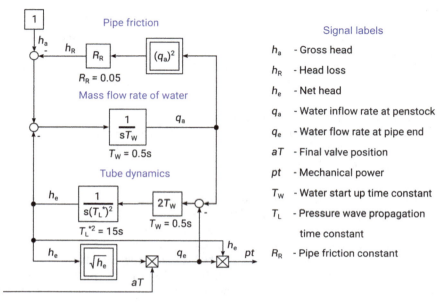

Figure 9.9 Model definition of the water column and turbine of a hydroelectric power plant

must be considered to calculate the fundamental frequency of the pipe vibrations properly.

Flow equation

$$\dot{q}_a = -\frac{1}{T_W} \cdot (h_a - R_R \cdot q_a^2 - h_e) \tag{9.22}$$

Pressure head equation

$$\dot{h}_e = -\frac{2 \cdot T_W}{(T_L^*)^2} \cdot (q_a - q_e) \tag{9.23}$$

Torricelli equation

$$q_e = aT \cdot \sqrt{h_e} \tag{9.24}$$

Mechanical power equation

$$pt = q_e \cdot h_e \tag{9.25}$$

Modified pressure wave time constant

$$T_L^* = \frac{2 \cdot \sqrt{2}}{\pi} \cdot T_L \tag{9.26}$$

With (9.22)–(9.26), the behaviour of the hydropower plant can be simulated to a good approximation for dynamic investigations.

266 *Intended and unintended islanding of distribution grids*

9.2.7 *Modelling of decentralised (renewable) generation*

As societies strive for sustainability and resilience in their energy systems, the need for accurate models of decentralised (renewable) generation units gains importance. Considering the best international practices, decentralised generation units can be modelled by three different approaches in the time domain [32,33]:

- Admittance-based static power generation models,
- RMS models for the simulation of electromechanical balancing processes, and
- EMT models that can represent electromagnetic transients within the time-domain simulation.

The choice of model type depends on the necessity of the analysis and the time frame in which the dynamic or transient behaviour is to be investigated. Whereas in the past grid operators mainly used the admittance-based static generation model, today the dynamic behaviour of renewable energy sources is the focus of stability studies. Therefore, these models must also be kept in various simulation tools, from grid planning to operational management. In simulations that consider electromechanical interactions to analyse, e.g., the frequency changes around the nominal value, voltage sags, or rotor angle stability, the modelling of the fast transient processes is not necessary. To consider the switching transients of the power electronics and the inverter control, detailed EMT models of inverter-coupled generators are required.

Today, most decentralised generation units are connected to the grid by inverters. These can be categorized as grid-following (GFL) inverters and grid-forming (GFM) inverters. GFL inverters synchronize with the external power system using a phase-locked loop (PLL), which estimates the voltage angle at the inverter's connection point. As most current grid-connected inverters use the grid-following approach, the correct PLL signal is crucial for inverter control and proper grid synchronization. In low inertia grids and grid areas with low SCR, inverters that operate in grid-following mode are susceptible to instability because their PLL controller may struggle to synchronize with the grid [1,34].

Regarding DER models for the medium- and low-voltage distribution levels, there is still a lack of universally accepted generic dynamic models. The latest efforts in this area have been undertaken by the Western Electricity Coordinating Council (WECC), but the models are constantly under modification, since the emergence of new grid code requirements influences the development of standardized models [35]. Consequently, the limited availability of DER generic models could also be attributed to these evolving regulatory demands. The DER_A model [35,36] developed by WECC can be applied for dynamic RMS simulations considering balanced network conditions (positive sequence). Detailed EMT simulation models of DER are available for specific purposes, e.g. short-circuit or FRT studies [37,38] but rarely contain all relevant control modes required to achieve full grid code compliance. Thus, validated power system simulation models for distribution systems depend on detailed black-box or grey-box vendor models. To take a WPP as an example, the individual models for each wind turbine type within the plant must be available. These

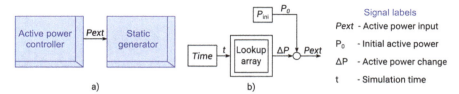

Figure 9.10 (a) Composite frame and (b) active power controller model definition for the static generator representing a WPP

are represented by transfer function block diagrams with comprehensive documentation including controller parameters. Depending on the study to be conducted, the models must allow the analysis of balanced or unbalanced time-domain behaviour (RMS or EMT), following symmetric or asymmetric incidents. Depending on grid code requirements the models must adhere to specified frequency, voltage and time ranges, covering all control types within the entire WPP. This can be further validated through appropriate tests. Best practices and a detailed description of general dynamic models of inverter-based devices, generation units and renewable energy systems are described in [9,39–41].

The WPP model used in the network represented in Section 9.3.1 is shown in Figure 9.10. The idea behind such a representation is that the WPP would always operate at its MPP, while the other power plants in the network would provide the required frequency regulation services. This is achieved with the highly simplified active power controller, shown in Figure 9.10(b). The signal (*Pext*) serves as the active power input to the static generator. This value is determined from the initialisation routine and equals the initial active power setpoint (P_0). If there is no change in the wind speed, the active power output of the WPP will stay constant at its initial value. The effect of wind speed changes is simulated using the lookup array block. This uses the recorded simulation time (*t*) as its input and outputs the corresponding change in WPP active power output as ΔP.

9.2.8 Hydrogen storage power plant

A steady and substantial energy buffer can ideally be provided by large-scale electrical energy storage. The concept for such a system, the hydrogen storage power plant (HSPP), has been devised at the University of Rostock [42] and is explained in the following section.

The internal HSPP structure is presented in Figure 9.11. Such a power plant has been designed to provide all the required ancillary services as currently done by conventional coal-fired plants. In the HSPP, following an active power imbalance in the three-phase network, the tasks of providing instantaneous, primary and secondary control reserve (IR, PCR and SECR) are accomplished by the three main storages, i.e. supercapacitor, battery and hydrogen storage, respectively [43–45]. Since these components operate with direct current (DC), the plant requires a grid-interfacing inverter. This inverter operates in grid forming mode, allowing the HSPP to ensure stable

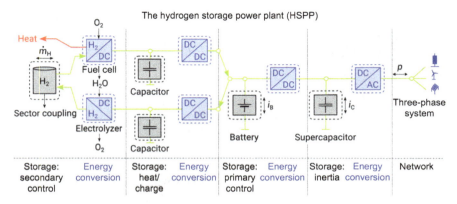

Figure 9.11 Working principle of the internal components of a HSPP

operation of a network even under high DER infeed and very few or no synchronous generators.

In an HSPP, a supercapacitor provides IR since the component can charge and discharge instantaneously with a high power gradient in response to network disturbances. In addition, its ability to undergo frequent charging and discharging cycles makes it an ideal device for inertia emulation [46]. The voltage of the supercapacitor is controlled to govern the PCR from the battery using the adjacent DC–DC converter. Contrary to the supercapacitor, the battery is suited for PCR provision since it is a cheaper form of storage with a higher energy density. Rapid charging or discharging of the battery is detrimental to its average lifespan, so the use of battery systems for providing IR is still unproven [47]. Hence, the combination of the supercapacitor parallel to the battery satisfies the required IR and PCR requirements.

The battery voltage is regulated to direct the SECR response from the hydrogen storage. If there is surplus energy in the network, e.g. due to increased DER infeed, the battery voltage rises. After exceeding an upper threshold, the electrolyser path (in Figure 9.11) is activated. The excess energy is then used to synthesize hydrogen and transfer it to the storage. In case of increased demand, if the battery voltage reduces beyond a lower threshold after providing PCR, the fuel cell path is activated. SCR provisions are then met by the fuel cell generating electrical energy from the stored hydrogen. The power flow, during each situation, is controlled by the respective DC–DC converter between the hydrogen storage and the battery. In this way, the HSPP adapts its active power output during positive and negative disturbances ensuring stability even in grids with high DER penetration. Details of the controller models linking the storages and converters of the HSPP are available in a registered patent [42].

The hydrogen produced by the HSPP via electrolysis can be stored in a liquid organic hydrogen carrier (LOHC) system. Such a system enables safe storage and transportation of hydrogen at a high energy density under ambient conditions using the currently available infrastructure [48]. In addition to being used in the HSPP, the

Dynamic modelling for distribution grid analysis in the time domain   269

stored hydrogen can play a vital role in sector coupling[‡] (Power-to-X) and industrial decarbonisation.

9.3 Case study on the provision of ancillary services in an island network with high DER infeed

To achieve the goal of carbon neutrality, the proportion of renewable and low-carbon technologies in the electricity mix has to be near 100% in 2050. One consequence of this climate target has been the drastic increase in the share of DERs within the power system. The integration of these DERs such as wind and solar power into distribution grids has posed significant challenges due to their intermittency and variability. This is particularly visible in islanded grids which operate independently of the main grid. Such DERs are gradually replacing coal-fired TPPs to reduce greenhouse gas emissions. However, these conventional plants consisting of synchronous generators are also responsible for providing frequency ancillary services in the form of IR, PCR and SECR, which maintain the active power balance in a network. The complete phase-out of such plants to decarbonize the energy system means that frequency regulation needs to be provided by alternative sources. It is difficult to establish the required constant energy buffer for this ancillary action solely by creating headroom on PV and WPPs. In practice, this would correspond to an operation of the intermittent DERs below their maximum power point (MPP) for long periods, which is a technically incompatible and economically unaffordable solution.

To maintain a stable and reliable power supply in such scenarios, alternative frequency regulation mechanisms are essential. Large battery energy storage systems (BESS) or HSPPs have emerged as promising solutions for ensuring the stable operation of power systems even with high infeed levels of intermittent renewable energy sources. These solutions utilize electrolysis to produce hydrogen during periods of excess renewable generation, which can be stored and later converted back to electricity through fuel cells. This energy would then be utilized during periods of high demand or low renewable generation.

The simulations performed in the following sections investigate the frequency-regulating characteristics of HSPPs in an islanded distribution grid. By utilizing RMS simulations the efficacy of HSPPs in maintaining grid frequency within acceptable limits is evaluated under varying operational conditions and renewable energy penetration levels. The key objectives of these studies include:

- Assessing the HSPP's frequency regulation performance.
- Investigating the response time and behaviour of the HSPP's internal components while providing frequency support.
- Analyzing the influence of varying renewable energy penetration levels on grid stability.

[‡]Sector coupling refers to integrating various energy sectors, like electricity, heating and transportation, to optimize energy use, enhance efficiency and reduce carbon emissions. This involves linking their systems through technologies such as power-to-heat, power-to-gas and electric vehicles (i.e. Power-to-X) to create synergies and increase flexibility in transitioning to a sustainable energy system.

- Identifying optimal operating strategies for BESS, HSPPs and DERs to enhance frequency regulation capabilities.

The next section begins with an explanation of the electrical network on which the RMS simulations are performed. The network contains four different types of power plants (i.e. TPP, HPP, WPP and HSPP). In the first scenario, the frequency-regulating properties of all these power plants are observed along with the consequent effect on the internal HSPP components (shown in Figure 9.11).

This is followed by a second investigation which shows the frequency governing capabilities of the HPP, WPPs and HSPPs during increased renewable infeed. The final case study replaces the HSPPs in the electrical network with BESSs and WPPs and implements the same network disturbance as in the first scenario to perform a quantitative comparison of the frequency stability. These comprehensive simulations provide valuable insights into the role of hydrogen storage in facilitating DER integration in islanded distribution grids while ensuring stability and reliability.

9.3.1 Example island network

To exhibit the frequency regulating characteristics of the HSPP in island grids, an example grid model with high DER and partial conventional infeed is designed. This is shown in Figure 9.12. The network consists of 25 equidistant nodes, each connected to either a power plant or a load. The nodes are interconnected via transmission lines, each 250 km long and at a voltage level of 110 kV. The line impedances are equal in magnitude with a resistance-to-reactance ratio of 0.1. This is a generalized grid structure that has been used in multiple research studies (including voltage-reactive power control). The purpose is to create a weak network and show that HSPPs ensure stable operation even under such conditions.

There are eleven power plants, of which five are HSPPs. Of the six other plants, four represent WPPs, and the other two each denote a conventional HPP and a coal-fired TPP. The remaining 14 nodes each house a load, where the active and reactive

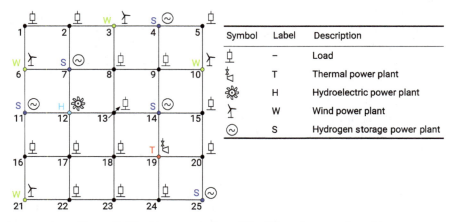

Figure 9.12   Representation of the 25-node example grid

Dynamic modelling for distribution grid analysis in the time domain 271

Table 9.3   Network summary and initial loadflow setpoints of the network elements

Network elements or quantities	Power setpoint (MW)	Total power (MW)
TPP (1)	5.00	5.00
HPP (1)	5.00	5.00
WPP (4)	5.25	21.00
HSPP (4)	5.00	20.00
HSPP at Node 25 (1)	5.13	5.13
Total generation	–	**56.13**
Loads (14)	4.00	56.00
Losses	–	0.13
Total consumption	–	**56.13**

power consumption is constant during the initial power flow. The loadflow setpoints for all these elements are represented in Table 9.3. The HSPP at node 25 is the slack generator and thus balances the power losses in the network. This particular setpoint is adopted so the active power outputs of the different power plants can be compared, as discussed later in Figure 9.14.

The network modelling and simulations are carried out in the software DIgSI-LENT PowerFactory. The base power value of the per-unit system is 10 MVA. The power plants supply the reactive power required by the network, but the associated results and control methods are not included since a simplified analysis is presented in this chapter.

Two separate investigations are initially performed to test the dynamic behaviour of the HSPP in combination with the other power plants. In the first scenario, a step increase is applied to instantaneously raise the power consumption of the load at node 13 of the 25-node example network. In the second case, the generation of all the WPPs is ramped by equal magnitudes and the response of the power plants to the presence of surplus power in the grid is analyzed. In addition, for comparative analysis, a third case is also included by replacing the HSPPs with BESSs and WPPs.

9.3.2   Frequency regulation and HSPP performance evaluation

The step increase applied at 200 s causes the power consumption by the load at node 13 (see Figure 9.12) to increase from 4 MW to 9.6 MW. This is represented in per unit values in Figure 9.13 with the apparent power base of 10 MVA. The magnitude of the load increase is equal to 10% of the total power consumption (56 MW) in the network. In response, the grid frequency reduces immediately at the inception of the load step, as shown in Figure 9.14. The RoCoF varies slightly for every power plant owing to their position in the network. The magnitude of this RoCoF is partly determined by the physical inertia of the turbines and generators in conventional power plants. The other factor governing the RoCoF magnitude is the virtual inertia due to the GFM of the HSPP inverter. Both the physical and virtual inertia of the corresponding power plants are responsible for their respective IR provision. Once the RoCoF magnitude starts to level off, the resulting frequency deviation is used to govern the PCR provision from the power plants. Finally, once the SECR starts after a few minutes of the disturbance, the frequency is gradually returned to 1 per unit.

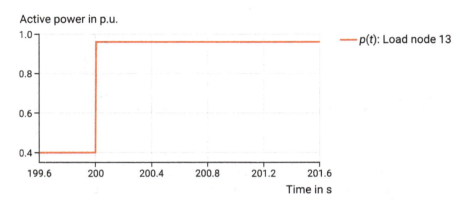

Figure 9.13 Step increase in the power consumption by the load at node 13

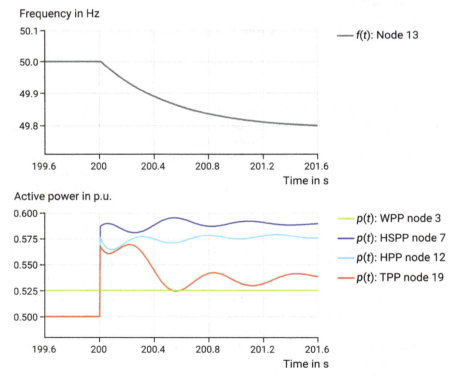

Figure 9.14 System frequency and plant outputs during the short time frame

The ancillary response of the four different types of power plants to the increase in power demand at node 13 is shown in Figure 9.14. In these investigations, since the WPPs are represented as converters operating at their MPP, they do not provide any frequency ancillary service. Figure 9.14 highlighting the IR provision of the different power plants exhibits that the HSPP at node 7 immediately reaches its peak output. This is because this HSPP has the lowest electrical distance to the changing load at node 13 amongst the displayed power plants in the figure. Consequently, the RoCoF at node 7 is higher than in node 12 or 19. Due to the lower electrical distance between nodes 12 and 13 than that between nodes 13 and 19, the HPP at node 12 experiences a higher rate of frequency change and provides a greater IR than the TPP.

This is more evident in Figure 9.15, which displays the supply of PCR by the different power plants. The power output of the HPP and TPP continues to rise steadily, with the HPP showing a greater rate of increase. Meanwhile, the output of the HSPP decreases steadily so that the overall increase in power generation balances the increase in power demand. Figure 9.16 demonstrates the SECR illustrating that the power output of the HSPP and TPP levels off at the same value. This is because all three power plants (i.e. HSPP, TPP and HPP) have identical time constant magnitudes for their secondary controller. However, the HPP has a slightly lower power

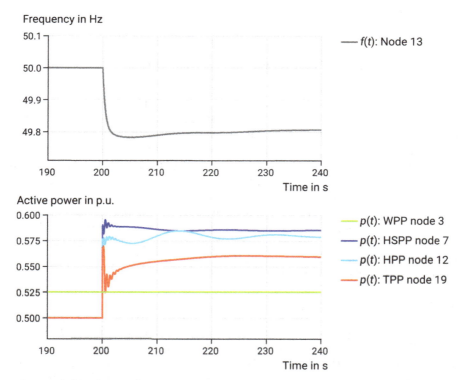

Figure 9.15 System frequency and plant outputs during the medium time frame

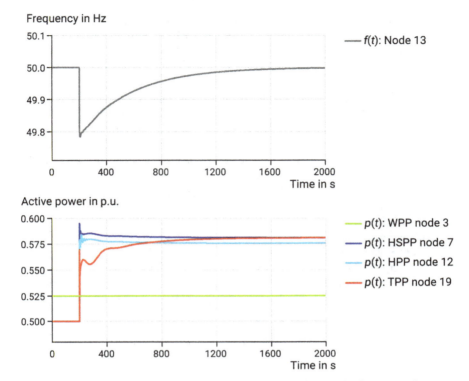

Figure 9.16 System frequency and plant outputs during the long time frame

output owing to the friction in its penstock walls. Since the utilized 25-node example grid is an island network (i.e. a single control area), the secondary controller is only responsible for restoring the system frequency to its nominal value. The power exchange between multiple control areas is not considered in this example.

Next, the behaviour of the internal components of the HSPP at node 7 is investigated to examine the power plant's ability to provide frequency ancillary services (i.e. IR, PCR and SECR). Figure 9.17(a) shows the change in the active power output of the HSPP at node 7. Figure 9.17(b) illustrates the resulting changes in currents out of the respective storages, while Figure 9.17(c) exhibits the voltage or mass levels of these three storages. These trends are presented in three different time scales in Figures 9.17–9.19 to highlight the regulating services provided by the HSPP and the storages associated with each of the control reserves.

- **HSPP Short Time Frame Response**

The graph in Figure 9.17(b) shows that, like the rotating turbines and generators in a TPP, the supercapacitor immediately starts to supply IR with the onset of the positive disturbance. This allows the HSPP to meet the increased network demand. As a result, the supercapacitor voltage decreases. To ensure that the supercapacitor can respond to further disturbances, the DC–DC converter between this storage

Dynamic modelling for distribution grid analysis in the time domain   275

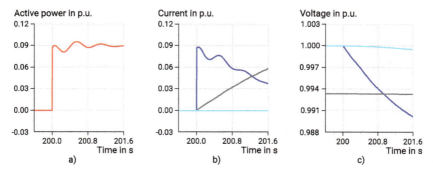

Figure 9.17 (a) Increase in power output, (b) change in current flow from the HSPP storages and (c) voltage levels of the storages during the initial short time frame

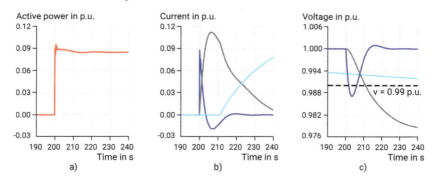

Figure 9.18 (a) Increase in power output, (b) change in current flow from the HSPP storages and (c) voltage levels of the storages during the medium time frame

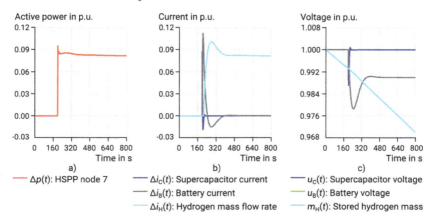

Figure 9.19 (a) Increase in power output, (b) change in current flow from the HSPP storages and (c) voltage levels of the storages during the long time frame

276   *Intended and unintended islanding of distribution grids*

component and the battery starts governing the power flow to supply the disturbed network demand. Part of this active power subsequently recharges the supercapacitor to its nominal value, as shown in Figure 9.18(c). This recharging phase is visible in Figure 9.18(b) when the supercapacitor current is negative.

- **HSPP Medium Time Frame Response**

For PCR provision, the DC–DC converter only uses the energy stored in the battery. Therefore, the battery current increases and its voltage decreases, as shown in Figure 9.18(b) and (c). In addition, the DC–DC converter limits the battery current gradient to lower the stress on the storage device and in the process improves its lifetime.

The battery voltage operates within a defined dead band under steady-state conditions. When the voltage surpasses the lower threshold of 0.99 p.u. in Figure 9.18(c), the DC–DC converter on the upper branch between the battery and the fuel cell in Figure 9.11, increases its power flow to the grid. This continues until it fully supplies the disturbed network demand on its own. Furthermore, the converter recharges the battery and raises its voltage to be within permissible limits of the dead band. It can perform these functions since it controls the adjacent fuel cell. As a result, it increases the power supplied from the fuel cell, according to the required power demand of the grid. This supply of SECR can be seen in the form of increased hydrogen mass flow in Figure 9.18(b).

- **HSPP Long Time Frame Response**

In the longer time frame represented in Figure 9.19(c), the consequent decrease in the stored hydrogen mass is shown. During steady-state operation, the network demand is fully supplied by hydrogen storage alone. The supercapacitor and battery currents return to zero, and their voltage levels are also restored to specified setpoints. The HSPP continues to output a constant active power owing to the steady rate of hydrogen mass flow, as shown in Figure 9.19(a) and (b). However, the HSPP is not only able to supply energy to the grid, like a TPP but can also store it. This is clarified in greater detail in the following investigation.

9.3.3   *Frequency regulation during high DER infeed*

Fossil-fuelled power plants will not be part of the future electricity grid due to the expected decarbonisation of the energy system. Hence, this investigation involves examining the dynamic behaviour of the power plants in the absence of the coal-fired TPP in node 19 of the 25-node example network. The modified island network is shown in Figure 9.20 with the updated loadflow values being represented in Table 9.4. The base apparent power value of the per-unit system is still 10 MVA.

A ramp is then applied to the power generation of the four WPPs in the network from 200 s to 800 s. The corresponding function is implemented in the lookup array block, shown in Figure 9.10. This portrays a situation when the power output of the WPPs would increase steadily due to an increase in the wind speed. As a result, the

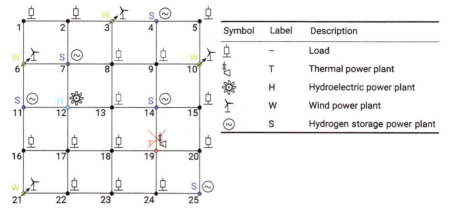

Figure 9.20 Representation of the 25-node island grid example for scenario II

Table 9.4 Network summary and loadflow setpoints for scenario II

Network elements or quantities	Power setpoint (MW)	Total power (MW)
HP (1)	10.00	10.00
WPP (4)	2.00	8.00
HSPP (4)	2.00	8.00
HSPP at Node 25 (1)	2.05	2.05
Total generation	–	**28.05**
Loads (14)	2.00	28.00
Losses	–	0.05
Total consumption	–	**28.05**

output of each WPP increases from 2 MW to 10 MW, shown in per unit values in Figure 9.21. At the same time, the power output of the HSPP and HPP decreases to maintain the balance between generation and demand. The output of the HPP levels off at 5 MW (0.5 p.u.) since it should function at least at 50% of its nominal power to maintain feasible operation. This is compensated by the five HSPPs, which then start to reduce their power output at a faster rate. This alteration in the active power reduction rate is depicted in Figure 9.21 by the gradient change of the HSPP curve once the HPP reaches its lower limit.

Due to this increase in WPP power generation, the grid frequency rises steadily, reaching its peak value when the ramp ends, as shown in Figure 9.21. After this time, the secondary controllers of the power plants take over and return the frequency to its nominal value of 1 per unit.

Such values are chosen for the WPP output ramp so that the power output of every HSPP reduces beyond zero and its energy-storing ability can be studied. In the long time frame, this excess energy available in the network is used by the HSPP to produce hydrogen via the electrolyser. The hydrogen is then stored for periods of

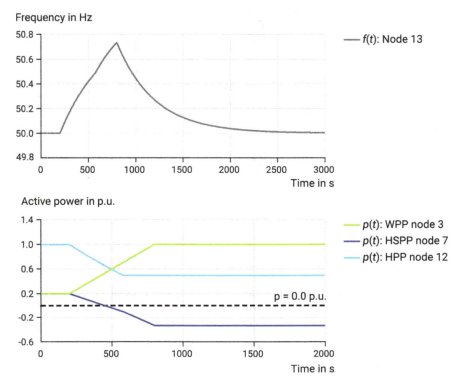

Figure 9.21 System frequency and plant outputs due to wind ramp

low renewable generation, dark doldrums[§] as well as sector coupling. This exhibits how HSPPs can ensure stability in a network with high DER infeed.

9.3.4 Frequency regulation without HSPPs

The final investigation involves analysing the frequency stability of the 25-node island network by replacing the HSPPs with alternative power plants. In this case, the HSPPs at nodes 4, 14 and 25 are replaced with WPPs operating at their MPP. The two HSPPs at nodes 7 and 11 are replaced with BESSs, which provide PCR. The dynamic model of the BESS is not explained in this chapter since it is available as an open-source example in the software DIgSILENT PowerFactory [49]. The TPP at node 19 is included again, so the results from this scenario can be compared with those in the first scenario (see Section 9.3.2). The updated grid structure is shown in

---

[§] Dark doldrums refer to periods of reduced DER infeed caused by simultaneous low wind speeds and limited solar irradiance. Such conditions present significant challenges for maintaining a consistent generation of electrical energy. Strategies such as energy storage, demand response and diversification of renewable sources are essential to counter the adverse impact of such situations. The hydrogen storage of an HSPP is dimensioned to withstand 2 weeks of dark doldrums.

Figure 9.22 and the corresponding loadflow values are presented in Table 9.5. The base apparent power value of the per-unit system is 10 MVA.

Next, the same load step as in the first scenario (Figure 9.13) is implemented to increase the power consumption at node 13 from 4 MW to 9.6 MW. The consequent system frequency and active power output of the different power plants (per unit) are displayed in Figure 9.23 during the long time frame. Since the WPPs operate at their MPP (i.e. without any reserve to provide frequency ancillary services) and the BESS only provides PCR, the IR originates only from the physical inertia in the TPP and HPP. This leads to a high RoCoF due to the load change, leading to a frequency nadir of around 49.6 Hz, as shown in Figure 9.23. This is much lower compared to the nadir of about 49.8 Hz, as presented in Figure 9.16. These results comply with the ones shown in Figure 9.1 and highlight the importance of IR as well as the susceptibility of the current grid to frequency events.

Figure 9.23 illustrating the active power outputs of the different plants show that the IR originates primarily from the HPP and TPP. The compressed time scale on the

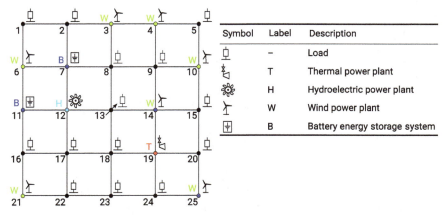

Figure 9.22 Representation of the 25-node island grid example for scenario III

Table 9.5 Network summary and loadflow setpoints for scenario III

Network elements or quantities	Power setpoint (MW)	Total power (MW)
TPP (1)	5.00	5.00
HPP (1)	5.00	5.00
WPP (6)	5.25	31.5
BESS (2)	5.00	10.00
WPP at Node 25 (1)	4.62	4.62
Total generation	–	**56.12**
Loads (14)	4.00	56.00
Losses	–	0.12
Total consumption	–	**56.12**

280 *Intended and unintended islanding of distribution grids*

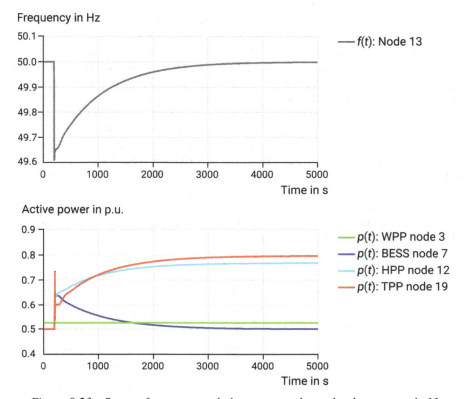

Figure 9.23 System frequency and plant outputs due to load step at node 13

x-axis gives the impression that the BESS also provides some IR but in reality, its response is much more delayed compared to the two conventional plants. Compared to the first scenario, which has seven power plants providing PCR (5 HSPPs, 1 HPP and 1 TPP), this case has only four such plants (2 BESSs, 1 HPP and 1 TPP). This also contributes to a lowered frequency nadir.

The SECR is supplied only by the two conventional plants. This factor along with the lowered frequency nadir results in a longer duration for the frequency restoration to the nominal value of 1 per unit. For simplification, the recharging of the BESS has been ignored in this investigation. If this was incorporated, either the outputs of the conventional plants would need to be raised or an increase in DER infeed (intermittent and uncontrollable) would be required. This further highlights the flaw in a system operating with plants that cannot deliver the full range of required ancillary services.

9.4 Summary

With the proliferation of renewable energy sources such as solar and wind power, distribution systems are facing unprecedented challenges due to the intermittent and

Dynamic modelling for distribution grid analysis in the time domain   281

variable nature of these energy resources. To address this escalating complexity, appropriate analytic approaches are necessary to facilitate the integration of renewable energy sources. In this evolving landscape of electrical power systems, dynamic simulation in the time domain has become increasingly important for the analysis of distribution systems. Such simulations consider time-varying phenomena and transient events, that are essential for a comprehensive understanding of the behavior of distribution grids under changing conditions. This chapter presents the importance of dynamic simulation in the context of power system stability and reliability, with a focus on distribution system operation. Basic information on modeling the energy system in the time domain is presented and several generic models, including conventional power plants, DERs and energy storage systems, are described. The chapter not only provides a basic understanding of dynamic modeling on different time scales and the difference between RMS and EMT simulations, but also contains relevant references for interested readers who need further guidance in the field of dynamic simulation of electrical power systems. A case study on the provision of ancillary services in an island grid with high DER feed-in, utilizing some of the previously described modeling approaches, completes the chapter.

References

[1] Milano F, Dörfler F, Hug G, *et al.* Foundations and challenges of low-inertia systems (invited paper). In: *2018 Power Systems Computation Conference (PSCC)*; 2018. p. 1–25.

[2] Prostejovsky AM, Brosinsky C, Heussen K, *et al.* The future role of human operators in highly automated electric power systems. *Electric Power Systems Research*. 2019;175:105883. Available from: https://www.sciencedirect.com/science/article/pii/S0378779619302020.

[3] ENTSO-E. Frequency ranges; ENTSO-E guidance document for national implementation for network codes on grid connection; 31 January 2018. Accessed: 2024-03-23. https://eepublicdownloads.entsoe.eu/clean-documents/Network%20codes%20documents/NC%20RfG/210412_IGD_Frequency_ranges.pdf.

[4] Dozein MG. *System Dynamics of Low-Carbon Grids: Fundamentals, Challenges, and Mitigation Solutions*. Department of Electrical and Electronic Engineering, University of Melbourne; 2021.

[5] Vorwerk J, Markovic U, Aristidou P, *et al.* Quantifying the uncertainty imposed by inaccurate modeling of active distribution grids. *Electric Power Systems Research*. 2022;211:108426. Available from: https://www.sciencedirect.com/science/article/pii/S0378779622005673.

[6] Hussain A, Kim CH, and Mehdi A. A comprehensive review of intelligent islanding schemes and feature selection techniques for distributed generation system. *IEEE Access*. 2021;9:146603–146624.

[7] Society IPE. Stability definitions and characterization of dynamic behavior in systems with high penetration of power electronic interfaced technologies. Tech Rep PES-TR77. May 2020.

282   *Intended and unintended islanding of distribution grids*

[8] Hatziargyriou N, Milanovic J, Rahmann C, *et al*. Definition and classification of power system stability – revisited and extended. *IEEE Transactions on Power Systems*. 2021;36(4):3271–3281.

[9] Yamashita K, Martinez V, van Cutsem T, *et al*. Modelling of Inverter-Based Generation for Power System Dynamic Studies; 2018. Cigré Technical Brochure 727.

[10] Kundur P. Power System Stability And Control. *EPRI Power System Engineering Series*. McGraw-Hill; 1994.

[11] Chua LO and Lin PM. *Computer-Aided Analysis of Electronic Circuits: Algorithms and Computational Techniques*. Englewood Cliffs, NJ: Prentice-Hall; 1975.

[12] Press WH. *Numerical Recipes: The Art of Scientific Computing*. 3rd ed. Cambridge: Cambridge Univ. Press; 2007.

[13] Butcher JC. *Numerical Methods for Ordinary Differential Equations*. 3rd ed. Chichester, UK: Wiley; 2016.

[14] Jin S, Huang Z, Diao R, *et al*. Comparative implementation of high performance computing for power system dynamic simulations. *IEEE Transactions on Smart Grid*. 2017;8(3):1387–1395.

[15] Driscoll TA and Braun RJ. *Fundamentals of Numerical Computation*. Siam Society for Industrial and Applied Mathematics, Philadelphia, PA; 2017.

[16] O'Rourke CJ, Qasim MM, Overlin MR, *et al*. A geometric interpretation of reference frames and transformations: dq0, Clarke, and Park. *IEEE Transactions on Energy Conversion*. 2019;34(4):2070–2083.

[17] IEEE Guide for Synchronous Generator Modeling Practices and Parameter Verification with Applications in Power System Stability Analyses. IEEE Std 1110-2019 (Revision of IEEE Std 1110-2002). 2020; p. 1–92.

[18] Watson N and Arrillaga. *Power Systems Electromagnetic Transients Simulation*. 2nd ed. Energy Engineering. Institution of Engineering and Technology; 2018.

[19] Dommel HW. Digital computer solution of electromagnetic transients in single- and multiphase networks. *IEEE Transactions on Power Apparatus and Systems*. 1969;PAS-88(4):388–399.

[20] Borghetti A, Caldon R, Mari A, *et al*. On dynamic load models for voltage stability studies. *IEEE Transactions on Power Systems*. 1997;12(1): 293–303.

[21] Renmu H, Jin M, and Hill DJ. Composite load modeling via measurement approach. *IEEE Transactions on Power Systems*. 2006;21(2):663–672.

[22] WECC Modeling and Validation Subcommittee. WECC Composite Load Model Specification; April 2021. Accessed: 2024-03-27. https://www.wecc.org/Reliability/WECC%20Comp%20Load%20Model%20Specification_final.pdf.

[23] Dandeno PL. Current usage and suggested practices in power system stability simulations for synchronous machines. *IEEE Transactions on Energy Conversion*. 1986;EC-1(1):77–93.

Dynamic modelling for distribution grid analysis in the time domain   283

[24] Milano F. *Power System Modelling and Scripting*. Power Systems. Springer Berlin, Heidelberg; 2010.

[25] Society IPE. IEEE Guide for Synchronous Generator Modeling Practices and Applications in Power System Stability Analyses. IEEE Std 1110-2019 (Revision of IEEE IEEE Std 1110-2003). 2019.

[26] Weckesser T, Jóhannsson H, and Østergaard J. Impact of model detail of synchronous machines on real-time transient stability assessment. In: *2013 IREP Symposium Bulk Power System Dynamics and Control – IX Optimization, Security and Control of the Emerging Power Grid*; 2013. p. 1–9.

[27] Dandeno PL, Hauth RL, and Schulz RP. Effects of synchronous machine modeling in large scale system studies. *IEEE Transactions on Power Apparatus and Systems*. 1973;PAS-92(2):574–582.

[28] Padiyar KR. *Power System Dynamics Stability and Control*. BS Publications; 2008.

[29] IEEE Power Engineering Society. Dynamic Models for Turbine-Governors in Power System Studies. Technical Report PES-TR1. 2013.

[30] IEEE Power Engineering Society. IEEE Recommended Practice for Excitation System Models for Power System Stability Studies. IEEE Std 4215-2016 (Revision of IEEE Std 4215-2005). 2016. p. 1–207.

[31] Weber H and Welfonder E. Dynamische Netzreduktion zur Modalanalyse von Frequenz-und Leistungspendelungen in ausgedehnten elektrischen Energieübertragungsnetzen. Electrical Engineering (Archiv fur Elektrotechnik). 1992;76(1):59–69.

[32] Lammert, G, Yamashita, K, Ospina, LDP, *et al.* International industry practice on modelling and dynamic performance of inverter based generation in power system studies. *CIGRE Science and Engineering*. 2017;8:25–37.

[33] Kyesswa M, Çakmak H, Kühnapfel U, *et al.* Dynamic modelling and control for assessment of large-scale wind and solar integration in power systems. *IET Renewable Power Generation*. 2020;14(19):4010–4018. Available from: https://ietresearch.onlinelibrary.wiley.com/doi/abs/10.1049/iet-rpg.2020.0458.

[34] Rocabert J, Luna A, Blaabjerg F, *et al.* Control of power converters in AC microgrids. *IEEE Transactions on Power Electronics*. 2012;27(11):4734–4749.

[35] Pourbeik P. PROPOSAL FOR DER_A MODEL; 2016, last revision 19.06.2019. Accessed: 2024-03-19. https://www.wecc.org/Reliability/DER_A_Final_061919.pdf.

[36] North American Electric Reliability Corporation (NERC). Reliability Guideline, Parameterization of the DER_A Model. Technical report; September 2019. Accessed: 2024-03-19. https://www.nerc.com/comm/RSTC_Reliability_Guidelines/Reliability_Guideline_DER_A_Parameterization.pdf.

[37] Qi J, Li W, Chao P, *et al.* Generic EMT modeling method of Type-4 wind turbine generators based on detailed FRT studies. *Renewable Energy*. 2021;178:1129–1143. Available from: https://www.sciencedirect.com/science/article/pii/S096014812100923X.

[38] Wang Y, Wen M, and Chen Y. A simplified model of Type-4 wind turbine for short-circuit currents simulation analysis. *IET Generation, Transmission & Distribution*. 2022;16(15):3036–3049. Available from: https://ietresearch. onlinelibrary.wiley.com/doi/abs/10.1049/gtd2.12496.

[39] IEC 61400-27-1:2020, Wind energy generation systems – Part 27-1: Electrical simulation models – Generic models; 2020.

[40] IEEE Recommended Practice for the Design and Application of Power Electronics in Electrical Power systems. IEEE Std 1662-2016 (Revision of IEEE Std 1662-2008). 2017; p. 1–68.

[41] E. Farantatos (EPRI). Model User Guide for Generic Renewable Energy System Models; July 2018. Accessed: 2024-03-28. https://www.epri.com/resear ch/products/000000003002014083.

[42] Weber H. Verfahren zur externen und internen Regelung eines schwungmasselosen Speicherkraftwerks. Deutsches Patent- und Markenamt; 2020. Patent No. DE102019124268A120200312. Available from: https://www.tbi-mv.de/ fileadmin/user_upload/patente/archiv/DE102019124268A1.pdf.

[43] Gerdun P, Ahmed N, Vernekar V, *et al.* Dynamic operation of a storage power plant (SPP) with voltage angle control as ancillary service. In: *2019 International Conference on Smart Energy Systems and Technologies (SEST)*; 2019. p. 1–6.

[44] Ahmed N, Gerdun P, and Weber H. Active power control based on hydrogen availability in a storage power plant. *IFAC-PapersOnLine*. 2020;53(2):12708–12713.

[45] Töpfer M, Ahmed N, and Weber H. Dimensioning the internal components of a hydrogen storage power plant. In: *2020 International Conference on Smart Energy Systems and Technologies (SEST)*. IEEE; 2020. p. 1–6.

[46] Zhang R, Fang J, and Tang Y. Inertia emulation through supercapacitor energy storage systems. In: *2019 10th International Conference on Power Electronics and ECCE Asia (ICPE 2019-ECCE Asia)*. IEEE; 2019. p. 1365–1370.

[47] Dozein MG and Mancarella P. Frequency response capabilities of utility-scale battery energy storage systems, with application to the august 2018 separation event in australia. *In: 2019 9th International Conference on Power and Energy Systems (ICPES)*. IEEE; 2019. p. 1–6.

[48] Teichmann D, Arlt W, Wasserscheid P, *et al.* A future energy supply based on liquid organic hydrogen carriers (LOHC). *Energy & Environmental Science*. 2011;4(8):2767–2773.

[49] DIgSILENT GmbH. DIgSILENT BESS frequency control. In: *DIgSILENT PowerFactory*; 2022.

Part III

Islanded grids in practice

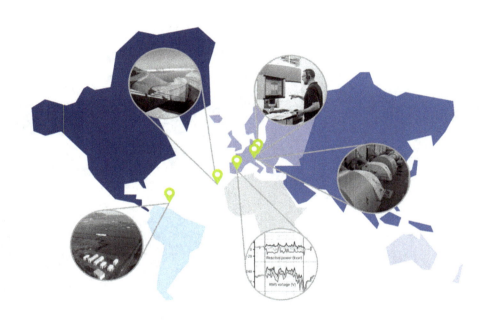

Chapter 10

Insular power systems

Andreas Knobloch[1], Mohamed Mostafa[1], João Abel Peças Lopes[2], Carlos Moreira[2], Leonel Carvalho[2], Hugo Morais[3], Lucas Pereira[4] and Michael Finkel[5]

Islands and other isolated power systems depend on thermal power generation from Diesel or other fuels to supply their electric loads. This type of power generation is a reliable and well-known established technology but brings a lot of undesired side effects such as exhaust gas pollution, noise and a lot of preventive maintenance demand [1,2]. Moreover, rising fuel costs and the cost of transporting fuel to remote islands are putting additional economic pressure on energy costs, making electricity generation from renewable energy sources an interesting competitor today.

Especially in remote locations, electricity from renewable energies (PV and wind) can often be produced at lower costs. To enable large renewable energy shares in those systems, a sophisticated and fully integrated power flow and energy management system is installed. In addition, more grid support functions have to be provided by each renewable component [3]. Large-scale grid-forming inverters can act as the backbone for genset-free grid operation and allow higher shares of renewable energy. The large-scale integration of inverter-controlled renewable generators into isolated power grids poses significant challenges for local grid operators as the proportion of conventional synchronous generators connected to the grid is gradually reduced or even eliminated. To mitigate the aforementioned challenges while enabling greater integration of renewables, isolated grid operators have established new grid code requirements for renewables and installed battery energy storage systems (BESS) to provide power frequency control. In isolated systems with close to 100% renewable integration, the BESS converter control mode can be based on grid-forming structures as these provide better performance to mitigate grid frequency fluctuations.

A rising number of projects are proving the concept to work and providing experiences about the impacts on grid operation. This chapter describes the experiences

---

[1]SMA Solar Technology AG, Business Segment Large Scale, Niestetal, Germany
[2]Faculdade de Engenharia da Universidade do Porto & INESC TEC, Portugal
[3]Instituto Superior Técnico, University of Lisbon & INESC-ID, Portugal
[4]Interactive Technologies Institute | LARSyS; Instituto Superior Técnico, University of Lisbon, Portugal
[5]Department of High Voltage Engineering and Electric Power Supply Systems, Augsburg Technical University of Applied Sciences, Germany

288 *Intended and unintended islanding of distribution grids*

with large grid-forming inverters on the Caribbean island St. Eustatius and the hybrid power system in Madeira.

10.1 Caribbean island St. Eustatius

10.1.1 Introduction

On remote islands without access to public electricity, a stable and secure power supply by renewable energy with at times up to 100% inverter penetration has long been a reality. Pioneering hybrid system projects like on the Caribbean island St. Eustatius [2,3] give an insight into hybrid generation control of future grids. The following selected results of extensive practical tests demonstrate how state-of-the-art utility-scale battery inverters with grid-forming capability can operate in harmony with conventional and renewable PV generators, but also in standalone operation [4].

The Island of St. Eustatius is part of the Antilles in the Caribbean Sea and has about 4,000 inhabitants. The electrical grid is based on a 12.45 kV medium voltage network – operated by Statia Utility Company (STUCO) – and covers a peak load demand of about 2 MW. Until 2016, the power generation was entirely based on diesel generators and more than 78% of STUCO's income was going directly into fuel-related operational costs [1,3].

The project St. Eustatius has been built in two separate project phases, 2016 and 2017, both planned and executed by the SMA Sunbelt Energy GmbH.

The first project phase was designed for a 23% PV energy share and consisted of a 1.9 MWp solar plant and a 1 MW battery storage system. After one year of operation, the estimated and actual performance of project-phase 1 was evaluated, showing very satisfying results. A PV energy share of more than 23% was achieved even though solar irradiation was about 5% less than expected. Ageing, efficiency and auxiliary consumption of the battery system were shown to be within the expected range. This project phase already led to diesel savings of more than 870,000 l/a and a reduction of CO_2 emissions of approximately 2300 t/a [3].

The second project phase of the project was designed to double the diesel savings and achieve 46% PV energy share. To achieve these goals, not only the solar plant and battery storage system had to be extended, but the operational strategy had to be modified too. During the design phase, it became obvious that only a part-time diesel-free operation during the day would result in the most cost-effective solution and allow further increases of PV energy share in the future [3].

Figure 10.1 depicts the St. Eustatius PV hybrid system setup. In total, it consists of 5.2 MVA of battery inverters, 5.77 MWh battery capacity, 3.85 MVA of solar inverters and a hybrid plant controller to supply the electrical grid with a peak demand of about 2 MW [2].

10.1.2 Droop-based grid-forming control

Figure 10.2 depicts the fundamental principles for droop-based grid-forming control battery storage inverters as an important part of the grid stabilising control system. Following a grid-forming control strategy, the inverter acts as a voltage

Insular power systems 289

Figure 10.1 St. Eustatius PV hybrid system. Left: PV storage system view from above. Right: System setup

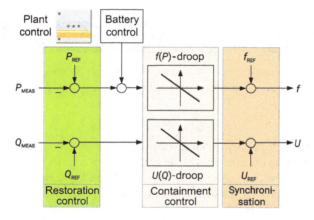

Figure 10.2 Droop-based grid-forming control structure [4]

source, where the output voltage is controlled and the current flow results by load or grid conditions. Voltage amplitude U and frequency f are determined by proportional droops, emulating the characteristics of synchronous machines. Therefore, active and reactive power measurements are used. Droop control [5] enables communication-free parallel operation of multiple gird-forming inverters with equal load share and fast containment reserve provision. In addition, the overlaid controller provides control action for the restoration of voltage amplitude and frequency to desired reference values with adjustable load share between different inverters (see Figure 10.5) [4].

10.1.3 System operation without genset inertia

The St. Eustatius hybrid power plant, with its conventional and inverter-based generation combined with battery storage and parallel operation of grid-forming and

290 *Intended and unintended islanding of distribution grids*

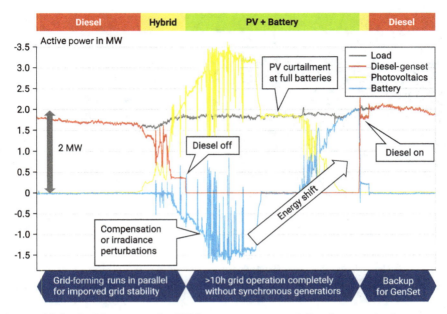

Figure 10.3 Load coverage by PV, battery storage and diesel genset in the course of the entire day on the island of St. Eustatius [4]

grid-supporting control, proves that a stable operation of low inertia power grids is feasible even without the need for conventional must-run-units [4].

Figure 10.3 shows the power flows in the PV hybrid system on a typical day. At night, diesel gensets cover the load demand. In the morning hours, the solar system reduces the power from the gensets until they reach a minimum load level. From this point, the batteries are charged until a sufficient energy amount enables to switch off the diesel engines. Hereafter, the load is completely supplied by PV plus storage without the power, energy and inertia of the diesel gensets. The batteries compensate for the difference between load and PV by charging or discharging the storage. When the batteries are fully charged, the solar power is curtailed to the load demand. In the evening, the power provided by the batteries is slowly increasing due to the reduced solar power (energy shift) and finally, the diesel gensets start again to provide the energy needed during the night. At night, grid-forming battery inverters still run in parallel for optimised genset operation and for uninterrupted backup purposes in case of a genset outage [6].

10.1.4 *Frequency stability at normal operation with large solar irradiance perturbations*

A big challenge for system frequency control at normal operation is often addressed to imbalances caused by large fluctuations of renewable power. Grid-forming battery

Insular power systems 291

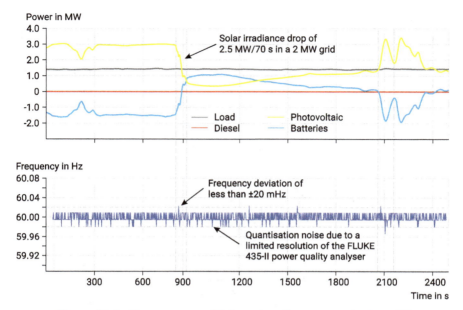

Figure 10.4   Frequency sensitivity to irradiance perturbations [4]

storage is able to compensate for the power imbalance immediately, without notable frequency deviations [3]. The system is capable of compensating huge and dynamic PV power fluctuations, typically caused by exceptionally fast-moving clouds in this region, within milliseconds. Grid frequency observations (see Figure 10.3) in stand-alone operation show that battery storage inverters compensate even power imbalances larger than load demand (solar power ramps of 2.5 MW per 70 s at 2 MW load) immediately and without a notable frequency deviation [6]. As shown in Figure 10.4, the system frequency remains close to its nominal value, with deviations of less than ±20 mHz.

10.1.5   Frequency stability and uninterrupted power supply at sudden genset outage

Power system splits, with a high power imbalance between generation and demand combined with a stepwise inertia decline in the remaining areas, represent the most critical scenarios for frequency stability in electricity grids. On the island of St. Eustatius, a power system split scenario was under investigation by a sudden disconnection of the complete conventional genset generation, which fully covered the load before failure (for detailed test setup, see [3]). Simultaneously, the complete genset inertia was disconnected.

Figure 10.5 depicts the system frequency, active power measurements and setpoints provided by the plant controller. The result indicates that grid-forming devices instantaneously take over the load with an equal share immediately, providing very fast containment reserves with dynamics similar to inertial response. After

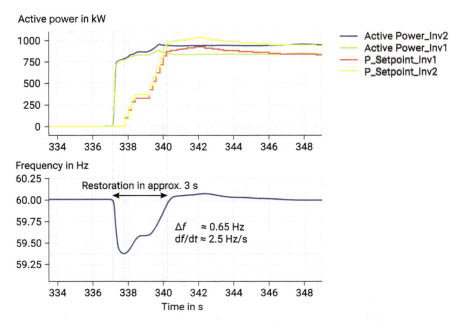

Figure 10.5 Frequency containment and restoration at unintended full genset disconnection (2 MW step) [4]

500 ms, the hybrid plant controller detects the diesel-off-situation and starts providing frequency restoration reserves by modifying the battery inverter's active power setpoints. In less than 3 s after genset disconnection, the frequency is restored back at nominal value. The maximum frequency deviation of the 60 Hz grid reached $\Delta f \approx -0{,}65$ Hz with a rate of change of frequency (RoCoF) of about $df/dt \approx -2{,}5$ Hz/s. The frequency nadir and the RoCoF depend on the inverter's control parameter settings as explained in [3,6].

10.1.6 Fast fault clearing and voltage stability after short-circuit faults

The provision of sufficient current in case of a short-circuit event in the medium-voltage grid is an important design criterion in inverter-based grids. Therefore, the verification of providing sufficient current for triggering the applicable protective devices was an integral part of the site acceptance tests. For both diesel-on and diesel-off-mode, the following scenarios were tested [3]:

- Single phase-to-ground fault
- Two-phase short circuit
- Three-phase short circuit

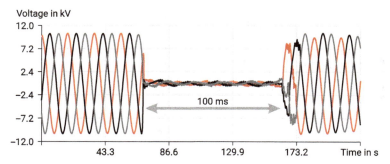

Figure 10.6 Fast fault clearing and voltage recovery after a 3-phase short-circuit fault [4]

All tests were passed successfully and demonstrated the ability of grid-forming battery inverters to achieve fast clearing times of approx. 100 ms and fast voltage recovery within 20 ms (see Figure 10.6) [4].

10.1.7 Fault current contribution

The AC current capability of commercial, cost-optimised battery inverter hardware is usually limited to almost nominal values. Facing the challenge in inverter-penetrated grids, appropriate grid-forming control strategies for overcurrent protection, as well as oversizing issues, are under discussion. For a current limiting inverter output voltage adjustment at short-circuit, three approaches have been investigated in [7,8]: virtual impedance, fall back current control and pulse width reduction. As shown in Figure 10.7, grid-forming battery storage inverters are able to provide an inherent, delay-free fault current with a maximum amplitude, where quality and accuracy depend on the individual overload capability and control design realisation. FRT rest results in [7,8] show that by an appropriate grid/plant system design (with usually a higher amount of inverter-based generation plus storage power compared to the peak load) and with appropriate grid protection units, there is no need for oversizing of individual inverters [4].

10.1.8 Power quality and operation

During the site acceptance tests, it could be shown that the system handled safely all addressed critical failure scenarios. This section focuses on quality indicators under normal operation conditions and data recordings after the first nine months of automatic operation.

The parallel operation of the Sunny Central Storage 2200 in grid-forming mode led to significant improvements in voltage harmonics and both frequency and voltage stability.

The measurements in [3] show that the grid frequency was always between 60.005 Hz and 59.992 Hz and thus is in a very small range of 13 mHz. This range is smaller than the frequency deviation allowed to activate the primary control reserve in the UCTE network [9].

294 *Intended and unintended islanding of distribution grids*

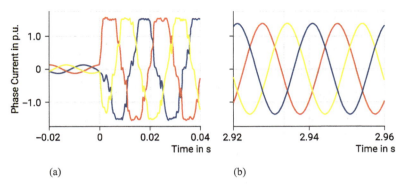

Figure 10.7 Fault ride-through current capability (in per unit) at operational limits of utility-scale grid-forming battery inverters: (a) maximum fault current contribution, with an increased distortion and (b) sinusoidal current contribution with a lower fundamental magnitude [8]

Additionally, the measured voltages (200 ms average) stayed in the range between 12.43 kV and 12.37 kV, which correspond to 99.84% and 99.36% of nominal voltage. As the last indicator for power quality, the total harmonic distortion (THD 40) of voltages on the medium voltage level remained in the range between 0.8% and 1.3% for all three phases.

Since its commissioning in November 2017, the plant has been in continuous operation. After one year of operation, the measured operational values were compared with the design targets. A solar energy share of 46–54% on a daily base was registered and thus has exceeded expectations. The most important KPI is the net energy fed into the grid. The combined system Phase 1 and Phase 2 had the design target of an energy yield of 6.4 GWh. The measured value of injected energy was 6.49 GWh; this reflects approximately 1,700,000 litres of Diesel saved and 4,600 tons of CO_2 within one year of operation [2].

10.2 Madeira Island

10.2.1 *About the Madeira Archipelago*

The Madeira Archipelago is formed by two main islands (Madeira and Porto Santo), which are permanently inhabited, and a set of small secondary islands – Desertas (with 3 islands, being only one seasonally inhabited) and Selvagens (with 3 islands, two of which also seasonally inhabited). Being located 1,000 km away from Lisbon, in the North Atlantic Ocean, the Madeira Archipelago is one of the seven European Outermost Regions.

The total extension of Madeira Island is 736.75 km^2, and its resident population is roughly 246,000 inhabitants, of which 105,000 (43%) live in Funchal, the region's capital. The total extension of Porto Santo is 42.5 km^2, and its resident population is roughly 5,100 inhabitants. However, it is important to remark that one of the key

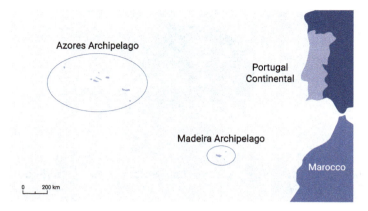

Figure 10.8 Madeira Archipelago

drivers of the local economy is tourism, with more than four million visitors in 2022, representing a considerable increase in the number of people living on both islands throughout the year.

Empresa de Eletricidade da Madeira SA (EEM) is the sole system operator (TSO/DSO) in the archipelago and oversees the production, transport and distribution of electricity, being responsible for managing the electrical systems in Madeira and Porto Santo and for carrying out the necessary investments to meet needs and guarantee the quality of electricity supply services. Regarding the transmission and distribution of electrical energy, EEM has the exclusivity of services, while, for production, the system is open to independent producers, namely private ones, who supply the grid with produced being remunerated though feed-in type tariffs.

10.2.1.1 Generation

Energy production in Madeira Island is guaranteed by 2 thermal plants (150 MW fuel and 56 MW gas), 11 hydro plants (76 MW, including two pumped storage power plants with 27.6 MW), 12 wind farms (63 MW), 1 solid waste plant (8 MW) and more than 800 solar PV plants (50 MW), most of them microgeneration [10,11]. The generation portfolio is currently supported by a 16.4 MWh/27.7 MVA battery energy storage system (BESS).

10.2.1.2 Transmission and distribution networks

The transmission network comprises substations (HV/MV) and transmission lines. In Madeira, the voltage level used in the transport network is 60 kV. In 2022, 32 substations were in service, of which 26 are intended to feed the MV network on the island of Madeira and 3 on the island of Porto Santo, with the remaining 3 intended exclusively for transport (energy transit between voltage levels 60 and 30 kV). The MV distribution network is designed to distribute energy from distribution substations. The most common operating voltage in the MV network is 6.6 kV, although

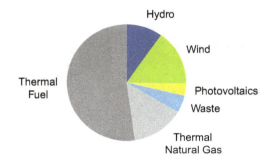

Figure 10.9 Madeira's energy mix by primary energy source (database: 2022, [12])

30 kV is also used in specific situations. In 2022, there are a total of 1678 secondary substations in Madeira Island, 1635 of which are at 6.6 kV and 43 at 30 kV.

10.2.1.3 Energy Mix

As shown in Figure 10.9, in 2022, the renewable component remained at 33.4% of the total energy emitted into the electric grid. Wind energy production registered an increase of 5.4% when compared to the previous year [12], mainly due to the impact of the recently installed battery system, which avoided the curtailment of such resources during several periods. On the other hand, the photovoltaic component showed a growth of only 0.1%, despite the significant increase observed in self-consumption units, suggesting that the resource was slightly lower than in the previous year.

10.2.2 Sustainable energy action plan

The Regional Government of Madeira is responsible, in general terms, among other attributions, for conducting the policy of the Region and adopting the necessary measures to promote economic and social development and to satisfy regional collective needs.

Madeira approved the first energy plan in 1989, with subsequent updates in 1992, 2002, 2012 and 2022. The current plan – Action Plan for Sustainable Energy and Climate of the Autonomous Region of Madeira (PAESC-RAM) – was approved in December 2022, defining the objectives and targets for the time horizons 2030 and 2050 in the areas of Energy and Climate, in accordance with Regulation (EU) 2018/1999 of the European Parliament and Council and with the Plano Nacional de Energia e Clima (PNEC), which will allow the Region to effectively monitor and report information on its contributions to the plan national.

With the implementation of the actions of this action plan, it is expected to reach, before 2030, a share of renewable energy resources of 55% in electricity production and 18% in primary energy demand, a 45% reduction in fossil fuel consumption and a 55% reduction in GHG emissions, compared to 2005. For 2050, more ambitious goals are defined, increasing the share of renewables in the electricity production

Insular power systems   297

mix by 95%. A reduction of 85% in fossil fuels consumption and GHG emissions is also strived for by 2050.

10.2.3   Generation expansion plan for the Madeira electric power system

As the economic activity in Madeira increases, energy consumption rises, accompanied by an increase in renewable energy generation, which demands a comprehensive generation expansion plan. To develop the generation expansion plan, a thorough assessment of the current and projected electricity demand is crucial. This evaluation considers factors such as population growth and increase in tourism activities, the main economic activities in the island. Detailed data analysis and forecasting techniques are employed to estimate future energy requirements accurately.

A primary objective of the generation expansion plan is to enhance the share of renewable energy sources within the Madeira Electric Power System. Embracing clean energy technologies not only reduces greenhouse gas emissions but also ensures long-term energy sustainability. Solar, wind and hydroelectric power are promising renewable sources that can contribute significantly to the decarbonisation of the generation portfolio. The Action Plan for Sustainable Energy of the Madeira Island, developed within the scope of the Pact of the Islands to which the Autonomous Region of Madeira adhered to in 2011, defines the energy policy framework in Madeira, namely regarding the incorporation of renewable electricity for the coming years. Ambitious targets have been defined for Madeira, such that 60% of the electricity consumed in the island should be of renewable origin before 2030, and by 2050, 95% of the electricity consumption should be from renewable sources. This will be achieved through an increase in installed capacity in hydro plants, wind farms and solar PV plants.

Madeira presents considerable hydroelectric potential due to its orography and water resources. The generation expansion plan focuses on the development of small- to medium-sized hydroelectric projects that can efficiently utilise local rivers and waterways. Careful consideration is given to minimise environmental impacts while maximising electricity generation. Hydro storage pumping plants are also central to this strategy. Harnessing the region's wind resources presents another valuable opportunity for renewable energy generation. Strategic placement of wind turbines can tap into the strong and consistent winds of Madeira. Conducting detailed wind resource assessments, identifying suitable sites and collaborating with wind energy developers are integral parts of the generation expansion plan.

Expanding solar power capacity involves the installation of photovoltaic systems across suitable locations. Solar farms and rooftop installations are foreseen for the next year in Madeira. The generation expansion plan aims to incentivise private investments in solar energy. The evolution of the installed capacity in special regime plants (generation using renewable sources developed by private investors) and generation owned by the utility are illustrated in Figure 10.10. From the analysis of Figure 10.10, one can conclude that the contribution from renewable plants increased considerably from 2014 to 2024 as a result of an increase of installed capacity in

298 *Intended and unintended islanding of distribution grids*

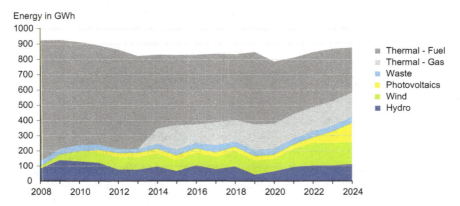

Figure 10.10 Evolution of the energy generated per technology (adapted from [10,11])

renewable power plants (25 MW in wind power, 62 MW of solar PV and 31.5 MW in hydro power).

Figure 10.11 shows typical load diagrams for the four seasons and the generation technologies used to meet demand. While generation from wind is relatively constant throughout the year, hydropower only makes a significant contribution in spring and autumn.

As renewable energy sources are time-variable in nature, energy storage systems play a pivotal role in balancing supply and demand. Incorporating battery storage facilities at strategic locations allows for effective energy management and provides strong support in frequency regulation. Thus, the generation expansion plan considers the deployment of energy storage systems to optimise the integration of renewable energy resources, which increased in 16.5 MW of hydro pumping storage together with 18 MW of batteries. A synchronous condenser of 15 MVA and inertia constant of 6 s is also planned to be installed in the near future to keep the stability of the system in scenarios with a low presence of synchronous units. A pure hydro pumping system was built with 15 MW generation installed capacity and 3 pumps of variable speed with 5.1 MW.

10.2.4 *Grid expansion plan for the Madeira electric power system*

Expanding the generation capacity necessitates the enhancement of the existing grid infrastructure. Robust transmission and distribution networks are vital to accommodate the increased electricity generation, ensure efficient power transmission and maintain grid stability. The generation expansion plan included the construction of new transmission lines (60 kV) and new substations to connect the newly developed renewable energy projects to the grid. This integration ensures seamless power flow, minimises transmission losses and improves grid resilience. At the same time, Madeira has an ambitious plan to supply ferries, cruise ships and merchant ships with electricity, when docked in the ports of Madeira (Funchal and Caniçal). This involves a large demand increase, which needs to be supplied by renewable power

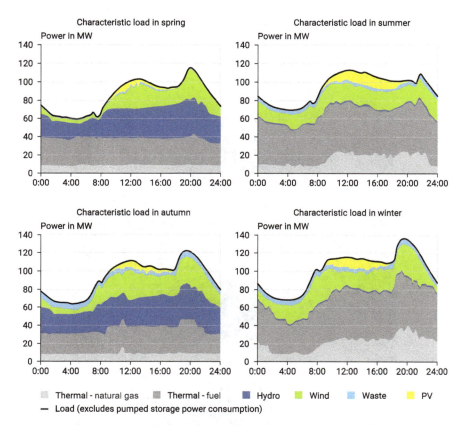

Figure 10.11 Typical load diagrams per season of the year and the generation technologies used to satisfy demand [10,11]

sources. New plans for the reinforcement of renewable generation, together with additional storage, are under study. This also requires an expansion and reinforcement of the transmission grid to supply the increased demand in the Funchal and Caniçal ports. The 60 kV infrastructure will be expanded for this purpose with new substations inside the ports. Figure 10.13 describes the structure of the main grid infrastructure at 60 kV and 30 kV.

10.2.5 Security of supply through reliability and generation adequacy assessment

Implementing 100% renewable-based electric systems in physical islands represents a major challenge. These isolated grids, disconnected from larger mainland networks, have long relied on fossil-fuelled generators to meet their energy needs and should have special requirements regarding the composition of their generation portfolio to ensure the security of supply. However, the imperative to reduce

300 *Intended and unintended islanding of distribution grids*

Figure 10.12   Upper reservoir of the Calheta power plant

greenhouse gas emissions and the prospect of using local primary energy resources, such as solar and wind energies, have prompted a radical transformation in these grids.

The variability of hydro, solar and wind generation is a major source of uncertainty in the long-term security of supply, especially in systems with limited storage capabilities. The elaboration of optimal expansion plans for islanded generation systems requires advanced models capable of addressing the time variabilities associated with renewable generation in different timeframes (e.g. daily, seasonal and yearly variations) and the operation strategies adopted to guarantee the safe, economical and continuous supply of electricity to end-users. Naturally, this planning exercise can only be carried out with high-resolution models capable of simulating the hourly operation generation systems, enabling the emulation of the operator's decisions throughout time, specifically in terms of unit scheduling and storage use. Furthermore, it is crucial to incorporate a comprehensive range of annual regimes for hydro, wind and solar sources into the analysis, as well as the unavailability of

Insular power systems 301

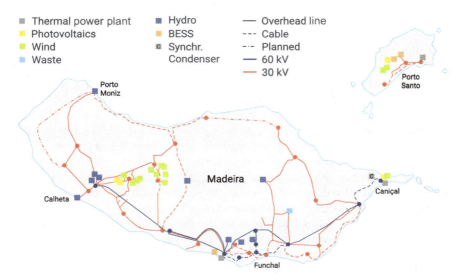

Figure 10.13 Transmission and distribution grid of Madeira (based on [12,13])

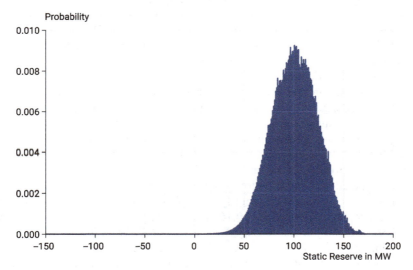

Figure 10.14 Static reserve probability distribution

the generation fleet stemming from planned and forced outages. Therefore, using an advanced modelling approach is of utmost importance as it effectively broadens the scope of the assessments by encompassing extreme conditions. This enhancement in modelling contributes significantly to the confidence in the calculated values for traditional risk indices, such as Loss of Load Expectation (LOLE) and Expected Energy not Supplied (EENS). As a didactic example for calculating the LOLE, Figure 10.14

302 *Intended and unintended islanding of distribution grids*

shows the probability distribution of static reserve states for a planning configuration of Madeira's generation system. The index LOLE results from accumulating the probabilities of negative reserve states.

Risk indices for each year of the planning horizon can be determined using either analytically based or simulation-based methodologies [14]. Notably, the application of Sequential Monte Carlo Simulation (SMCS) [15] stands out as a powerful tool for comprehensively addressing the temporal dependencies among stochastic variables. Conceptually, risk indices can be incorporated into the planning exercise as an independent criterion that requires minimisation, akin to the objectives of cost minimisation or renewable energy maximisation. Alternatively, they can be treated as a threshold that must not be exceeded. Consequently, a candidate plan, specifying the units to be installed during each year of the planning horizon, must be evaluated concerning the trajectories of the risk indices over the years.

Bearing this in mind, a planning study for the generation system of Madeira Island over the horizon of 2020–2030 was developed, aiming to find duly timed investment solutions in new generation capacity. The alternatives included renewable-based options, namely, a new hydro plant with pumped storage in 2024 and a geothermal plant in 2025, and fossil-fuelled options, namely, new thermal units of 16.5 MW, which, contrarily to the pumped hydro and geothermal options, can be installed in any year of the planning horizon. A predefined trajectory accounting for wind and solar capacity growth was considered, along with a decommissioning plan for existing thermal power plants. Four load trajectories (LC1 – 0% load growth, LC2 – 0.5% load growth, LC3 – 1.5% load growth and LC4 – 3.2% load growth) were defined to reflect varying consumption levels.

It is worth mentioning that the pumped hydro plant considered in this study is presently already in operation. While this planning exercise also focused on the contribution of the geothermal plant, this option was only considered mainly due to its capability to displace fossil-fuelled generators on the basis of the load diagram, allowing for a greater share of renewables in the total energy consumption of the island. Moreover, static energy storage based on battery systems was not considered due to its prohibitively high cost at the time for daily and/or seasonal storage applications as compared to pumped hydro. Notice that a renewable-based system requires firm generation to guarantee the security of supply in periods with low wind and solar. As an example, Figures 10.15 and 10.16 illustrate the variability of renewable capacity in weeks 10 and 36, respectively, for the year 2030, for a given conjugation of yearly wind, hydro, and solar regimes. The technologies represented in these plots have almost no degree of controllability, with the exception of run-of-river hydro, which can store water for a few hours. It is clear that there might be periods in the year where these resources can exceed the total load but others where other units must operate, namely, units that can offer firm power.

Accordingly, three main objectives for the sizing problem were defined as:

- To minimise the maximum LOLE in the planning horizon;
- To maximise the renewable energy use in 2030;
- To minimise the present value of capitalised costs in 2017, accounting for investment, operational, production and emission expenses discounted at a given rate.

Insular power systems 303

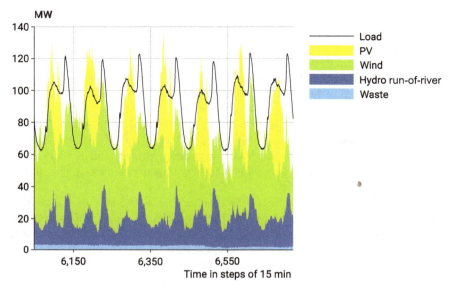

Figure 10.15   Illustration of renewable generation in week 10 of 2030

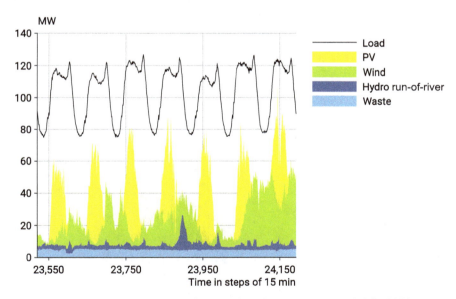

Figure 10.16   Illustration of renewable generation in week 36 of 2030

These objectives are conflicting in their nature. Naturally, there is no optimum solution for this class of problems but a set of efficient, non-dominated solutions, i.e. those belonging to the Pareto Front. The generation of non-dominated solutions relied on the Non-dominated Sorting Genetic Algorithm-II (NSGA-II) [16] algorithm.

The risk indices were determined based on yearly analysis of the generation system availability, including detailed operation of the pumped storage units, with their specific roundtrip efficiency, for maximising the use of renewables. Chronological series from 2014 to 2016 for hydro, wind, solar and waste-to-energy were considered, whose combination led to 81 dissimilar yearly regimes. Forced outage rates for each unit were computed based on past records.

Water pumping for storing surplus renewable energy was only permitted during off-peak periods, denoted as the period outside the window between 17h30 and 22h30. As an illustration, Figure 10.17 shows the yearly operation of the pumped storage units with a resolution of 15 minutes, along with the total energy stored. Noticeably, there is a seasonal variability, with more hydro energy available towards the end of the year.

Yearly planning solutions consisted of the number of 16.5 MW thermal units to be installed while the new pumped-hydro plant, which is composed of two turbines of 15 MW and five pumping units of 5.5 MW, and the three new geothermal units of 10 MW each would be commissioned in 2024 and 2025, respectively. NSGA-II was able to find 301 efficient solutions. Table 10.1 illustrates an efficient planning solution in the Pareto Front, while Table 10.2 shows the evolution of the LOLE over the years in Option C and Option D. Notice that there are no new investments before 2023. Moreover, 2028 is the critical year with the greatest LOLE. Including a new pumped-hydro plant reduces the LOLE, but it is insufficient to reduce the number of new thermal units required to be installed. The most effective measure to improve the security of supply is installing the geothermal plant, which allows a significant decrease in the generating systems' LOLE.

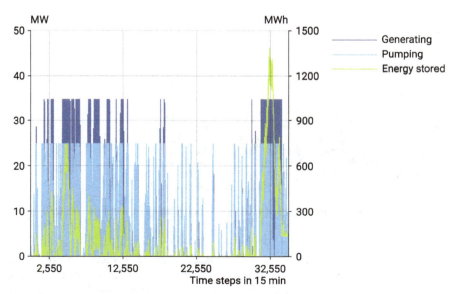

Figure 10.17 Illustration of pumped storage operation along the year

Insular power systems   305

Table 10.1   Overview of an efficient planning solution

Year								Option A	Option B	Option C	Option D
								Wo[1]/Geo-thermal Wo/New Pumped Hydro	Wo/Geo-thermal W[2]/New Pumped Hydro	W/Geo-thermal Wo/New Pumped Hydro	W/Geo-thermal W/New Pumped Hydro
2023	2024	2025	2026	2027	2028	2029	2030				
# Thermal Units								Max LOLE (h/yr)	Max LOLE (h/yr)	Max LOLE (h/yr)	Max LOLE (h/yr)
5	0	1	1	0	0	1	0	142,492	127,364	13,447	11,814

[1] Wo: Without
[2] W: With

Table 10.2   Evolution of LOLE over the planning horizon under different load scenarios

Year	Option C				Option D			
	LC1 LOLE (h/yr)	LC2 LOLE (h/yr)	LC3 LOLE (h/yr)	LC4 LOLE (h/yr)	LC1 LOLE (h/yr)	LC2 LOLE (h/yr)	LC3 LOLE (h/yr)	LC4 LOLE (h/yr)
2020	0.009	0.014	0.048	0.299	0.009	0.014	0.048	0.299
2021	0.009	0.015	0.070	0.629	0.009	0.015	0.070	0.629
2022	0.009	0.019	0.100	1.281	0.009	0.019	0.100	1.281
2023	0.062	0.131	0.544	5.323	0.062	0.131	0.544	5.323
2024	0.051	0.103	0.598	7.678	0.044	0.089	0.513	6.698
2025	0.059	0.149	0.741	11.040	0.051	0.109	0.645	9.722
2026	0.009	0.023	0.175	4.240	0.008	0.018	0.147	3.714
2027	0.009	0.023	0.225	7.516	0.008	0.025	0.195	6.616
2028	0.009	0.033	0.300	**13.447**	0.008	0.025	0.260	**11.814**
2029	0.001	0.005	0.079	6.386	0.001	0.005	0.069	5.611
2030	0.001	0.005	0.107	11.754	0.001	0.005	0.093	10.342

Only the solutions with LOLE less than eight hours per year in load scenarios LC1, LC2 and LC3 were considered, which resulted in the final planning solutions represented in Table 10.3. Notice there is a high risk in high-load scenarios when only five thermal units are installed. Moreover, if no geothermal is installed, then the percentage of renewable penetration in 2030 is less than 50%, again in an extreme load growth scenario.

In conclusion, incorporating risk indices, such as the LOLE, into the decision-making process, apart from cost and renewable penetration, provides a quantitative analysis of the risks of each planning decision has under various externalities, such as

306 *Intended and unintended islanding of distribution grids*

Table 10.3 Final planning solutions

Sol.	Therm. units #	Pumped hydro	Geo- therm.	LOLE (h/yr)				% Renewable 2030			
				LC1	LC2	LC3	LC4	LC1	LC2	LC3	LC4
1	5	No	Yes	0.31	0.82	6.19	**252.00**	84.7	80.1	70.0	55.4
2	8	No	Yes	0.04	0.12	0.50	10.32	84.7	80.1	70.0	55.4
3	5	Yes	Yes	0.76	1.68	7.89	**229.66**	87.3	82.4	71.9	56.9
4	8	Yes	Yes	0.04	0.11	0.48	9.19	87.3	82.4	71.9	56.9
5	7	Yes	No	0.70	1.54	6.93	**150.06**	67.3	63.6	55.7	**44.1**
6	10	Yes	No	0.04	0.11	0.47	10.81	67.3	63.6	55.7	**44.1**

load growth, ultimately leading to better-informed decisions, which can be adjusted along time to mitigate potential adverse outcomes.

10.2.6 System dynamic performance – the need of synchronous inertia and the role of battery energy storage

The large-scale integration of converter-interfaced renewable-based generators in isolated power systems is posing tremendous challenges to local system operators as the share of conventional synchronous generators connected to the grid is being progressively reduced or even eradicated. An immediate consequence consists of renewables curtailment, especially during valley loads, as a result of the system operator requirements demanding a minimum number of fossil-fuelled plants to be in operation to ensure minimum spinning reserve, appropriate voltage control and minimum short circuit power. Aiming to mitigate these challenges, while allowing more renewable generation integration, insular system operators have been defining new grid code requirements for renewable generation units [17] (Section 10.2.7) and installing BESS for the provision of fast power-frequency regulation services [18], as it is the case of Madeira Island, where a 27.7 MVA/15 MW/16.4 MWh lithium-ion BEES has been in full commercial operation since mid-2022 being capable to be operated in grid-following or grid-forming modes depending on the system operator decision.

Although many local system operators have been installing BESS for fast power-frequency regulation, it has also been observed that in operation scenarios with very high instantaneous shares of converter-interfaced renewables, the integration of additional synchronous inertia must be used as a complementary solution to improve frequency dynamics in the face of fault induced disturbances risking key frequency indicators such as frequency nadir and the RoCoF. This is also the case with the Madeira Island Electric Power System, where a larger part of the reversible hydro power plants was recently refurbished to be capable of operating in a synchronous condenser mode. Recent installations were also constructed with this capability (ex: Calheta III hydro station). In addition, the installation of a 15 MVA synchronous condenser (inertia of $H = 6$ s in the machine base power) for voltage/reactive power support, as well as for providing additional grid inertia, is already planned to be in operation for 2025.

One of the key issues regarding the dynamic security of this network is related to the need to avoid under-frequency load shedding, which can be triggered either by the RoCoF or by the under-frequency excursion. During the voltage sags, most of the power associated with converter-interfaced units is significantly affected by the fault ride-through reaction of the converter-interfaced sources, which is then followed by its ramp-limited active power recovery after the voltage is restored [19]. Consequently, there is a critical concern regarding the delayed post-fault active power recovery gradient of modern wind turbines with respect to short-circuits in isolated power systems. In fact, during high instantaneous wind power generation levels, the magnitude of the mismatch between generation and consumption following fault clearance and delayed recovery of active power injection from wind generators could be larger than the traditional extreme case of the largest generator tripping, hence risking frequency stability.

As an illustrative example, a day-time operation scenario in spring with 120 MW total load (95 MW of system natural load and 25 MW of pumping power) together with a generation scenario consisting of 60 MW of wind, 55 MW of solar PV, 5 MW of the waste-to-energy plant is considered. A particular characteristic of this scenario is the consideration of a fleet of hydro generators operating in synchronous condenser mode, adding about 230 MW.s of inertia (rotational kinetic energy) to the grid. Moreover, a BESS with 25 MVA/20 MW connected to a grid-following type power converter and providing fast power-frequency regulation is in operation in this scenario. The critical security constraint for the Madeira system is avoiding automatic under-frequency load-shedding situations. In particular, it is considered that load shedding occurs at least under one of the following conditions take place:

- Rate of change of frequency ($\Delta f / \Delta t$, RoCoF) lower than -1.5 Hz/s
- Absolute frequency value lower than 49 Hz.

A symmetrical three-phase short-circuit located in one of the longest transmission lines of this island's high voltage (60 kV) network was simulated. The fault is cleared within 100 ms together with the disconnection of the line, also leading to the disconnection of some generation plants without fault ride-through capability. This refers to older wind and PV units, adding up to 20 MW of a combined generation loss, as well as fixed-speed hydro pumps (disconnected by means of undervoltage protection relays), partly counterbalancing this deficit with the disconnection of about 8 MW of load. Figure 10.18 presents the simulation results of this base case condition regarding the frequency of some synchronous units.

A sensitivity analysis with respect to the BESS power converter sizing is also presented, as well as the influence of integrating additional rotational inertia. Because of the considered network fault, the BESS power converter activates the fault ride-through mode and its active power response capacity in the initial moments is significantly affected: as a result of the voltage sag, the BESS converters reach the current limit, and the injected active power is proportional to the amplitude of

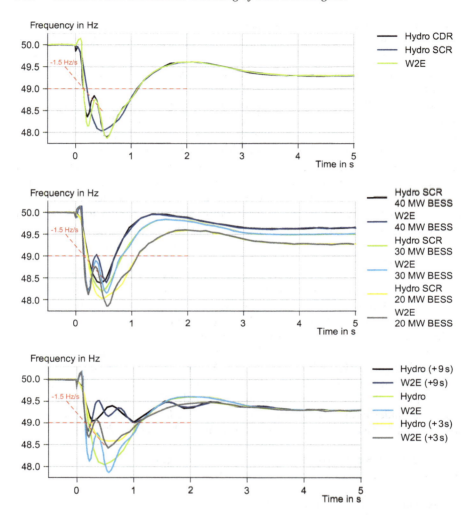

Figure 10.18 Generators frequency in different operation conditions: base case (top), sensitivity to BESS power converter sizing (middle) and sensitivity to additional rotational inertia integration (bottom). The dashed red line indicates the RoCoF threshold for under-frequency load-shedding activation

the voltage sag. Consequently, the influence resulting from the capacity increase in the BESS power converter is negligible regarding the fast dynamics observed in the initial frequency drop. The improvements provided by the additional inertia are remarkable, as the initial frequency drop is significantly sustained.

Considering the future scenario where the BESS in the Madeira Island is to be duplicated, as well as the installation of a synchronous condenser (15 MVA,

6 s inertia constant in the machine MVA base), it is addressed the impact in frequency stability resulting from the operating the BESS power converters in a grid-forming mode (GFL). Moreover, while considering different operating scenarios, Figure 10.19 provides a summary of the key frequency indicators (nadir and RoCoF) computed with respect to the grid Centre of Inertia (CoI) considering different operation modes of the BESS power converters (GFM versus GFL – grid following), as well as different number of synchronous condensers activated (SC – the 15 MVA synchronous condenser; SH – the synchronous condensers from Socorridos and Calheta III hydro power stations).

Grid-forming power converters are advanced power electronic interfaces that play a crucial role in modern power systems. Unlike traditional grid-following converters that synchronise with an existing power grid, grid-forming converters have the unique capability to define grid frequency and voltage independently. By acting as virtual synchronous machines, grid-forming power converters can contribute to seamlessly integrating renewable energy sources and battery storage systems into the grid, enhancing overall system resilience and flexibility. Their ability to contribute to grid stability makes them vital components in the transition towards a more sustainable and decentralised energy landscape. Nevertheless, the results evidence that a proper management of GFM and rotational inertia is still vital in the Madeira Island

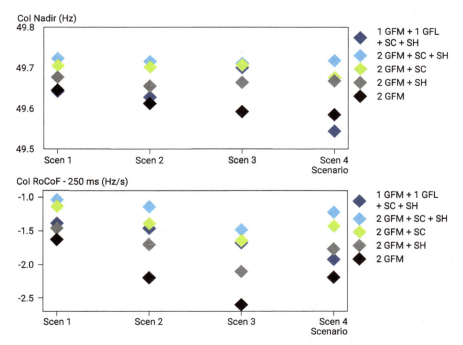

Figure 10.19 Key frequency indicators evolution in several scenarios and considering different operation modes for the BESS power converters and a number of activated synchronous condensers

310 *Intended and unintended islanding of distribution grids*

power grid to comply with the security rules, particularly in the scenarios without thermal power generation connected to the grid. It is also notorious that the preferred combination of synchronous condensers in operation depends on the operating scenarios, hence demanding specific attention.

10.2.7 Madeira Grid Code – a short overview

As already discussed in Chapter 4, the share of Converter Interfaced Renewable Energy Sources (CI-RES) impacts the way power systems are operated. When the integration of CI-RES plants in the network is relatively low/moderate, they are not required to provide ancillary services aiming to support grid operation, since the resulting impacts of their integration can be handled by the existing operation practices. However, the large-scale integration of inverter-based generation required the definition of a new set of connection requirements with the objective of mitigating the adverse impacts it brings with respect to power system dynamic performance. CI-RES generators must actively participate in the provision of ancillary services. The design of specific grid codes or the adaptation of widely used requirements to the specific characteristics of autonomous power systems is a pathway being largely adopted to evolve towards higher integration of renewables in insular systems while aiming to mitigate the resulting adverse impacts of this transition [20]. These connection requirements are the main concern, taking into account future operation scenarios with 100% infeed from renewables in a system with a high diversity of resources, as it is the case of the Madeira Island power system.

The grounding framework for the identification of the requirements to be applied to new RES-based generation facilities in the Madeira Island is based on the European grid code [21], attempting to thoroughly adapt and adjust some of its norms to the specific characteristics of the Madeira power system [17]. From the assessed operating scenarios driving the definition of the grid code, it becomes notorious the risk of fast frequency variations due to reduced inertia operating regimes, hence leading to the specification of a RoCoF withstand capability of 4 Hz/s (measured in a moving window of 250 ms) for all the new generating units aiming to secure the operation of the system and to improve the frequency ride-through capability. Moreover, all units must be capable of withstanding frequency deviations in the 47–53 Hz range for at least 20 s and in the range 47.5–52 Hz without any time restrictions.

Following the conceptual model existing in the European grid code, the classification of each generation site is based on its installed capacity within a set of categories and applicable connection requirements per category, as summarised in Table 10.4.

A special class of power facilities, with a power rating of up to 2.5 kW and intended to foster self-consumption in a domestic environment is also considered (class A-special). These units must comply with frequency requirements.

The under-voltage ride-through requirement is transversal to existing grid codes. Nevertheless, during the voltage sag, there exists the possibility of giving priority to reactive current injection in CI-RES to active current injection. In the particular case of Madeira Island, it was found that active current control should be prioritised (reactive current control can also be included if total current is below nominal values),

Table 10.4 Installed capacity limit per generation type and applicable connection requirements (cf. Chapter 4)

Generation system type	Type A	Type B	Type C	Type D
Installed capacity limit	<0.1 MW	<1 MW	<5 MW	≥5 MW
Requirements				
Low-voltage ride-through	–	X	X	X
High-voltage ride-through	–	X	X	X
Reactive current control	–	–	X	X
Active current control	–	–	X	X
Post-fault active power recovery gradient	–	X	X	X
LFSM-O[1]	X	X	X	X
LFSM-U[2]	–	–	X	X
Active power ramps	–	–	–	X

[1] Limited frequency sensitive mode for over-frequency.
[2] Limited frequency sensitive mode for under-frequency.

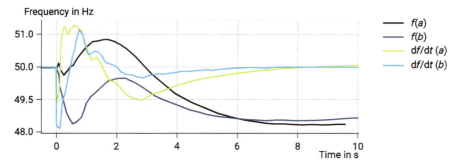

Figure 10.20 *A transmission line short-circuit disturbance in an off-peak scenario, comparing the outcome for frequency f and df/dt with (a) and without (b) priority to active current injection*

which is a key requirement to contain voltage-dip-induced frequency events in low rotational inertia scenarios. The evaluation of the effects resulting from this requirement was performed by simulating a network fault and comparing case (a) with zero active current infeed during the voltage dip against case (b), where priority is given to set active current equal to the pre-fault value, being the reaming current capacity of the CI-RES used to modulate reactive current injection as a function of the terminal voltage. This behaviour is considered only in type C and type D generators and the resulting impact in terms of the CoI frequency is depicted in Figure 10.20. The results demonstrated that the active current injection requirement constitutes a complementary alternative to the synchronous inertia increase. This justifies the need for voltage-sensitive active current control as its benefits are fundamental to contain the system's response to the operator's reference disturbance.

When integrating large amounts of renewable generation, one possibility to counterbalance grid inertia deficit is to emulate synthetic inertia through control solutions in CI-RES. For instance, in wind generators, this can be achieved by taking advantage of the accumulated kinetic energy in the machine's rotor. Within the scope of the Madeira Grid code, wind generators were taken into account for the evaluation of inertial responses. Furthermore, two different approaches were considered. First, a conceptual approach for the provision of synthetic inertia based on the filtered value of the grid frequency time derivative. An alternative approach was also evaluated, focusing on the state-of-the-art behaviours found in some wind generator manufacturers, defining a power versus time envelope for the active power deployment [22], as depicted in Figure 10.21 In this case, the active power envelope defines the capability of providing an active power step increase for a few seconds, followed by a recovery period where power drops below the pre-fault value.

To evaluate the relevance of this requirement, a disturbance consisting of a sudden generation disconnected was considered, as well as three situations for comparison purposes: normal operation (without synthetic inertia provision), conceptual approach to inertia provision based on an equivalent inertia of $H = 1.7\,s$, and the state-of-the-art implementation based on envelope from Figure 10.21. The state-of-the-art response provides significant benefits as the active power increase is fed to the grid, significantly containing the frequency deviation following the disturbance. However, the power decrease following initial power deployment (necessary due to the recovery stage of the inertial response) leads to a large frequency dip (Figure 10.22), which the system cannot recover from without violation of predefined security criteria. Hence, as none of the considered methods provided significant benefits for the system's dynamic security, the Madeira Island grid code does not further consider this functionality as a connection requirement.

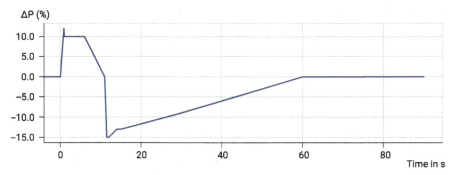

Figure 10.21 Predefined active power characteristic for the considered state-of-the-art inertia emulation capability in wind generators

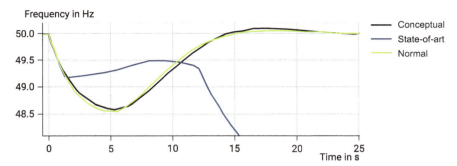

Figure 10.22   CoI frequency in the face of different options for synthetic inertia provision from types C and D wind farms

10.2.8   Using storage at the secondary substation for voltage control

A considerable amount of renewable generation in Madeira is being connected to the distribution grids, bringing operational problems that need to be properly managed. Beyond the previously discussed frequency and congestion management, voltage constraints may arise from the intensive utilisation of small-scale renewable generation, primarily photovoltaic (PV) [23]. In the case of Madeira, PV installations on rooftops for self-consumption account for around 8 MW, with a strong tendency to increase in the coming years as a result of continuous government incentives.

Voltage issues in low-voltage (LV) grids are further compounded by emerging technologies such as electric vehicles (EVs). When photovoltaics (PVs) and EVs are combined, voltage fluctuations become pronounced throughout the day, presenting new challenges for their effective operation. Another significant obstacle to the optimal management of LV grids is the absence of monitoring devices [24]. These challenges are pervasive across all systems; however, they are particularly crucial in islands and rural areas. The extensive distances covered by cables, coupled with a limited number of consumers, exacerbate the impact. In isolated systems, an additional challenge arises, primarily due to variations in voltage levels at upper levels, impacting the voltage profiles of LV grids (where the slack bus is not set at 1 p.u.).

An example of the approaches under study to address such issues in Madeira Island is described next, where battery storage systems are used to help control voltage in an LV rural distribution grid. For this use case, an LV grid depending on a secondary substation with a transformer having an apparent power of 250 kVA, passing from 6,600 V to 400 V is considered. This substation is one of Madeira's LV substations with a high amount of PV generation, with a total capacity of 36 kWp (14% of the feeder capacity), distributed over nine DGs, three of which have three-phase installations. This grid serves around 100 consumers. Figure 10.23 shows an illustration of the LV grid being studied, with the approximate location of the PV generators [25].

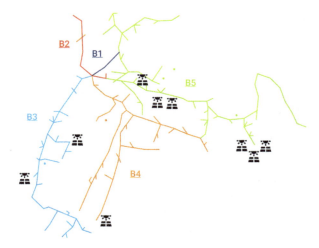

Figure 10.23   Overview of the rural grid that was considered in this use case

The grid was modelled using PandaPower, following an initial model done with DigSILENT in the scope of the Horizon 2020 SMILE project [26]. The load and production profiles used were also collected from the SMILE project for households in the vicinity [27].

To test the voltage control using active and reactive power control, some tests were done. The tests were divided into two groups: domestic and community storage. The conditions of the simulation were the following:

- Production: 125 kWp;
- Demand: 165 kW (peak), 80 kW (off-peak) and 100 kW (average);
- Season: Spring
- Day: Saturday
- Period of simulation: 11:00–18:00

In Figure 10.24, the total load and production under these conditions are presented. The result of the simulation without storage is presented in Figure 10.25(a). As expected, the nodes at the edge of the grid experience a much higher voltage level (close to 1.1 p.u.).

One of the developed experiments consisted of two use cases for reactive power management: (b) considering community storage at the secondary substation (4 BESSs), (c) 15 residential BESSs distributed across the grid. The results of the simulations are presented in Figure 10.25(b) and (c), showing that in both cases, it was possible to reduce the voltage levels in most of the nodes in the edge of the grid to under 1.05 p.u.

Ultimately, the primary finding from this analysis is that BESSs are effective in voltage control within LV grids. Their efficacy, however, hinges on two crucial factors: first, their location; second, their control functions.

Regarding the first point, it is concluded that BESS at the residential level can alleviate part of the problem by directly addressing voltage issues at the source. This

Insular power systems 315

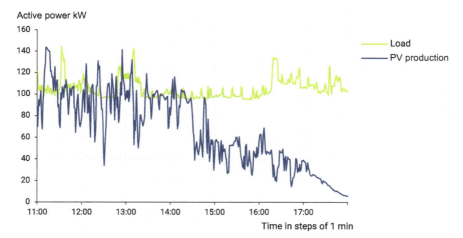

Figure 10.24   Load and PV production in the reference period: 11:00–18:00

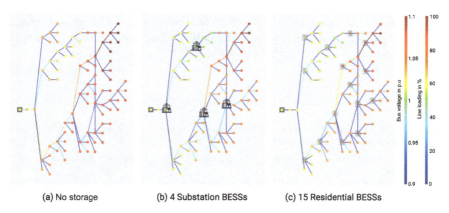

Figure 10.25   Simulation results for reactive power control: secondary substation vs residential storage

is particularly true when customers have both PV systems and EVs in the same installation. However, strategically placing small BESS units within the grid can also ensure the proper functioning of the overall system. This approach has the added advantage of providing visibility in the grid, facilitating improvements in state estimation algorithms used in control centres. BESS installed at the secondary level can be valuable, primarily for mitigating the impact of fluctuations in higher voltage levels on LV grids (ensuring the slack at 1 p.u.). As for the second point, the balance between active and reactive power, as well as the respective control functions, should be adjusted based on network characteristics (overhead or underground).

In summary, given the evolution and expansion of networks, the use of small BESS installations within the grids presents an interesting solution, allowing for the

316 *Intended and unintended islanding of distribution grids*

recalibration of control functions during operation. Nevertheless, the chosen BESS systems should facilitate this level of flexibility.

10.2.9 Conclusions

The generation expansion plan for the Madeira Electric Power System represents a forward-looking approach to meet the region's increasing electricity demands sustainably. By prioritising the integration of renewable energy sources, enhancing grid infrastructure, and fostering a favourable regulatory environment with concern for the security of supply and dynamic security, Madeira can achieve a reliable, robust, clean and resilient power system that benefits both its residents and the environment.

Very important was the design of a specific grid code for the Madeira power system such that renewable energy conversion systems comply with a set of demanding technical requirements, to assure the stability of the system.

Based on these results, several actions can be taken in the near future to enhance the grid infrastructure. Of particular interest is second-life storage that can be deployed at specific nodes of the distribution grid (e.g. at the edge or near points with higher levels of PV penetration) to deliver fast-response services such as voltage and frequency regulation, hence reducing the overall cost of the systems when compared to using fist-life storage or promoting grid reinforcements.

References

[1] Rojas EG, Sadri H, and Krueger W. Experience on MW-Sized Hybrid PV, Battery Storage and Genset System; Case Study of St. Eustatius Island. *36th European Photovoltaic Solar Energy Conference and Exhibition.* 2019; p. 1–3.

[2] Schömann O, Bülo T, Hardt C, *et al.* Experiences with Large Grid-Forming Inverters on Various Island and Microgrid Projects. *4th International Hybrid Power Systems Workshop* | Crete, Greece | 22–23 May 2019. 2019; p. 1–5.

[3] Schömann O, Bülo T, Hardt C, *et al.* Experiences with Large Grid Forming Inverters on the Island St. Eustatius, Portability to Public Power Grids. *8th International Workshop on the Integration of Solar Power into Power Systems* | Stockholm, Schweden | 16–17 October 2018. 2018; p. 1–6.

[4] Knobloch A, Hardt C, Falk A, *et al.* Grid Stabilizing Control Systems for Battery Storage in Inverter-Dominated Island and Public Electricity Grids. 13 ETG/GMA Fachtagung "Netzregelung und Systemführung", 18-19092019, Berlin. 2019; p. 1–6.

[5] Engler A. Regelung von Batteriestromrichtern in modularen und erweiterbaren Inselnetzen [dissertation]. Universität Gesamthochschule Kassel. Verlag Dissertation.de, Berlin; 2002.

[6] Knobloch A and Premm D. PV storage hybrid system for 100% solar power on the remote island of St. Eustatius. In: Kraiczy M, editor. Report IEA-PVPS T14-14:2021, PV as an ancillary service provider – Laboratory and

field experiences from different IEA PVPS countries. IEA PVPS TCP; 2021. p. 47–49.

[7] Knobloch A and Führer O. Transstabil-EE, Subproject: Control Technologies for Solar Parks and PV Inverters [Final Report]. SMA Solar Technology, Niesetal: SMA; 2019.

[8] Duckwitz D, Knobloch A, Welck F, Becker T, Gloeckler C, and Buelo T. Experimental Short-Circuit Testing of Grid-Forming Inverters in Microgrid and Interconnected Mode. *NEIS 2018; Conference on Sustainable Energy Supply and Energy Storage Systems, Hamburg, Germany*, 2018; p. 84–89.

[9] UCTE. UCTE Operation Handbook – Policy 1: Load Frequency Control and Performance [Handbook]. UCTE, Brussels, Belgium: UCTE; 2009. Available from: https://www.entsoe.eu/fileadmin/user_upload/_library/publicatio ns/entsoe/Operation_Handbook/Policy_1_final.pdf.

[10] EEM. Plano de Desenvolvimento e de Investimento na Rede de Transporte e Distribuição em at e mt DA RAM (PDIRTD-RAM), Período Regulatório 2022–2024 [Report]. Avenida do Mar e das Comunidades, Funchal: Empresa de Electricidade da Madeira (EEM); 2021. Available from: https://www.erse. pt/media/cuskjng5/pdirtd-ram-2022-2024.pdf.

[11] EEM. Relatório EEM 2022 – Caraterização da Rede de Transporte e Distribuição em AT e MT - Direção de Estudos e Planeamento [Report]. Avenida do Mar e das Comunidades, Funchal: Empresa de Electricidade da Madeira (EEM); 2022.

[12] EEM. Electricidade da Madeira: 2022 Annual Report [Annual Report]. Avenida do Mar e das Comunidades, Funchal: Empresa de Electricidade da Madeira (EEM); 2022. Available from: https://www.eem.pt/media/1070838/ en_eem_annualreport_2022.pdf.

[13] AREAM. Caracterização da Rede Eléctrica da Região Autónoma da Madeira, Report of the project TRES [Project Report]. Funchal: Agência Regional da Energia e Ambiente da Região Autónoma da Madeira (AREAM); 2010. Available from: http://proyectotres.itccanarias.org/files/TRES-Caracterizao_ da_rede_electrica_da_RAM.pdf.

[14] Billinton R and Allan RN. *Reliability Evaluation of Power Systems*. New York: Springer Science+Business Media New York, Originally published by Plenum Press, New York in 1986; 1996.

[15] Billinton R and Li W. *Reliability Assessment of Electric Power Systems Using Monte Carlo Methods*. New York: Springer Science+Business Media New York, Originally published by Plenum Press, New York in 1994; 1994.

[16] Deb K, Pratap A, Agarwal S, *et al.* A Fast and Elitist Multiobjective Genetic Algorithm: NSGA-II. *IEEE Transactions on Evolutionary Computation.* 2002;6(2):182–197.

[17] Beires P, Moreira CL, Lopes JAP, *et al.* Defining Connection Requirements for Autonomous Power Systems. *IET Renewable Power Generation.* 2019; p. 3–12.

[18] Psarros GN, Karamanou EG, and Papathanassiou SA. Feasibility Analysis of Centralized Storage Facilities in Isolated Grids. *IEEE Transactions on Sustainable Energy.* 2018;9(4):1822–1832.

318   *Intended and unintended islanding of distribution grids*

[19]   O'Sullivan J, Rogers A, Flynn D, *et al.* Studying the Maximum Instantaneous Non-Synchronous Generation in an Island System—Frequency Stability Challenges in Ireland. *IEEE Transactions on Power Systems.* 2014;29(6):2943–2951.

[20]   Rodrigues EMG, Osório GJ, Godina R, *et al.* Grid Code Reinforcements for Deeper Renewable Generation in Insular Energy Systems. *Renewable and Sustainable Energy Reviews.* 2016;53:163–177. Available from: https://www.sciencedirect.com/science/article/pii/S1364032115009107.

[21]   European Commission. Commission Regulation (EU) 2016/631 – Establishing a Network Code on Requirements for Grid Connection of Generators [Commission Regulation]. Brussels, Belgium; 14 April 2016.

[22]   Attya AB, Dominguez-Garcia JL, and Anaya-Lara O. A Review on Frequency Support Provision by Wind Power Plants: Current and Future Challenges. *Renewable and Sustainable Energy Reviews.* 2018; 81:2071–2087. Available from: https://www.sciencedirect.com/science/article/pii/S1364032117309553.

[23]   Bayer B, Matschoss P, Thomas H, *et al.* The German Experience with Integrating Photovoltaic Systems into the Low-Voltage Grids. *Renewable Energy.* 2018;119:129–141. Available from: https://www.sciencedirect.com/science/article/pii/S0960148117311461.

[24]   Grilo A, Casaca A, Nunes M, *et al.* A Management System for Low Voltage Grids. In: *2017 IEEE Manchester PowerTech*; 2017. p. 1–6.

[25]   Hashmi MU, Horta J, Pereira L, *et al.* Towards phase balancing using energy storage. arXiv preprint arXiv:200204177. 2020.

[26]   Ponnaganti P, Bak-Jensen B, and Pillai JR. Battery Energy Storage System based Voltage and Frequency Control of An Island Distribution Network. CIGRE 2020 e-session, Active distribution systems and distributed energy resources (C6). 2020;(C6-211_2020).

[27]   Pereira L, Cavaleiro J, and Barros L. Economic Assessment of Solar-Powered Residential Battery Energy Storage Systems: The Case of Madeira Island, Portugal. *Applied Sciences.* 2020;10(20). Available from: https://www.mdpi.com/2076-3417/10/20/7366.

Chapter 11

LINDA projects – droop-based practical examples

Christoph Steinhart[1], Tobias Lechner[2], Sebastian Seifried[2], Dominik Storch[2], Michael Finkel[2] and Georg Kerber[3]

The core objective of all LINDA projects is to develop a transferable emergency supply concept for particularly critical consumers in the event of a prolonged and widespread power blackout based on grid-coupled generation (e.g. grid-following photovoltaic (PV) generation and combined heat and power (CHP) plants) as an energy source without major adaptions.

The central element here is a grid-forming unit that uses the existing grid codes on the basis of a droop control system in such a way that further distributed generation (DG) units can be stably integrated into the islanded emergency supply grid without requiring the establishment of blackout-proof communication. The design of the droop control and the reaction of the DGs to this control concept are presented in more detail. Several LINDA projects that have already been carried out with practical field tests serve as examples of the successful testing of the basic concept in a real environment. This chapter also gives an insight into how to deal with practical challenges such as frequency stability during load steps and optimisation potential.

11.1 Motivation and aims

11.1.1 Motivation

The increasing degree of electrification leads to a growing addiction to supply with electrical energy in nearby all areas of life. In typical power failure scenarios, it is assumed that the electrical power supply can be fully restored within a few hours. To bridge this time, battery systems and emergency diesel generators are used in critical infrastructures such as hospitals, water supply or communication systems. Due to their limited power and energy reserves (e.g. fuel reserves for diesel generators or battery storage capacity), these emergency power systems are not adequately

[1]SWM Infrastruktur GmbH & Co. KG, Munich, Germany
[2]Department of High Voltage Engineering and Electric Power Supply Systems, Augsburg Technical University of Applied Sciences, Germany
[3]Electrical Power Engineering and Networks, Munich University of Applied Sciences, Germany

320 *Intended and unintended islanding of distribution grids*

designed for a prolonged and widespread power outage. A study by the German government in 2010 [1] has shown that long-term and widespread blackouts can have a serious impact on society and even lead to a national crisis. The comprehensive and appropriate supply of necessary goods cannot be ensured, which leads to a threat to public safety. Furthermore, digitalisation and sector coupling are leading to a massive increase in the dependency of all areas of life on electrical energy.

Beyond that, the fluctuating and uncontrollable feed-in of renewable energy sources makes the grid restoration after a blackout more complex (cf. Chapter 5).

These DG units automatically connect to the grid and start their infeed when a stable grid voltage is available again, at the grid restoration process. This can lead to a power flow from the distribution grid to the main network, which can destabilise the grid restoration process [2]. With increasing complexity and recovery time until normal network operation is restored, damages caused by the non-availability of electrical energy are rising [1]. Due to this fact, emergency supply systems gain significance.

Against this background, the research project LINDA (German acronym for Local Island Power Supply and Accelerated Grid Restauration with Distributed Generation Systems in Case of Large-Scale Blackouts) was initiated with the aim of developing and testing an emergency supply system for critical infrastructures based on DG-fed islanded grids. These should be able to operate over a long period of time in the event of prolonged power outages. With the provision of electrical energy for critical infrastructure by decentralised island grids, the potential harm for society from a blackout would be immensely decreased.

11.1.2 Main objectives

LINDA's core objective is the development of a cost-efficient backup concept by using the existing infrastructure and facilities and avoiding retrofitting and reparameterisation as far as possible. Therefore, the basic control concept uses the existing $P(f)$ and $Q(U)$ functions of the DG according to the German grid codes for balancing the active and reactive power (cf. Chapter 4). Only one leading power plant (LPP), which provides the black-start capability and is the network forming unit, gets a special, island-optimised controller parametrisation, which is activated in case of use. By using the system variables frequency and voltage as control parameters, no additional (fail safe) communication infrastructure is required. This leads to the backup concept with low investment and amendment requirements and high transferability to other grids.

11.2 Basic concept

11.2.1 Basic requirements: SCR and inertia

For a stable operation of the island grid, sufficient power of grid-forming capabilities and inertia is required. Since grid-following capability requires magnetic short circuit power, the ratio between grid-forming and grid-following capacities is a point of interest. Therefore, the short circuit ratio (SCR) is commonly used as a comparative value. As shown in [3], typical grid-following inverters

LINDA projects – droop-based practical examples 321

(Category 1: System-supporting) can require an SCR of at least 3 to guarantee a stable network operation. Otherwise, inverters of categories 2–4 (2: Extended system-supporting; 3: Grid-forming; 4: Extended grid-forming) may be required. A classification of grid-forming and grid-following inverters is also described in [3]. The concrete SCR value up to which an inverter can be operated stably depends on its structure and control. An SCR value below 3 does not fundamentally rule out the stable operation of grid-following inverters, but proof of stability requires detailed consideration.

It should be noted that often only the LPP can be influenced, and the behaviour of the DG inverters is given because the LINDA concept shall work without extensive adjustment on the existing DG infrastructure in the grid section. So, a sufficient SCR is one basic criterion for a planned island grid.

11.2.2 Leading power plant and load management

In case of a major power failure, DG units are used to establish electrical islands. At least one DG with black-start capability and a reliable minimum performance act as a LPP, which is the grid-forming generation unit. Furthermore, the LPP is responsible for the balance of active and reactive power, which is coherent with the control of frequency and voltage. Examples of suitable generation technologies for these tasks are hydroelectric power plants, gas engines or turbines, diesel engines, grid-forming inverters in combination with storage systems, and so on.

If the voltage is established, the locally available DG units will reconnect to the (islanded) grid and generate additional infeed. As there is typically no special battery-buffered communication infrastructure suitable for control in the distribution grid, the power output of the DG units in the medium- and low-voltage grid is not directly controllable. Therefore, the control of the power output of the DG is managed by utilising the frequency droop characteristic required by German and European grid codes (cf. Figure 4.8). The principal structure of the described islanded grid is shown in Figure 11.1.

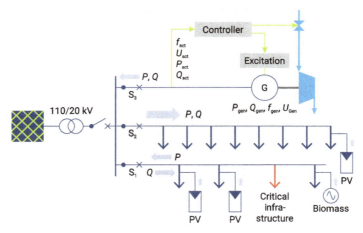

Figure 11.1 Schematic overview of the analysed and tested islanded grid (adapted from [7])

This can be realised by implementing corresponding $P(f)$- and $Q(U)$-droop curves in the LPP. The principle is shown in Figure 11.2.

An additional load management system can disconnect parts of the isolated network depending on the actual feed-in of the fluctuating renewable energy sources and the power demand via underfrequency load shedding and thereby prevent a collapse of the system. In times of higher available feed-in power, additional parts of the grid can be connected. The disconnection of single loads or network strings in low-frequency situations, regarding to their priority allows a power balance with a prioritised supply of critical infrastructure in times of low supply capacity. Further information is published in [7].

The research project LINDA successfully proved that the developed concept of islanded grid operation based on the $P(f)$ and $Q(U)$ droop characteristics for DGs implemented in the German grid codes is feasible. Hence, no communication between the DGs and the loads in the islanded grid or a microgrid management system is necessary. This results in an increased transferability of the concept to other grid areas. One objective of the follow-up project LINDA 2.0 is to further develop and transfer the LINDA concept to mobile generators used for the auxiliary supply of low voltage (LV) grids during maintenance work of the distribution system operator (DSO). This allows testing and improving the concept in many different LV grids. Additionally, important knowledge for the power grid restoration, the frequency and voltage-dependent behaviour of LV grids, can be gained [8].

DSOs often use mobile generators in LV grids for maintenance without a shutdown of LV grids. In this case, any infeed of DG has to be prevented to keep the island stable. Therefore, the frequency is raised to 51.7 Hz to disconnect all DGs (mainly PV systems) via the rules required by the grid codes (e.g. [4]). The DG capability cannot be utilised in this operation. Therefore, a hybrid mobile generator consisting of a battery energy storage system, a grid-forming inverter and an additional range extender (diesel generator set) is under development (see Figure 11.3, right). The range extender will only be set in operation when the battery charge is no longer sufficient for the islanded grid operation. This hybrid mobile generator allows the infeed of DGs during islanded grid operation and thus reduces operation costs

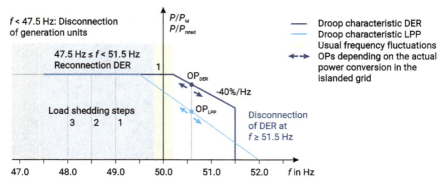

Figure 11.2 Droop concept for active power control (adapted from [7])

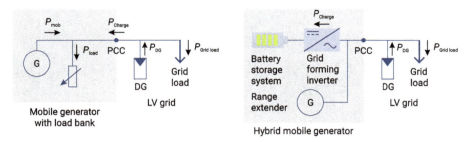

Figure 11.3 Simplified single-line diagram of the modified mobile generator (left) [10]* and the hybrid mobile generator (right) [11] operation in the LV grid

and CO_2 emissions. Based on the initial LINDA concept, this further development allows reverse power flows (infeed exceeds load in the grid) that are desired to charge the battery. In order to allow reverse power flows in the islanded grid operation, the LINDA concept has to be further developed and improved [9].[†]

To gain the experience and knowledge necessary for developing the hybrid mobile generator, a conventional mobile generator (diesel generator set) was equipped with an additional load bank (see Figure 11.3, left) to absorb reverse power flows. Field tests with this mobile generator with load bank showed that the higher the infeed of DGs, the more sensitive is the island grid system stability in the case of load and generation changes. Hence, a curtailment of the DGs by increasing the grid frequency also increases the grid stability [10].

A frequency droop characteristic $f(P)$ that increases the islanded grid frequency with an increasing active power output of the DG in the islanded grid was developed. Since the measurement-based determination of the infeed of the DG in a LV grid is only possible with great effort, the residual load at the point of common coupling (PCC) of the hybrid mobile generator is used as an input parameter. This residual load is called the charging power P_{Charge} and equals the difference of the infeed (P_{DG}) and the load ($P_{Grid\ load}$) in the grid. A positive charging power equals a charge and a negative charging power to a discharge of the battery [11].

11.2.2.1 Design of frequency droop

For the design of the frequency droop characteristic, the aggregated grid behaviour ($P(f)$) of the LV grid, as well as the installed load and capacity of DG, is necessary to evaluate if stable quasi-stationary operating points can be obtained. This evaluation can be done in a closed-loop simulation (see Figure 11.4). DG in the German LV grid are mostly PV systems that can be divided into two groups regarding their $P(f)$-behaviours in the case of an increasing grid frequency. They either reduce the

---

*This paper was presented at the *21st Wind & Solar Integration Workshop* and published in the workshop's proceedings.
[†]Text reprinted with permission from CIGRE, Energy Balance Tool for the Operational Planning of Hybrid Mobile Generators – Islanded Grid Operation with the Infeed of Distributed Generation Systems, © 2023.

324 *Intended and unintended islanding of distribution grids*

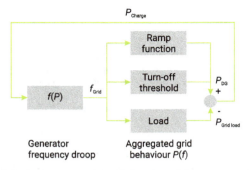

Figure 11.4 Simplified representation of the closed-loop control simulation for the design of the frequency droop [11]

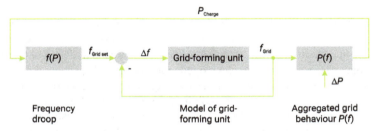

Figure 11.5 Simplified representation of the overall network simulation of the islanded grid operation in the LV grid

active power output in the form of a ramp function or have defined turn-off thresholds [11]. A detailed description of the categorisation of PV systems according to their commissioning date can be found in [8].

11.2.2.2 Overall network simulation with designed frequency droop

With a derived frequency droop, the model of the grid-forming unit and a model of the aggregated grid behaviour ($P(f)$), an overall network simulation (also dynamic) of the islanded grid operation with mobile generators and the infeed of DG can be performed according to the model shown in Figure 11.5. Figures 11.4 and 11.5 describe the approach for developing a frequency droop to balance the active power in the islanded grid. This approach can be used accordingly to develop a voltage droop to balance the reactive power in the grid as well.

11.2.3 Practical approach for the derivation of a frequency droop characteristic

As an example, the derivation of the frequency droop characteristic used for the operation of the hybrid mobile generator in the LV grid is described in this section.

For this practical approach, grid data of the DSO involved in LINDA 2.0 (aggregated installed PV system power in the LV grid) and a publicly available database – Marktstammdatenregister (MaStR) [12] – of all PV systems installed in the German grid (commissioning date defines the $P(f)$-behaviour of the individual

DG, see [13]) are statistically analysed to derive the aggregated $P(f)$-behaviours of three representative grids with high, medium and low PV infeed. The characteristic-parametrisation of the hybrid mobile generator must be robust in terms of variations of the $P(f)$-behaviour of individual islanded grids since the generator is supposed to be operated in many different LV grids without any major adaptions. The determined $P(f)$-behaviour represents the estimated behaviour of all PV systems in Germany. A method to determine the detailed individual $P(f)$-behaviours of LV grids with the help of the [12] and the postcode of the islanded grid is described in Section 11.3. One commonly used conventional mobile generator in the rural grid in the southern Germany area has a nominal apparent power of 275 kVA. To replace it with the hybrid mobile generator, LV grids with a transformer size of up to 400 kVA are analysed.

Flexible loads (e.g. electric vehicle chargers) and storage systems reduce their active power demand and change the operation mode from charge to discharge in the case of frequencies below 49.8 Hz [4] to support the grid stability. Both are neglected in the following investigations since their penetration is not yet significant in the analysed LV grids. The frequency-depended behaviour of LV loads is not considered as it is nearly negligible according to [14]. Loads are modelled as constant values in the simulation, and in extreme cases, there is no load in the LV grid, resulting in the maximum possible charging power. Since flexible loads and storage systems are neglected, only frequencies above 49.8 Hz are considered as DGs curtail their active output power in overfrequency situations. A detailed description of the statistical analysis and the derivation of the three grids as well as the practical approach for the derivation of a frequency droop characteristic is published in [11]. The aggregated $P(f)$-behaviours of the three representative grids and the derived frequency droop characteristic are displayed in Figure 11.6.

The frequency droop characteristic is designed to operate in constant frequency operation at 50 Hz (blue background in Figure 11.6) to only curtail the PV infeed by increasing the frequency when the aggregated PV system power is higher than 85 kW. This allows to operate the representative grid with medium PV infeed without increasing and thereby curtailing the DG, even if there is no load in the grid. When the charging power exceeds 85 kW ($P_{\text{Charge droop start}}$), the mobile generator switches to the frequency droop operation (green background) and increases the frequency from 50 to 51.5 Hz with increasing charging powers up to the maximum permissible value ($P_{\text{Charge max}}$). The maximum permissible charging power determines the gradient of the droop characteristic. This maximum value depends on the inverter and battery system capabilities. Above that value, the stationary operation of the hybrid mobile generator is impermissible (red background). Steady-state operating points with positive charging power values above 51.5 Hz are unattainable (grey background), as all DG in the grid must be disconnected at that point according to the rules implemented in the grid codes. With a maximum charging power of 200 kW, the frequency is increased with a rate of 1.3 Hz per 100 kW (equals to an increased rate of 38% of the maximum charging power per Hertz) above charging power values of 85 kW. In the grids with medium and low PV system infeed, the resulting grid frequency is 50 Hz, as intended with both droop characteristics. With high PV infeed, the quasi-stationary operating points are at 50.8 Hz (approx. 132 kW). The

326 Intended and unintended islanding of distribution grids

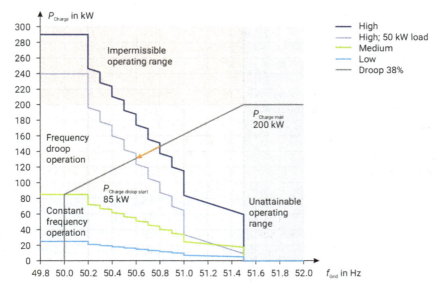

Figure 11.6 Resulting quasi-stationary operating points of the aggregated grid behaviours in combination with the generator frequency droop characteristics [11]

operating point change when adding a load to the grid with high infeed can also be seen in Figure 11.6 (indicated by the orange arrow) [11].

As the example in Figure 11.6 shows, the aggregated PV system power in the grid can be higher than the maximum permissible charging power of the hybrid mobile generator due to the high level of possible curtailment of DG. Therefore, the PV system power in the island grid is not the dimensioning factor of the inverter size; it is the peak load in the grid.

The basic concept described influences the behaviour of DG´s and loads for balancing power. In the following, the focus is on the frequency-dependent behaviour of these units according to the droop concept.

11.3 Behaviour of DGs according to the droop concept

11.3.1 Background

The installation of PV systems in German grids started around 1990 with the 1,000-roof-top program, which was funded by the German government. Since then, the standards for generation systems connected to distribution grids have been continuously developed to guarantee stable grid operation. The grid code published in 2006 allows a feed-in from PV systems between 47.5 and 50.2 Hz. By 2010, installed PV capacity had grown to 18 GWp. A frequency increase above 50.2 Hz would result in a lot of generation power disconnection and an unmanageable generation loss. As a result, a regulation (in German: Systemstabilitätsverordnung SysStabV) came into

LINDA projects – droop-based practical examples   327

effect in 2012 that required plants with a commissioning date between 2001 and 2011 to be upgraded. The regulation requires the implementation of a ramp function according to the then valid grid codes (e.g. [4]) or the shift of the switch-off threshold according to an equally distributed scheme between 50.2 and 51.5 Hz. Therefore, the behaviour of decentralised energy sources during frequency deviations in distribution grids cannot be estimated generally. In the following two sections, an approach to estimate the behaviour of DGs connected to a local network area during frequency deviations based on public data is given. Furthermore, a general overview of the German standards of the last years is provided.

11.3.2   Estimation of the frequency-dependent behaviour of PV systems

To determine the characteristic curve of the grid-leading power station according to Figures 11.2 and 11.6, the behaviour of DGs connected to the islanded grid must be approximately known. As described in Section 11.2.3, the MaStR provides information about the local administrative unit, the installed power, and the commissioning date for all DGs in Germany. According to the commissioning date and installed power, the behaviour of each DG can be assigned to a specific grid code, which was obligatory at the date of commissioning. The historical development of the German grid codes for DGs in LV grids is shown in Figure 4.8.

Using the information provided by the database, a categorisation can be done, and the grid behaviour can be estimated for disconnection and resynchronisation. For DGs that fall under the [15], a range is specified in which they can disconnect from the grid. The worst case is when they disconnect abruptly at 50.2 or 51.5 Hz. The process to estimate the behaviour of the DGs is shown in Figure 11.7.

The database has the restriction that only the local administrative unit is known, but not the substation to which it is connected. In practice, the quality of the result of this methodology depends on the existing database of the DSO.

11.3.3   Measured behaviour of a mixed PV system population

During several field tests, the aggregated frequency-dependent behaviour of PV systems connected to low-voltage grids was analysed by applying a frequency sweep between 50.0 Hz and 51.7 Hz. The frequency-dependent behaviour of PV systems can be extracted with the following assumptions:

*   The change of active and reactive power of loads is much smaller than the infeed of DGs
*   Disconnection of PV systems at 51.5 Hz
*   No disconnection of PV systems below 50.2 Hz

Figure 11.8 shows the frequency sweep between 50.0 Hz and 51.7 Hz. The left axis of the left diagram shows the frequency, and the right axis the measured active power. The difference in the residual load for the active power between 50.2 and

328   *Intended and unintended islanding of distribution grids*

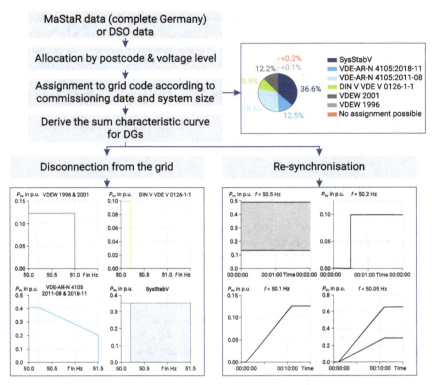

Figure 11.7   *Procedure for evaluating MaStR data and resulting diagrams for grid disconnection and resynchronisation as well as the percentage allocation to the standards on a sample dataset of a village in southern Germany in 2022 [13]*

51.7 Hz is shown in the diagram shown in the middle. To compare multiple samples, the residual load is normalised to the median of the power values measured for the frequency below 50.2 Hz. The right figure shows the normalised curve of the residual load for one sample of an LV grid. Figure 11.7 shows a relatively large proportion of PV systems that are operated on the grid according to a grid code with a defined switch-off threshold. Therefore, a clear power reduction at 50.2 Hz can be seen in Figure 11.8, and the frequency-dependent active power reduction up to 51.5 Hz is less pronounced.

In Figure 11.9, multiple samples of the LV grid in comparison with the estimation of Section 11.3.2 are shown. The MaStR only gives information about the PV systems in the local administrative unit and no information about the exact location is given. Therefore, the aggregated grid behaviour of one particular LV grid can deviate. If the composition of PV systems in the considered LV grid according to the grid codes is similar to the composition of the local administrative unit, the behaviour can be reproduced.

LINDA projects – droop-based practical examples 329

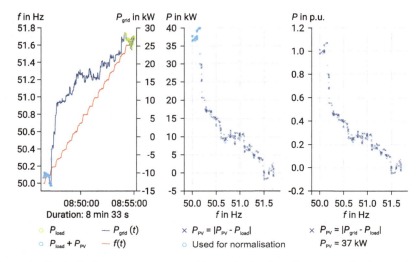

Figure 11.8 Left: Frequency sweep with marked values for normalisation and PV power calculation; centre: amount of calculated PV power, right: normalised PV power [13]

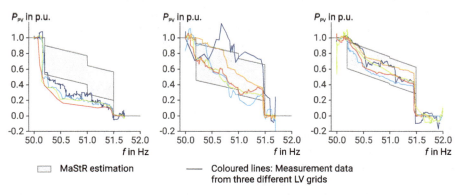

Figure 11.9 Results of several field tests in three low LV grids in comparison with the MaStR estimate (cf. [13])

11.4 Interaction between load and frequency

11.4.1 Influence of load behaviour on frequency stability

The load behaviour influences the frequency stability in grid operation. On the one hand, rapid changes in the grid load lead to frequency deviations until the power imbalance is eliminated again. Therefore, the power curve during the process of load connection or load disconnection is relevant for the occurring frequency fluctuations. On the other hand, certain loads react to frequency fluctuations with a change in power. With the load reaction, a distinction must be made between the

330 *Intended and unintended islanding of distribution grids*

momentary reserve – the input or output of rotational energy due to mass inertia – and the load self-regulation effect. As shown in [16] and Chapter 2, the frequency-dependent load behaviour of different load types differs considerably. This results in regional and seasonal differences in the frequency-dependent load behaviour in the interconnected grid [16]. In local islanded grids with critical infrastructures, the load composition and, thus, the frequency-dependent load behaviour can differ significantly from the usual load mix in the interconnected grid. The frequency-dependent load behaviour is particularly pronounced in grid areas with a high proportion of directly coupled electrical machines. Against this background, the behaviour can be considerably more pronounced in a water supply system, for example, than in a typical low-voltage network with household customers, where the behaviour is often negligible [14]. It should be noted that new inverter loads behave differently than the existing load.

The frequency-dependent load behaviour for new inverter loads like charging devices of battery electric vehicle (BEV) is specified in the grid code [17]. For power infeed of the charging device (battery of the vehicle is discharged), it must satisfy the specified behaviour of [4]. DC- and inductive charging devices >12 kVA follow the same specifications as storage systems. So far, no grid code or specification prescribes a frequency-dependent behaviour of AC-charging systems, but [17] notes, that in future versions, the same behaviour as for DC charging systems is devised. For heat pumps, no specification has been given so far. In addition to the frequency-dependent load behaviour itself, especially the frequency stability of load switching is an important issue for island networks. This can be positively influenced by the following load bank concept.

11.4.2 Increase in frequency stability: load bank concept

In small island networks, the available rotating mass is often very limited, and the control speed of the LPP is limited by design (e.g. hydropower plant). On the other hand, when connecting network strings or discrete loads of critical infrastructures, the load jumps that occur cannot be limited arbitrarily. Therefore, unfavourable ratios of load jumps to rotational inertia and fast controllable generation power can occur in island grids. Against this background, a concept for optimising frequency stability during switching operations is presented.

The balancing of the power difference after a load change is usually done with the power plant control. Mechanical processes are often responsible for a change in power plant output, which is the reason for the time lag between the load step and power adjustment to the new setpoint. The focus of this section is on the investigation of a concept in which the frequency stability during switching operations in island grid operation is to be increased by active load counter switching. The basic idea is to use a controllable load to perform a second contrarian load switching when a first, original switching action triggers dynamic frequency deviations in the island grid. The active load counter switching compensates for the power difference and thus avoids the frequency deviation. A load bank, for example, is well suited for this task, as the total load can be divided into small steps during pre- or post-conditioning,

LINDA projects – droop-based practical examples   331

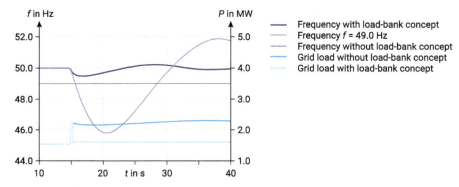

Figure 11.10 Principle of action of an active counter switching on frequency stability [16]

resulting in smaller frequency deviations than the original load step. The concept is investigated using a dynamic simulation model.

Figure 11.10 shows an example of an isolated grid, which is supplied by a 6 MVA hydropower plant. The initial load step has a power of 700 kW and leads to a frequency deviation of more than 4 Hz. The figure shows the effect of an active load bank (630 kW; time lag 250 ms) counter switching. With this measure, the frequency deviation can be limited to 0.5 Hz, although the load bank does not have exactly the same power (90%) as the initial load step and a time offset occurs between the switching actions. These deviations were deliberately chosen to not specify ideal conditions for the concept. As can be seen in the figure, this concept can massively increase frequency stability and avoid triggering underfrequency load-shedding stages (f = 49.0 Hz).

In [16,18], the concept is analysed in detail with different load types (pump, fan, ohmic load), switching lags (time between initial and contrary switching) and ratios of load bank to initial load steps. It is shown that this concept is robust and tolerates deviations from ideal conditions. Especially for LPPs with comparatively high rotating mass and low control speed, it allows the integration of significantly larger loads without triggering frequency thresholds (load shedding, protection).

A repeated automatic reaction of the load bank to the frequency gradient promises further optimisation potential. Using a pump ramp-up, it was shown that a RoCoF control can again significantly reduce the maximum frequency fluctuation compared to a single triggering of the load bank.

The great advantage of this concept is that it is relatively easy to implement, and the installation of a load bank is comparatively inexpensive, low-maintenance, and durable. Especially in critical infrastructures, discrete load jumps such as the activation of a pump of a drinking water supply can be stably integrated into a small emergency supply island network, which, without countermeasures, would possibly lead to the collapse of the islanded network.

332 *Intended and unintended islanding of distribution grids*

11.4.3 Challenge: RoCoF measurement

In [19], the previously described concept of contrary load switching was tested in a laboratory test. To trigger a contrary load adaption using a certain RoCoF threshold, the frequency estimation must be done in real time. Real time in this context means faster than the frequency answer of the dynamic grid behaviour to track the current grid frequency. The major restriction for real-time RoCoF estimation was the frequency estimation algorithm. The zero-crossing algorithm gives unprecise estimation results at disturbed grid voltages. Therefore, an Enhanced Phased Locked Loop and a Second Order Generalised Integrator with Frequency Locked Loop were tested regarding performance during voltage disturbances as an alternative to the zero-crossing algorithm. The advantage of both frequency estimation methods is that the RoCoF can be extracted from the structure itself. Both estimation methods show deviations of the estimated frequency during voltage distortions like harmonics, voltage drops, or phase steps. As a final result, it can be stated that if the large load to be switched is known, a controlled contrary load switch should be conducted [20].

11.5 Practical examples of the LINDA concept

This section describes the LINDA projects as practical examples of the LINDA concept. The basic concept was developed as part of the first LINDA project and also tested in practice. The concept was further developed in the subsequent LINDA projects, and its area-wide applicability was investigated. In both projects, the concept was successfully tested using different hydropower plants as well as drinking water supply facilities. Furthermore, the functionality of the basic control strategy was validated. In the following LINDA 4 H_2O project, the transferability on a broad scale was analysed using biogas plants as LPPs.

11.5.1 First LINDA project – key issues and field test sight

The findings of the research project LINDA are based on a dynamic simulation model of all relevant components of the grid area as well as staggered field tests. The simulation model was optimised and validated with the help of the measurement data from the field tests.

The project was implemented by the local DSO, hydro-power-plant operator and manufacturers, a manufacturer of a biomass power plant and scientific institutes.

Key issues:

- Establish a droop-based control concept which uses the $P(f)$ and $Q(U)$ functions of the DG according to the German grid codes for stable power control in the isolated network
- Develop an emergency electric energy supply for a drinking water supply system
- Optimise the hydropower plant controller for large load steps of the hydro pumps in relation to the available generation power of LPP
- Field Tests for the investigation of:

LINDA projects – droop-based practical examples 333

- o Stable interaction between hydropower and PV plants with regard to the droop control concept and load jumps
- o Stable interaction between hydropower and biogas plants with regard to the droop control concept and load jumps
- o Frequency-dependent behaviour of a historically grown PV system population according to different grid codes

LINDA field test sight:

The selected grid area offers good conditions for islanding tests because of the following conditions:

- Leading power plant: The Feldheim run-of-river power plant provides black-start capability and allows the establishment of an island network. A recently renewed digital turbine controller allows a simple parameter change between classic interconnected grid operation and island grid operation. As a result, optimised control parameters could be developed and used for island grid operation, which massively increases frequency stability, especially during load changes [21,22].
- Rain hydropower plant, which supports the LPP and allows synchronisation tests.
- Biogas plants, which allow testing of a stable interaction of different generation units via droop concept in the stand-alone grid.
- 185 PV plants with an installed capacity of approx. 1.8 MWp. Since the plant population has grown historically, the plants differ in terms of manufacturers, outputs, years of commissioning, and standard behaviour. This allows the testing of the control concept with a mixed-generation structure.
- A large drinking water supply system as an example of a critical infrastructure.

11.5.2 LINDA 2.0

In the first research project LINDA, a concept for an emergency power supply using local islanded grids, including DGs, was developed and successfully tested in the real grid. This island grid concept does not require a microgrid management system or communication between the LPP, the DGs, and the loads in the grid since it is based on the $P(f)$ and $Q(U)$ droop characteristics for DGs implemented in the German grid codes (e.g. [4]). This results in an increased transferability to other grid areas. The main objectives of the follow-up research project LINDA 2.0 are therefore [9][‡]:

- Transfer the LINDA concept to another grid with a hydropower plant and a waterworks.
- (Partially-)automate the LINDA concept to minimise the deployment of personnel to form an emergency power supply in the case of a blackout.
- Transfer and further develop the concept to the regular operation of mobile generators in LV grids to gain knowledge on forming and operating islanded power

[‡]Text reprinted with permission from CIGRE, Energy Balance Tool for the Operational Planning of Hybrid Mobile Generators – Islanded Grid Operation with the Infeed of Distributed Generation Systems, © 2023.

334 *Intended and unintended islanding of distribution grids*

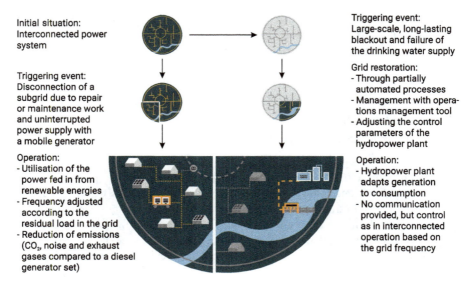

Figure 11.11 *Schematic overview of the LINDA 2.0 project*

supplies and restoring the grid. More information on this development is given in Section 11.2.3.

In summary, the field test areas were particularly well suited because they have black-start capable power plants, diverse DG facilities with sufficient capacity, a synchronisation capability with the interconnected grid and critical infrastructure. Further information can be found in [23].

The staggered field tests have shown that the LINDA concept works stably in practice and is transferable to other combinations of hydropower plants and drinking water supplies. Figure 11.11 provides a schematic overview of the research project LINDA 2.0. The series of field tests with the mobile generators showed that the concept is also applicable in low-voltage grids, even if the generation in the islanded grid exceeds the load and reverse power flows occur.

11.5.3 LINDA 4 H$_2$O: analysis of further application examples for the emergency power supply of critical infrastructures

While the main focus of the other LINDA projects is on the emergency power supply of large drinking water supply systems with hydropower plants, LINDA 4 H$_2$O is analysing the transferability to other, more typical supply constellations.

The focus is, therefore, on the emergency power supply of typical Bavarian drinking water supply systems by using biogas plants in islanded grid operation. The basic concept is illustrated in Figure 11.12.

The drinking water supply in Bavaria is organised in a very decentralised manner. Therefore, the idea is to analyse if more than 5,000 biomass units in Bavaria can be used to create emergency power supply islands. The average of the more than 1,200

LINDA projects – droop-based practical examples 335

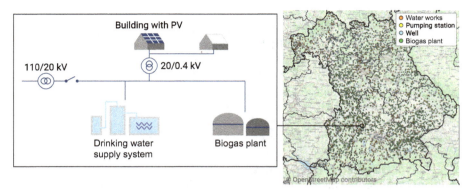

Figure 11.12 Overview diagram of the LINDA 4 H_2O research project, map data from OpenStreetMap: https://www.openstreetmap.org[§]

Bavarian drinking water supply systems with their own water production through wells supplies almost 8,300 inhabitants and has a total electrical power demand of about 97 kW [24].

The average Bavarian biomass unit, on the other hand, has an installed capacity of 270 kW. To implement an islanded grid emergency power supply with a biogas plant as a grid-forming unit, the corresponding plant must be black-start capable and also be able to act as a LPP in the islanded grid. It must be assumed that the majority of biogas plants have to be retrofitted to meet these requirements. However, in most plants, the prerequisites for retrofitting at moderate cost are good, although the type and scope of the retrofits can vary between the individual manufacturers and series.

A spatial proximity between drinking water supply systems and biogas plants can also often be observed. The results of the nearest neighbour analysis show that in over 50% of the wells and treatment plants, the nearest biogas plant is located within a linear distance of less than 3 km [25].

11.5.3.1 General rollout possibility of the islanded grid emergency power supply concept for drinking water supply systems

To determine the general potential of islanded grid emergency power supply of Bavarian drinking water supply systems with biogas plants, relevant data on the corresponding power plants and electrical consumers are first collected. The information on the biogas plants comes from the publicly accessible 'Marktstammdatenregister' of the German Federal Network Agency, while data from the Bavarian State Office for Environment is used for the evaluation of the water supply systems.

Based on the resulting database, in which, among other things, the locations and the power demand of the water supply systems as well as the locations and the installed power of the biogas plants are stored, the minimum spanning tree for cabling of all relevant consumers and generators along the existing road network is

---

[§]Information from OpenStreetMap is made available under the Open Database License (ODbL). The full legal text of this license is available from https://opendatacommons.org/licenses/odbl/.

336   *Intended and unintended islanding of distribution grids*

formed using the Kruskal and Dijkstra algorithms. This allows realistic modelling of the islanded grid, which is to be set up for the emergency power supply.

This model then serves as the basis for checking the technical feasibility of the islanded grid operation, whereby, among other things, load flow calculations are carried out automatically for each supply constellation. For all technically feasible power supply constellations, economic conditions are then determined, and comparisons are made with a conventional emergency power supply using diesel generators. Finally, this allows conclusions to be drawn on the economic competitiveness of the islanded grid emergency power supply. To increase the validity of the analysis, the investigations are carried out for several scenarios and supply variants.

It can be summarised that the technical potential of this emergency power supply solution is high. Low-voltage island grids are possible for about 30% of the Bavarian drinking water supply systems, while medium-voltage island grids can even supply up to 80% of the water supply systems. Economic competitiveness, on the other hand, is only present in a minority of cases and – depending on the scenario and supply constellation – is given for 2%–12% of the water supply systems.

Although the islanded grid emergency power supply solution is only economically competitive in a minor proportion of cases, it has significant non-monetary benefits. The most significant advantage of the islanded grid is the possibility of a substantially longer and largely independent emergency power supply. Another important advantage is that transport, handling and storage of the water-polluting substance diesel in drinking water protection areas is not necessary (compared to an emergency power supply using diesel generators) [25].

11.6   Concluding remarks

With the help of the implemented droop characteristics, the $P(f)$ and $Q(U)$ behaviour of the DGs specified in the standards is utilised, and the provision of power is controlled without additional communication infrastructure. The advantage of this concept is that it can be integrated comparatively easily and cost-effectively into the existing grid infrastructure, as costly retrofitting with blackout-proof communication infrastructure is avoided.

The results of the LINDA 2.0 and LINDA 4 H_2O research project show that the basic LINDA concept can not only be used in individual cases but that it can also be transferred to typical supply constellations of critical infrastructures and generation plants and, from a technical point of view, could be applied in many cases to increase the security of supply. The practical suitability of the LINDA concept was also demonstrated in the numerous field tests carried out.

References

[1]   Petermann T, Bradke H, Lüllmann A, Poetzsch M, and Riehm U. Hazards and Vulnerability in Modern Societies – Using the Example of a Large-scale Outage in the Electricity Supply. Summary, 2010. *Büro für Technikfolgen-Abschätzung beim Deutschen Bundestag (TAB)*. doi:10.5445/IR/1000137664

[2] Kapatanovic T. Übertragungsnetzsicherheit bei steigender Volatilität; Herausforderungen und Ausblick, *8. Int. Energiewirtschaftstagung der TU Wien*, Vienna, Austria, 2013.

[3] VDE FNN Info: Future requirements for power system stability, *Forum for Network Technology & Network Operation in the VDE* (VDE FNN), November 2022

[4] VDE-AR-N 4105. Power Generating Plants in the Low Voltage Network (VDE-AR-N 4105), *Technical Connection Rule*, VDE Verlag GmbH, Bismarckstraße 33, 10625 Berlin: Forum Netztechnik/Netzbetrieb im VDE (FNN); 2019.

[5] VDE. Technical Connection Rules for Medium-Voltage (VDE-AR-N 4110) [Technical Connection Rule]. VDE Verlag GmbH, Bismarckstraße 33, 10625 Berlin: Forum Netztechnik/Netzbetrieb im VDE (FNN); 2018.

[6] European Commission. Commission Regulation (EU) 2016/631 – Establishing a Network Code on Requirements for Grid Connection of Generators [Commission Regulation]. Brussels, Belgium; 14 April 2016.

[7] Steinhart ChJ, Finkel M, Gratza M, Witzmann R, Kerber G, and Schaarschmidt K. Local Island Power Supply and Accelerated Grid Restoration with Distributed Generation Systems in the Case of Large-Scale Blackouts. *CIRED Workshop* 2016, 14–15 June, Helsinki, Finland, paper no 0136.

[8] Seifried S, Lechner T, Stenzel D, *et al.* Determination of the Voltage and Frequency Dependent Behaviour of Low Voltage Grids – Test Procedure for a Modified Mobile Generator, *11th Solar & Storage Integration Workshop*, Berlin, Germany, 2021.

[9] Lechner T, Bogendörfer F, Seifried S, *et al.* Energy Balance Tool for the Operational Planning of Hybrid Mobile Generators – Islanded Grid Operation with the Infeed of Distributed Generation Systems, *CIGRE International Symposium Cairns* 2023, Cairns, Australia.

[10] Lechner T, Seifried S, Timmermann J, *et al.* Island Grid Operation of a Modified Mobile Generator – Test and Optimization in a Living Lab with High PV Penetration, *21st Wind & Solar Integration Workshop*, The Hague, Netherlands, 2022.

[11] Lechner T, Seifried S, Timmermann J, *et al.* Frequency Droop Characteristic for Grid Forming Battery Inverters – Operation in Islanded Grids with the Infeed of Distributed Generation System, *CIRED 2023 International Conference & Exhibition on Electricity Distribution*, Rome, Italy, 2023.

[12] Bundesnetzagentur: Marktstammdatenregister (MaStR), available from: https://www.marktstammdatenregister.de/MaStR.

[13] Seifried S, Lechner T, Storch D, *et al.* Vermessung des frequenzabhängigen Verhaltens von Niederspannungsnetzen im Inselnetzbetrieb und Nachbildung mit Daten des Marktstammdatenregisters, *13. Internationale Energiewirtschaftstagung an der TU Wien (IEWT 2023)*, Vienna, Austria, 2023.

338 *Intended and unintended islanding of distribution grids*

[14] Palm S, Schegner P, and Schnelle T. Measurement and Modeling of Voltage and Frequency Dependences of Low-voltage Loads, *2017 IEEE Power & Energy Society General Meeting*, Chicago, IL, 2017, pp. 1–5.

[15] Verordnung der Bundesregierung. *Verordnung zur Gewährleistung der technischen Sicherheit und Systemstabilität des Elektrizitätsversorgungsnetzes (Systemstabilitätsverordnung – SysStabV)*, Berlin: Bundesanzeiger Verlagsgesellschaft mbH, 2012.

[16] Steinhart ChJ, Lokale Inselnetz-Notversorgung auf Basis dezentraler Erzeugungsanlagen mit Fokus auf die Frequenzstabilität, Dissertation, *Technical University of Munich*, Germany; 2020.

[17] VDE-AR-N 4100. Technical rules for the connection and operation of customer installations to the low voltage network (TAR low voltage), *Technical Connection Rule*, VDE Verlag GmbH, Bismarckstraße 33, 10625 Berlin: Forum Netztechnik/Netzbetrieb im VDE (FNN), 2019.

[18] Steinhart ChJ, Gratza M, Fischer M, *et al.* Optimierung der Frequenzstabilität bei Lastschaltungen im Inselnetz, 10. *Internationale Energiewirtschaftstagung an der TU Wien*, IEWT, Vienna, Austria, 2017.

[19] Seifried S, Lechner T, Distl T, Finkel M, Steinhart ChJ, and Witzmann R.Erhöhung der Frequenzstabilität in Inselnetzen mit Hilfe einer ROCOF geregelten Lastbank, *ETG Kongress*, Wuppertal, 2021.

[20] Seifried S, Fischer S, Storch D, Lechner T, Finkel M, and Witzmann R. Implementation and Test of Frequency Estimation Methods for RoCoF-based Load Switching in Islanded Grid, *CIRED 2023 International Conference & Exhibition on Electricity Distribution*, June 2023, Rome, Italy.

[21] Gratza M, Witzmann R, Steinhart ChJ, *et al.* Frequency Stability in Island Networks: Development of Kaplan Turbine Model and Control of Dynamics, *PSCC 2018, Power Systems Computation Conference*, 11–15 June 2018, Dublin, Ireland.

[22] Gratza M Steinhart ChJ, Witzmann, R, and Finkel, M. Parametrierung eines dynamischen Kaplan Turbinenmodells anhand von Messdaten, *16. Symposium Energieinnovation, Energy for Future, Wege zur Klimaneutralität*, 12–14 February 2020, Graz, Austria.

[23] Abschlussbericht zum Verbundvorhaben LINDA, Verbundprojekt Lokale Inselnetzversorgung und beschleunigter Netzwiederaufbau mit dezentralen Erzeugungsanlagen bei großflächigen Stromausfällen, Projektlaufzeit: 01.08.2015 – 31.07.2018, Förderkennzeichen: 0325816A-H, 31.10.2019.

[24] Storch D, Finkel M, and Witzmann R. Charakterisierung bayerischer Trinkwasserversorgungsanlagen und Approximation des zugehörigen elektrischen Leistungsbedarfs, *DVGW energie | wasser-praxis 09/2022*, S. 32-39, 2022.

[25] Storch D, Seifried S, Lechner T, Finkel M, and Witzmann R. Notstromversorgung von Wasserversorgungsanlagen mit Biogasanlagen im Inselnetz – Analyse der technischen und betriebswirtschaftlichen Umsetzbarkeit, *13. Internationale Energiewirtschaftstagung an der TU Wien*, Vienna, Austria, 2023.

Chapter 12
Practical aspects of bottom-up grid restoration

Herwig Renner[1]

Grid restoration after a major blackout, affecting a wider area, is a challenging task. It is the process to return the system to normal operation after a local or regional blackout. Each power system has system-specific characteristics that should be considered when planning restoration. Therefore, the identification and sufficient consideration of all relevant aspects in the specific power system restoration makes both power system restoration and the planning of restoration actions challenging tasks. Power system characteristics will be significantly different during the restoration process, compared to normal operation. System strength is reduced, frequency and voltage will vary in a wider range and several assets may not be available. System-wide blackouts are rare events which means that restoration processes are seldom executed in practice, usually, there is a lack of experience for the operational staff.

Regarding the restoration strategy, typically, top-down approaches are used, giving the lead and organisation of grid restoration to the transmission system operator (TSO). In case of available stable voltage from the TSO, the distribution system operator's (DSO) focus is on the consecutive sectorial reconnection of consumers. The operation of islanded distribution grids is not foreseen in this concept. The disadvantage of the top-down strategy is that it requires a strong and stable voltage source.

The alternative is the bottom-up strategy with a (small) islanded grid [1–4]. The initial energy is provided by a black-start generator which performs the initial network restoration. At this stage, the system is rather weak with a low short-circuit level and limited generation capacity. Step by step the island is expanded by connecting local load and generation. Several islanded grids can be formed in parallel, thus allowing a faster restoration. One by one those grids are synchronised, forming larger sub-systems until the complete power system is energised. Sometimes this strategy is used as a backup strategy in case for whatever reason the TSO is not able to provide a stable system. A precondition for this strategy is the technical and organisational ability of the DSO's staff to start and operate an islanded grid. A coordinated approach of all stakeholders – TSO, DSOs and power plant operators – might provide the fastest and the most stable grid restoration since many processes can be performed

---

[1] Graz University of Technology, Institute of Electrical Power Systems, Austria

340   *Intended and unintended islanding of distribution grids*

in parallel. This approach provides increased flexibility and resilience, considering the variety of possible disturbance characteristics.

12.1   Key issues for bottom-up grid restoration

For a successful bottom-up grid restoration, besides technical requirements legal, regulatory and logistic requirements have to be fulfilled. Also, in the case of grid restoration tests, those aspects must be considered in the planning phase. In the following list, the focus is put mainly on technical issues. The following list is mainly based on the

Black-start capability: It must be possible to start up a power plant without an external power supply to achieve black-start capability. A secure supply for the control equipment and auxiliary facilities of the generator such as excitation, governor, valves or forced bearing lubrication must be available. This functionality can be provided by emergency generators, often self-excited and driven by a small hydro turbine or a diesel engine. However, it should be mentioned that securing a power plant into house load – as is usually the goal in the case of grid collapse – does not account for black-start capability [1].

Control requirements: The generator unit must be able to cope with the requirements on voltage and frequency deviation in island operation. Usually, the voltage control meets the dynamic requirements even in cases of larger load changes due to the small-time constants of the excitation system and voltage controller in the range of some tenths of seconds [5]. However, the high rate of change of frequency (RoCoF) in island grids during grid restoration due to the rather small rotational energy is a challenge for active power and frequency control [1]. Dynamic limits are given by thermodynamic processes in the case of thermal units. Hydraulic units, especially storage power plants, are well-qualified for fast control. Limits, reducing the control speed, are given by transients in the plant's waterway (penstock, surge tank). In the case of Pelton turbines, those restrictions can be avoided by activating deflector control with nozzle pre-opening. In that mode, the nozzle is operated with a larger flow rate than needed. The surplus of water is deviated from the Pelton bucket by the deflector, allowing fast power control with head space in both directions. Due to the losses during that operation mode, the operation time is limited.

Automatic selection of the number of active nozzles, which ensures optimum efficiency during normal operation, should be avoided. This toggling will likely introduce frequency fluctuations, jeopardising the frequency stability of the islanded grid.

Strategy for energising the grid: In principle, there are two methods of energising the island network. In the first method, one or more machines are accelerated to nominal speed and voltage and synchronised to a busbar. Then, the remaining lines and transformers are connected step by step. The inrush current from large transformers could be problematic in a weak grid but can be avoided by using point-of-wave switching devices. A disadvantage of the method is that in case of an insulation fault of equipment, switching on this fault could jeopardise the stability of the islanded grid. In the alternative method, the configuration of the grid is established in a

Practical aspects of bottom-up grid restoration 341

de-energised state and the entire grid is slowly ramped up in terms of voltage. Possible short circuits are detected at the beginning of the process at low voltage [6].

Minimum load: Grid restoration requires at least one generator capable of operating from minimum to maximum load without limitations. Active power control during low loading conditions must guarantee a precise and fast response to load changes. These requirements are generally met well by Pelton turbines. However, if a unit is operated at a minimum load, this minimum load has to be provided reliably. From an organisational point of view, loads under the direct control of the power plant operator or grid operator are preferred. A good example of that kind of load are storage pumps. In addition to providing a minimum load to the generator, the torque-speed characteristics of the pump introduce some inherent support to stabilisation of frequency (self-regulating effect of load). If the stabilising load is provided by a third party, it must be available at any given time and the remote connection must be secure in cases of a blackout for a predefined time, which should reasonably be around the autonomy time of relevant switching stations of 24 hours [1].

Maximum load steps: It is important to know which amount of load can be connected during certain steps of the restoration process. As a rough rule of thumb, the load step should not exceed 10% of the total rated power of all machines participating in frequency control in the island. In practice, islanding tests as described below can be used to determine maximum load steps. One should consider that the load at the time of restoration might significantly differ from the pre-outage load. Amongst other effects, the so-called cold-load-pickup should be taken into account. The cold-load-pickup phenomenon refers to loads that interact with some kind of storage and kick in synchronously after an outage. The most prominent examples are the thermal loads of heating and cooling systems [7].

Protection: Short-circuit capacity in islanded grids will be significantly lower than in normal operation. Currents of remote faults can be below the pickup threshold of relays. Therefore, it might be necessary to adapt the settings of protection relays during islanded operation. In resonant grounded grids, the availability of Peterson coils in the islanded grid must be ensured. Settings of over- and under-frequency protection must be checked to avoid undesired tripping of equipment during the test. Especially the under-frequency protection of storage pumps, which might be used as a load during the test, can have rather tight limits. In normal operation, they are part of the load-shedding scheme and are usually among the devices tripped in the early stage of a frequency collapse.

Reactive power compensation: Especially in the case of larger islands the reactive power must be balanced. Uncompensated lines or cables act as large capacitive loads, appropriate reactive power compensation devices must be included in the restoration plan. Again, the protection settings of the compensation device must be checked carefully. Undesired tripping leading to capacitive load may result in the self-excitation of a synchronous generator, a well-known phenomenon, leading to power frequency over-voltage [8].

342   *Intended and unintended islanding of distribution grids*

Resonance and ferro resonance: In the weak grid with low short-circuit capacity and high grid impedance, low-frequency resonance might occur. Since the load in that kind of grid is rather low, the damping of the resonance is poor, which can result in high voltage distortion. Ferro resonance could also be an issue [9].

Consideration of distributed generation: Distributed generation in medium voltage and low voltage grids is not within the controllability of the grid operator. Operation of that kind of generation follows grid codes, which define the behaviour in case of over- and under-frequency as well as over- and under-voltage. Automatic frequency-dependent disconnection and reconnection of distributed generation is a risk for frequency stability. A strategy to avoid this problem is the operation of the system at a frequency above the nominal frequency, thus forcing distributed generation to be disconnected during the restoration process. However, with the steadily growing share of distributed generation, it will have to play an active role in future grid restoration strategies [4].

Resynchronisation: Resynchronisation requires a coordinated convergence of the frequencies of the grids to be connected. The exchange of information within the participating companies about actual and reference frequency and control settings after synchronisation is essential. The location of the connection must be selected carefully, undesired tripping due to (transient) overload during the synchronisation process must be avoided.

Interference with other grids: Especially during tests for grid restoration, the operation of grids with different frequencies is unavoidable. Due to capacitive or inductive coupling, mutual interference between the grids is possible. For instance, in the case of double circuit lines, with both circuits connected to different islands, inductive and capacitive interference will likely occur. Voltage fluctuations, oscillating with the frequency difference of the two islands (beat frequency), will be observable. Due to the non-ideal transposition of the circuits, negative sequence and zero sequence voltages can be introduced as well.

Exit procedure: In case of unforeseen serious problems, a scenario for a safe exit and shutdown of the test setup must be prepared. Equipment damage should be avoided and personal safety must be always guaranteed.

Communication: Reliable and secure communication channels between the participating partners, ideally independent of public channels, must be available. Maintaining discipline in communication is a must. Only clear definitions and unambiguous use of technical terms allow clear and unmistakable exchange of information. A common basis for discussion between grid operators and power plant operators must be found.

Training: The grid restoration concepts have to be practised regularly by all participating parties under preferably realistic conditions. Dynamic grid simulators can be used to test individual tasks with several participants to locate and improve weak points in the grid restoration concept. Besides the simulation, black-start tests

Practical aspects of bottom-up grid restoration   343

at the qualified power plants and grid areas are necessary to verify the requirements mentioned above. Additionally, a real black-start test leads to the operation of equipment beyond normal conditions and thus can help to detect hidden failures in the system [1].

12.2 Rules for grid restoration in Austria

The provisions on network restoration within the European ENTSO-E network are subject to the ENTSO-E Network Code on Emergency and Restoration at the European level [10] and the corresponding national regulations. According to national regulations in Austria, the TSO is responsible for grid restoration. However, agreements between TSO and DSOs exist, allowing the latter to start a bottom-up grid restoration in parallel to the TSO's activities. Finally, the synchronisation of individual islands is coordinated by the TSO.

12.3 Example of practical test of bottom-up grid restoration

The performance of practical tests is one pillar in the framework of bottom-up grid restoration. In the following, the setup and results from such a test in southern Austria are presented. In the 110 kV grid of Carinthia and southern parts of the 220/380 kV transmission grid of Austria, island grid and black-start trials have been regularly carried out since 2005 by the local distribution network operator Kärnten Netz GmbH (KNG), the TSO Austrian Power Grid (APG) and the power plant operators KELAG and Verbund Hydro Power (VHP) in cooperation with Graz University of Technology. The aim of the test was the verification of the black-start functionality, the operation of a real islanded system beyond normal frequency ranges, the comparison of different governor settings and control strategies and finally the successful resynchronisation of the islanded grid. The staff was trained to operate and communicate in exceptional situations and as a side effect, the detection of failures, hidden during normal operation, was possible. Some selected results from the test are described in the following sections.

12.3.1 Test schedule

The test was scheduled early in advance with the following points to be considered:

- Availability of the grid elements to be used, considering expected weather, load flow and possible maintenance activities.
- Availability of the power plants to be used, considering generation schedule, market prices and level of water reservoirs.
- Availability of staff in grid operator's and power plant operator's control centre.

The test itself will usually take some hours, depending on the successful execution of the subtasks. The test setup is given in the following:

344   *Intended and unintended islanding of distribution grids*

- Setup of communication of the participating bodies
- Adjustment protection relay settings (if necessary)
- Preparation of 110 kV switching state for the test
- Black-start of a hydropower plant and energisation of the test system
- Connection of storage pump to provide a base load of the test system
- Synchronisation of additional generation units (local and remote) to increase the inertia of the test system
- Performance of load variations to test the frequency control capabilities
- Synchronisation of a large pump storage power plant and transferring frequency control to this plant
- Performance of larger load variations by activating a large storage pump
- Resynchronisation of the test system to the interconnected main grid
- Intentional fall-back in islanded operation with power surplus or deficit to test the frequency control capabilities
- Performance of short circuit and earth fault tests to verify protection functionality under low fault level conditions
- Resynchronisation with the main grid and controlled shutdown of participating plants
- Restoration of the normal 110-kV switching state

In the following sections, some of the tasks described above are highlighted. Procedures and lessons learned from grid restoration and islanded operation tests in Carinthia/Austria are presented. The focus of the island grid tests was set on the improvement of the frequency control in load switching in the early stage of grid reconstruction, as well as the synchronisation of local grid islands.

12.3.2   Grid configuration during test

For the test, a part of the 110-kV grid was separated from the interconnected system. In total, a circuit length of approximately 140 km was involved in the test.

The participating power plants, a storage plant (PP1) and a pump storage plant (PP2), have black-start capabilities, while the run-of-river power plant PP3 is synchronised to the already established island grid to enhance system inertia and maintain voltage. The fixed-speed hydro pump located at PP4 is used as a load during the islanded operation. The schematic grid structure is shown in Figure 12.1.

Table 12.1 lists the relevant parameters of the units participating in the test. An important parameter is the rotating energy, which significantly contributes to frequency stability and limits the frequency gradients.

$$W_{kin} = H \cdot S_r \cdot \left(\frac{\omega}{\omega_r}\right)^2 \tag{12.1}$$

The inertia constant H is defined as

$$H = \frac{1}{2}\frac{J}{S_r} \cdot \omega_{mr}^2 \tag{12.2}$$

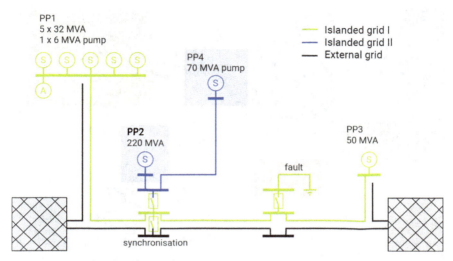

Figure 12.1 Schematic representation of the test grid (110 kV) and participating power plants

Table 12.1 Participating power plants

Power plant	Type	Turbine	Rated power S_r	Rotating energy W_{kin}
PP1	Storage	5 Pelton units and 1 pump	5×36 MVA 6 MVA pump	Total 450 MWs
PP2	Storage	Pelton unit	220 MVA	720 MWs
PP3	Run-of-river	Kaplan	50 MVA	89 MWs
PP4	Storage pump	Francis	70 MVA	308 MWs

The ω^2 term is responsible for the fact that a Pelton turbine with a low moment of inertia and high mechanical rated speed can contribute more to the rotating energy than a Kaplan turbine with a high moment of inertia and low mechanical rated speed.

12.3.3 Black-start and provision of stable initial configuration

The starting point is the hydropower plant PP1 with a total of five generation units with Pelton turbines and a small storage pump as shown in Figure 12.2. The plant is capable of starting without external auxiliary power, station supply is covered by internal sources.

The machines are synchronised to each other at a common 110-kV bus bar. All turbine governors operate in droop speed control. The reference speed for three

346 *Intended and unintended islanding of distribution grids*

Figure 12.2   Hydropower plant PP1 with machines PP1-1 to PP1-5

machines (leading units) is set to 50.0 Hz while the other machine's reference speed is set to 48.0 Hz. Thus, units PP1-1, PP1-2 and PP1-3 determine the system frequency, driving the other three machines into motor operation with their nozzles completely closed.

This initial configuration has the following characteristics:

- All five units in PP1 contribute to system inertia.
- Units PP1-4 and PP1-5 provide a minimum load (machine losses) for the system.
- In case the frequency drops below 48.0 Hz, for instance, after connecting a large load, all machines will actively support system frequency.

12.3.4   Start of storage pump in PP1

To test the frequency control capability of the start configuration, the storage pump of power plant PP1 is used as load. The pump has a rated power of 5 MW and is driven by an asynchronous motor. The start-up procedure takes place without an inrush current limiter. Before the pump is started, PP1-4 and PP1-5 run as motor

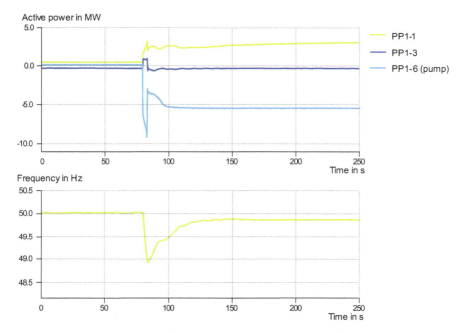

Figure 12.3   Time course of active power and frequency during start-up of the pump

loads providing additional inertia while PP1-1, PP1-2 and PP1-3 maintain the frequency as generators. Figure 12.3 shows an example of the power curves of one unit participating in frequency control (PP1-1), one unit providing spinning reserve (PP1-4) and the pump (PP1-6). Running up the pump with the ball valve closed leads to a load step of 6 MW followed by a ramp-shaped load increase to 9 MW.

It can be seen that due to the negative frequency gradient after the start of the pump, all units participate in the power balancing due to their inertia. At the terminals, in addition to the power from the driving turbine, a part of ΔP_{kin}, originating from the machine's inertia, can be observed. Since in the first moment, the turbine's reaction is zero, ΔP_{kin} corresponds to the negative load change ΔP, thus satisfying the active power balance (losses neglected). Hence, ΔP_{kin} can be calculated by

$$\Delta P_{kin} = \frac{dE_{kin}}{dt} = -2 \cdot H \cdot S_r \frac{\dot{\omega}}{\omega_0} = -2 \cdot H \cdot S_r \cdot \frac{RoCoF}{f_0} \tag{12.3}$$

In Figure 12.4, the focus is put on the kinetic power provided by PP1-1. The green line indicates the total power as measured at the generator terminals. The blue line shows the part coming from the turbine due to the governor's action.

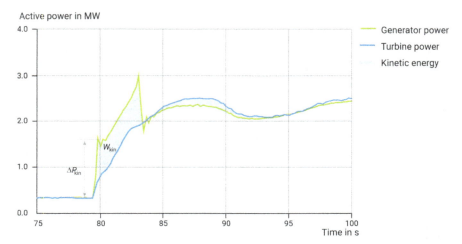

Figure 12.4 Time course of active power to demonstrate the effect of kinetic energy

The difference, the light blue area in Figure 12.4, is the kinetic energy, delivered during the negative frequency gradient and absorbed during the following positive frequency gradient.

After reaching the rated speed, the pump's power demand decreases to 3.5 MW and then slowly increases to the steady-state power of 5.5 MW when the ball valve opens. Subsequently, the output power of the frequency-controlling machines PP1-1, PP1-2 and PP1-3 is increased, while PP1-4 and PP1-5 again go into motor operation. The frequency nadir is 48.95 Hz, and the steady-state frequency is 49.89 Hz, determined by the governor's droop setting s.

$$\frac{\Delta f}{f_0} = s \cdot \frac{\Delta P}{S_r} \tag{12.4}$$

The steady-state power demand of 5.5 MW is covered by three units participating in frequency control, which results in a measured frequency deviation of 0.11 Hz, matching the droop setting of 4%.

After establishing a reasonable load in the system, the run-of-river powerplant PP3 is synchronised. It is operated with a constant flow rate, covering a constant part of the system load. Its contribution to system inertia is limited due to the low mechanical speed. However, it helps maintain the voltage in that area of the islanded grid.

12.3.5 Synchronising island and fall-back with power imbalance

In order to test the system's response for further positive and negative load steps, the following procedure was chosen during the test: The island grid, operating with the

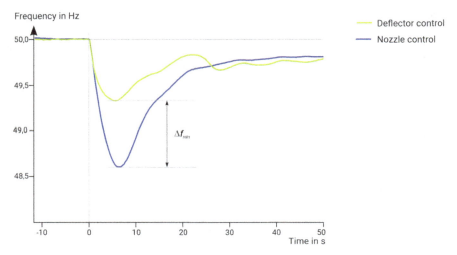

Figure 12.5   Comparison of frequency deviation during a load change with deflector control and conventional nozzle control

pump as load, is synchronised with the interconnected grid via a bus coupler circuit breaker. Subsequently, a power surplus or a power deficit is created in the test grid via the reference values of the turbine governors of the machines involved. When the breaker is opened, the test grid falls back into island operation and the power deficit respectively the power surplus leads to a frequency reduction respectively to a frequency increase. In this way, the reaction of the island grid to positive and negative load changes can be tested in a reproducible and controlled manner. Within the framework of the islanding test presented here, a comparison was made between classic nozzle control and deflector control with nozzle hold-back. The frequency curves of the two tests were superimposed for comparison in Figure 12.5. The higher control dynamics with nozzle hold-off lead to a significantly smaller frequency drop.

12.3.6   Synchronising with a second island

A further test was the synchronisation of two islanded systems and handing the frequency lead to one unit. Resynchronisation requires a coordinated convergence of the frequencies of the grids to be connected. When in droop speed control mode, the speed reference value must not match the actual steady-state frequency. The actual operating point (loading) of the machine responsible for speed control must be considered. Also, the rate of change of the reference speed setting must not be too fast to avoid over/undershooting the frequency goal. The point of connection must be carefully selected. Undesired tripping of the connection line due to transients and oscillations must be avoided.

350 *Intended and unintended islanding of distribution grids*

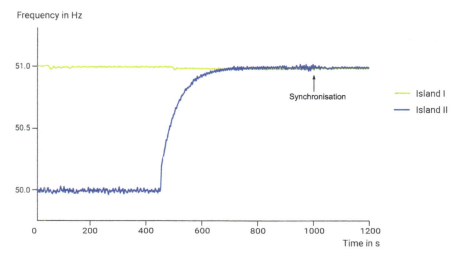

Figure 12.6 Frequency adjustment and synchronisation

After triggering the synchronisation sequence, the switching action itself will usually be controlled by a synchro-check device, actuating the circuit breaker. The synchro-check device ensures that the difference of voltage magnitude, frequency and phase angle on both sides of the breaker are within specified limits, allowing a smooth switching and avoiding transient stress for the machines.

For this purpose, a second islanded grid was started with the power plant PP2. Storage pump PP4 was used as load. The target frequency was intentionally specified as 51 Hz to test synchronisation at frequencies deviating from the rated frequency. As can be seen in Figure 12.6, the frequency of the first islanded grid is successfully adjusted to the frequency of the second grid and a smooth synchronisation is achieved. At that time, responsibility for frequency control is transferred to the power plant PP2.

12.3.7 Grid protection during islanding tests

At the end of the islanding test, a short-circuit test was carried out. For this purpose, the islanded grid was intentionally switched to a grounded busbar via a circuit breaker in a 110-kV substation. The distance of the power plants to the fault location is about 80 km. The fault current reaches approximately 1 kA and the voltage in the substation close to PP2 drops to approximately 50% of the rated voltage (Figure 12.7). This is sufficient for the distance relay to pick up and trip the faulty line as expected. The fault clearing time is 400 ms and neither loss of synchronism nor frequency instabilities were observed at the power plants.

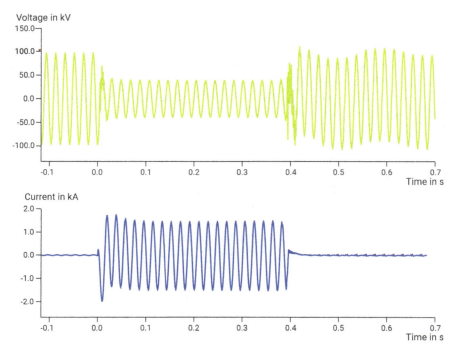

Figure 12.7   Voltage and current during a three-phase fault in the islanded grid

12.4   Conclusions

In addition to simulation-based grid restoration tests, real tests are essential for a successful grid restoration strategy. It is the only way, personnel can practice the operation of the grid and power plant under abnormal conditions like reduced inertia, reduced short-circuit capacity and frequency deviating from 50 Hz. Sometimes those tests even reveal hidden failures in the wiring or parameter settings of devices.

The tests carried out show that with the existing black-start capable power plants, it is possible to establish stable grid islands with a frequency maintained in a tolerable range. Minimum generator configuration, optimum control strategy and maximum allowable load connection can be derived from the test results. Synchronisation of different islands, even at frequencies not equal to 50 Hz, was tested successfully.

However, this puts demands on the operating team of the power plants in terms of operational management, which is difficult to master without training. The complexity of synchronisation increases with the size of the island – i.e. the number of generators in the island grids – and should therefore be carried out as early as possible during the process of grid restoration.

References

[1] Polster S, Schürhuber R, Renner H, *et al.* Best practice grid restoration with hydropower plants, *20th International Seminar on Hydropower Plants*, 14–16 November 2018, Vienna/Austria.

[2] Schmaranz R, Polster J, Brandl S, *et al.* Blackout – key aspects for grid restoration, *CIRED 2013*, Stockholm.

[3] Brandl S, Schmaranz R, Hübl I, *et al.* Evaluation of islanded grid operation tests and dynamic modelling, *CIRED 2011*, Frankfurt.

[4] Shen C. Bottom-up Network Restoration Based on Distributed Generation, Dissertation Kassel University, 2018.

[5] Machowski J, Bialek JW, and Bumby JR. *Power System Dynamics: Stability and Control*, 2nd ed. Chichester: John Wiley & Sons Ltd, 2008.

[6] Nikkilä A-J, Kuusela A, Weixelbraun M, Haarla L, Laasonen M, and Pahkin A. Fast restoration of a critical remote load area using a gradual voltage build-up procedure, *IET Generation, Transmission & Distribution*, 2020;14(7):1320–1328.

[7] Hachmann C, Becker H, and Braun M. Cold load pickup model adequacy for power system restoration studies, *MDPI Energies* 2022;15:7675.

[8] Nikkilä A-J, Kuusela A, Laasonen M, Haarla L, and Pahkin A. Self-excitation of a synchronous generator during power system restoration, *IEEE Trans. Power Syst.*, 2019;34(5):3902–3911.

[9] Nikkilä A-J, Kuusela A, Harjula A, Rauhala T, and Haarla L. Ferroresonance and subsequent sustained parallel resonance occurrence during power system restoration: analyses for system operation, *CIGRE Science & Engineering*, 2021;21:70–92.

[10] ENTSO-E, ENTSO-E Network Code on Emergency and Restoration, no. 714, 2015.

Chapter 13

Unintended islanded grids

Enrique Romero-Cadaval[1], Eva González-Romera[1], Francisco-José Pazos-Filgueira[2] and Michael Finkel[3]

For a long time, it was assumed that unintended islanded grids lose stability very quickly on their own and that islanded grids with a predominant feed-in via inverters in particular cannot occur. However, the fact that even large PV systems are capable of stabilising unintended islanded grids within minutes has been repeatedly demonstrated in a medium-voltage distribution grid operated by the Spanish grid operator Iberdrola.

13.1 Introduction

Nowadays, the presence of distributed generation in power systems is usual. However, strong supporting policies have caused that, in some countries, its introduction was not a progressive process. In 2007 and 2008, more than 3,000 MW of PV generation was connected to the Spanish power system. Almost 90% of this power was produced by large plants connected to MV or HV networks, resulting in a new and not mature situations that have not been sufficiently tested,

During these years, it was clearly seen that the commissioning of large PV plants arises several problems, which can be basically classified as:

1. Islanding operation during network maintenance or in a fault state.
2. Over-voltages in PV plants (as already described in [1]) that could cause the damage in the PV plant and in other customers connected to the same MV feeder.

In 2011, Iberdrola documented these problems, as they detected them in their MV and HV networks associated with the islanding behaviour of PV plants. In [2], the results of field tests in PV plants were also described to prove that islanding events are possible even when the balance of active and reactive power is not perfect

---

[1]Electrical, Electronics and Control Engineering Department, University of Extremadura, Spain
[2]i-DE Redes Eléctricas Inteligentes (Iberdrola Group), Bilbao, Spain
[3]Department of High Voltage Engineering and Electric Power Supply Systems, Augsburg Technical University of Applied Sciences, Germany

354 *Intended and unintended islanding of distribution grids*

and several inverters, either of the same brand or different brands, were involved, as it will be described later in Subsections 13.2 and 13.3.

The tests (13.4) concluded that, since voltage characteristics in the island were within normal limits, it was not possible to detect islanding behaviour with traditional protection methods or settings. Moreover, active methods without coordination between inverters of different brands proved to be ineffective. Another significative conclusion is that laboratory tests did not properly represent the on-field situations and results will be even worse when the requirements to guarantee transmission system stability are applied.

These conclusions were the reasons for launching the project PROINVER (Conversion and protection solutions in power electrical system scenarios with high penetration level of distributed energy resources) in 2011 with the participation of inverter, protection relay and communication system manufacturers, as well as laboratories and research centres. The main aim of the project was to define the specifications for developing new protection systems to solve the detected problems (as described in 13.5), to be either implemented in the inverters or based on relays and communications.

As a first step of this project, it was necessary to understand the reasons for the failure of protection systems embedded in inverters, since the conclusion of field tests suggested an interaction between the transient behaviour of the inverters and the loads connected to the network during islanding events. To this aim, two different analyses are carried out in Subsections 13.6 and 13.7, considering both passive methods and Sandia Frequency Shift (SFS, see Section 7.3.7) Islanding Detection Method:

1. Sensibility analysis of the constants used in SFS. Association with different kinds and numbers of inverters, equivalent quality factor in the point of common coupling, and presence of active loads (electric motors), have been assessed to identify how detection time is affected.
2. Modelling of a real network. It includes the simulation of the interaction among inverters with different anti-islanding configurations and power control parametrisations. The performance of SFS has been studied and assessed in the presence of active and reactive power balance between generation and loads in the network.

13.2 Operational experience on islanding events caused by large photovoltaic plants

Several cases of islanding behaviour during MV network maintenance or faults have been detected (Section 7.2 has introduced problems that arise with unintentional islands), giving rise to different kinds of negative consequences. Islanding events during maintenance works, in the absence of a fault, are the most severe cases and they happen when the anti-islanding systems fail, causing a quasi-permanent islanding situation. On the contrary, in case of faults, not only the system's ability to detect the islanding but also the disconnection time is relevant.

13.2.1 Dangerous work conditions

The following event, which took place during the maintenance of an MV/LV substation, describes an example of increased risk due to the failure of anti-islanding protection. According to the first two rules of the safety golden rules, the maintenance staff had to isolate the MV/LV substation from sources of supply and lock the isolation devices. These operations were made by opening, firstly, the MV feeder breaker at the HV/MV substation and, after that, opening the closest disconnector to the MV/LV substation. Since the MV feeder breaker was open, it was expected to open the switch with no load.

In this case, the workers considered that voltage absence verification in the MV feeder was not necessary. On the one hand, because isolation from voltage sources is a previous step to voltage verification, according to safety legislation in force (see Figure 13.1). On the other hand, because they did not have to work in that place, but on equipment located away from the isolation point. When the disconnector, which is designed to operate only with no load, was opened, an electric arc took place, which could damage the disconnector with the subsequent risk for the workers.

13.2.2 Impossibility of network operation or maintenance

The aforesaid event makes clear the need for additional voltage absence verification, going one step beyond the mandatory safety rules, for any kind of MV switching. In fact, when maintenance staff included this additional measure in similar cases, they verified the presence of voltage at different points of the MV network for a long time. Cases up to 40 minutes, or repetitive events after opening and closing the MV feeder breaker, have been reported.

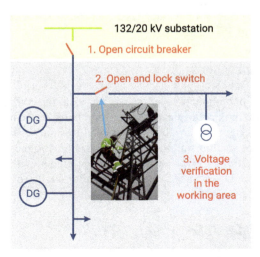

Figure 13.1 First three steps stablished in the safety procedure for maintenance in the MV network

356   *Intended and unintended islanding of distribution grids*

The problem in this case is that, consequently, maintenance works could not be carried out until the voltage sources (one or several PV plants) were identified. Even after identifying the voltage source, isolating the installations and locking the isolation devices in the open position, as required by safety rules, is a difficult matter. The reason for this additional problem is that network maintenance staff do not have access to the breaker and isolation devices within PV plants. Besides, PV plants can be dispersed kilometres away from the working area.

13.2.3   Failure of network automation

Another problem of sustained islanding behaviour is that voltage in the islanding network loses synchronism with the network. If breaker control includes synchronism or voltage absence verification for the closing operation, the breaker cannot close until the islanding event disappears, so the normal service cannot be restored, and further actions must be taken to identify the voltage sources (PV plants) and to disconnect them.

13.2.4   Damaged inverters

In those cases where the breaker control does not include voltage absence verification, the breaker closes without synchronism. This creates an over-current, irrelevant for the network, but that may be harmful for the inverter. This situation is very likely to happen in case of an earth fault followed by a fast reclosing, especially in networks with a limited earth-fault current. The reclosing time ranges from 0.4 to a few seconds, depending on local practices, so if the PV plant protection system is not fast enough, the breaker will close while the PV generators are still connected.

To prevent damage in the electronic switches (IGBTs) of inverters, extremely fast over-current protection is needed. This protection is implemented by many manufacturers, but some cases of fatal damages have taken place affecting up to 100% of the inverters of a large PV plant.

13.3   Technical reasons

The development of photovoltaic inverters was based, at the beginning, on the principle that they are generators of irrelevant power, compared to the network. Thus, most of the previous experience up to 2007 was related to small installations. The same technology tested for small installations was applied (due to the rapid growth of the PV generation in Spain in 2007 and 2008) to large plants, without any adaptation, reaching a generation power in many feeders similar, higher or even several times higher than the load power.

13.3.1   Behaviour of photovoltaic inverters

Inverters integrate control and protection algorithms (see Chapters 2 and 3), but they work without coordination among inverters of the same or of different plants. In addition, there is no standardisation between brands.

Unintended islanded grids   357

Inverters installed in the former plants do not try to keep any particular voltage, but to take out all the energy supplied by the PV panels, they produce a voltage slightly higher than the network voltage. Their control loop tries to keep the sinusoidal current that represents the available power with the given voltage, making the necessary corrections in very short times, typically tens or hundreds of microseconds. If the current is lower than the target, the inverter increases the voltage (and the opposite).

National regulations usually establish the voltage and frequency protection settings, and anti-islanding protection is introduced in IEEE Std 1547-2018 without standardising any detection method. Thus, the method is decided by each manufacturer, which could be a passive method, an active one, or a combination of both.

In any case, the anti-islanding system is usually tested in labs in conditions that could not be representative of actual networks since the test only includes a single inverter and a simplified load. On the contrary, in real networks, and especially in MV networks, several inverters (of the same or of different brands, and that could have the same or different islanding methods) are working in parallel.

13.3.2   *Behaviour of an islanding network*

A network working in islanding mode implies a balance between generation and consumption, both in terms of active and reactive power (details about the occurrence of unintentional electrical islands are available in Section 7.1). It is frequently assumed that this kind of balance is hardly reached if PV plants do not have power regulation and they do not provide reactive power.

However, the field test has proved that starting from a rough balance at the beginning of the event, the exact balance of active power is reached by means of consumption modification. Consumption depends on the voltage and the loads behave approximately as constant impedances in the field tests carried out by Iberdrola. In this situation, changing the voltage, generation and consumption match (even phase by phase) could happen and the islanding situation can be possible. If the generation is clearly lower than consumption, the voltage will decrease and the generation will trip, and if the generation is clearly higher than consumption, the voltage increases so the generation trips. An extreme case of this latter situation takes place when generation is several times higher than load. In this case, voltage increases so much that damages are possible depending on voltage magnitude and duration [1].

Regarding reactive power, in the field tests, it could be seen that inverters were not able to keep a stable power factor during islanding behaviour. The addition of all the different reactive powers produced by the different inverters and the transient behaviour of loads made the balance possible.

13.4   Field test

Two field tests in MV systems were carried out in substations where the dangerous conditions were previously reported. The test process started from a situation in which generation was higher than consumption. To roughly balance the power, some inverters were manually disconnected until the current by the substation MV circuit breaker decreased to a few amperes. Then, the feeder circuit breaker was opened. In both tests, the islanding event was reproduced, producing islanded networks of 700 kW and 2.5 MW, respectively.

Some parameters were more unstable during islanding than when connected to the main grid, but rms voltage, distortion and unbalanced magnitudes remained within normal limits. Figure 13.2(a) shows the active power of one inverter, which decreased constantly according to the declining irradiance of the evening. It can also be seen in this figure that the reactive power of individual inverters suffered a strong fluctuation and that voltage also fluctuated, but within limits.

Frequency also fluctuated but not as much as expected (Figure 13.2(a)), since consumption-generation balance is not done by frequency adjustment, as with traditional synchronous generation, but by means of voltage modification. Although there were some frequency changes, as can be seen in Figure 13.2(b), their magnitude was too small or short to be detected with the rate of change of frequency (RoCoF) settings compatible with system stability constraints.

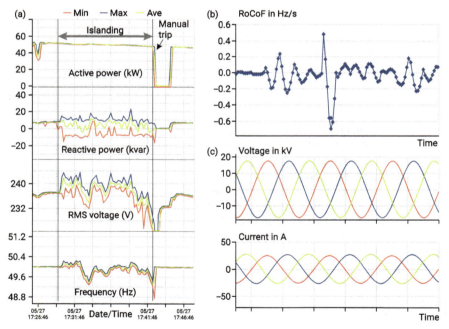

Figure 13.2 (a) Measured magnitudes during islanding operation: active power, reactive power, rms voltage and frequency (fluctuation can be seen as the difference between max.-blue and min.-red values). (b) Worst case of RoCoF in consecutive cycles. (c) Voltage and current at point of coupling to the MV network

In addition, the almost perfect waveforms shown in Figure 13.2(c) prove that an islanding identification based on voltage or current quality cannot be easily found.

Other conclusions obtained during field tests are the following:

- Network configuration allowed testing different combinations of generators. In one of them, only dozens of inverters with the same impedance measurement method supplied the islanding system. Those inverters failed to detect the islanding situation, probably affected by the interaction between inverters.
- Active anti-islanding methods, based on frequency shift, proved to be able to detect the islanding, but depending on the other generators that they had in parallel. When all or most of the PV inverters were of the same model, they worked satisfactorily but the time needed to detect the islanding condition was progressively longer when other inverters, with passive and impedance methods, were connected in parallel, reaching the case that were not able to detect it at all.

13.5 Potential solutions

Islanding events have to do with the transient behaviour of the inverters and their interaction with the loads connected to the network (and different islanding detection methods could be applied, as discussed in Section 7.3).

In the previous section, field tests proved that:

- Islanding events are possible, providing that the difference between generation and consumption is not excessive (about 10%).
- Islanding conditions can last a long time. The interaction inverter loads tend to find a stable point by voltage variations and not so much by frequency as expected.
- The failure of anti-islanding systems has to do with the interaction between several inverters, either of the same brand or different brands.
- The actual field situation is not represented by laboratory setups. Different inverter brands, from several countries, and implementing different anti-islanding methods have failed.
- If voltage characteristics in the island are within normal limits, it is not possible to detect islanding behaviour with traditional protection settings that allow the normal behaviour of the plant.
- Additionally, the requirements to improve fault ride-through capabilities, necessary to preserve the system stability in case of a voltage dip, as well as frequency insensitivity to events in transmission networks, complicate even more the detection of the islanding conditions, especially with passive methods.

In this context, active islanding detection methods (Section 7.3.), as a potential solution, require coordination between different brands, so that they can work in parallel without inferring each other. Meanwhile, test methods using several inverters in parallel, as shown in Figure 13.3 (an approach for simple islanding scenarios has been discussed in Section 7.1.4), would help to verify inverters' capabilities closer to real conditions. However, the use of some active methods could be a constraint for the penetration of inverter-based generation. Thus, frequency shift has proved to

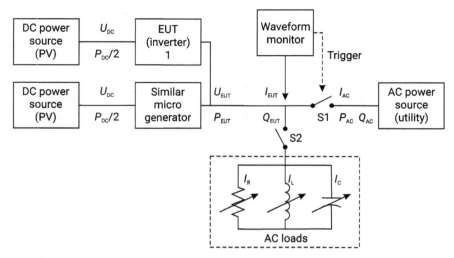

Figure 13.3 Test with several inverters in parallel (based on IEC 62116)

be an effective method, but if applied massively, it could lead to loss of stability in case of major events in the transmission network. Therefore, coordination between protections included in the inverters (either active or passive methods), protections of the point of coupling (relays) and communication-based protections (inter-tripping or telecontrol) is necessary.

Therefore, requirements from transmission networks (delayed or insensitive trip to prevent instability) and distribution networks (fast and certain trip, to prevent damages, or affecting power quality) are opposite to each other. A possible solution could be the use of requirements segmentation, as follows:

- Generation plants connected to MV or HV should give priority to transmission needs and have fault ride-through capabilities, passive protections (internal or at the point of coupling) and communication-based protections. These communications systems could be, in general, telecontrol to allow safe operation in case of maintenance and inter-tripping in some network configurations.
- Communication requirements for LV must be necessarily lower, so traditional inter-tripping or telecontrol should not be applied. Fast active methods should be used when telecontrol is not available, giving priority to safety instead of fault ride-through capabilities.

However, smart meters including an internal switch can be used as a simple and inexpensive telecontrol. Their use can allow reducing the amount of generation without fault ride-through capabilities or other characteristics that may affect system stability, thus permitting a higher penetration of dispersed generation. All inverters should be designed to withstand asynchronous reclosing and, in addition, to meet strict limits of temporary over-voltages in case of disconnection from the network.

13.6 Inverter sensibility analysis

During this stage of the work, a sensitivity analysis for the different parameters or causes that could affect the anti-islanding detection method was carried out. This analysis was done through simulation by determining the detection time measured from the time when the islanding situation starts to the time when this situation is detected, usually when the frequency goes out of the established interval (50 ± 1 Hz).

The first analysis tries to determine the influence of different K constants used for determining the chopping factor cf typical of SFS methods (see Section 7.3.7), according to Figure 13.4 and (13.1) [3]

$$cf = \frac{2t_s}{T} = cf_0 + K(f_m - f_{ref}) \tag{13.1}$$

(where f_m and f_{ref} are measured and reference values of frequency).

This first analysis also considers different quality RLC load (Figure 13.5) factors, Q, given by (13.2) at the fundamental frequency f_1:

$$Q = \frac{R}{L\,2\pi f_1} \tag{13.2}$$

Capacitor C is selected for the resonance frequency of this load to be equal to f:

$$C = \frac{Qf_1}{2\pi R f_0^2} \tag{13.3}$$

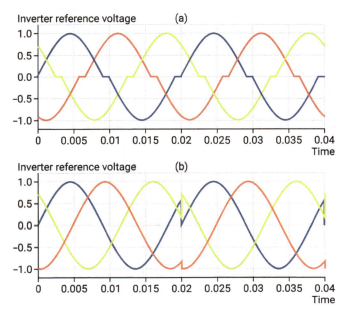

Figure 13.4 Two ways of implementing the SFS method ($cf = 0.1$): (a) type A defined in [3], and (b) type B used in [4]

362 Intended and unintended islanding of distribution grids

The analysis has been done using an average model for the inverter and assuming the systems shown in Figure 13.5, where the inverter or group of inverters supply 100% of the power demanded by the load. It is also considered that the resonant frequency of the RLC is $f_0 = f_1 = 50\,\text{Hz}$. This situation has been considered because it has been described as the worst case for SFS islanding detection method [5].

The detection times obtained by a set of simulations are shown in Figure 13.6 when a single inverter is supplying 100% of the load connected at the PCC. In this figure, one can see that there is a dependence of this time on the constant K and factor Q, which is not easy to determine, and this time also highly dependent on the type of inverter considered. But, at the same time, it could be concluded that to have detection problems in the proposed system, the factor Q has to be quite high, up to 4, which is very difficult to achieve in real systems.

As a second stage of this analysis, a group of two inverters was considered instead of a single inverter. The results are shown in Figure 13.7 for two cases. These simulations allow us to conclude that the detection time (and the operation of the

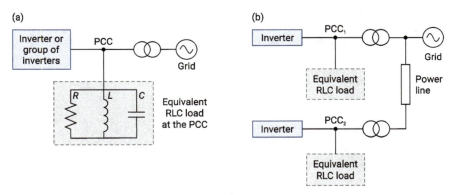

Figure 13.5   Systems under study: (a) one PCC system and (b) two PCC system

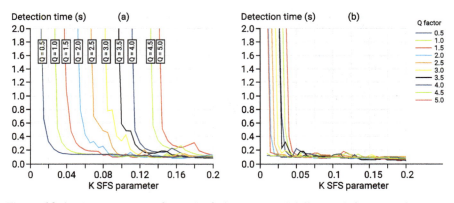

Figure 13.6   Detection time for a single inverter and different Q factors, when varying the parameter K: (a) inverter type A and (b) inverter type B. (Types are defined according to Figure 13.4.)

Unintended islanded grids 363

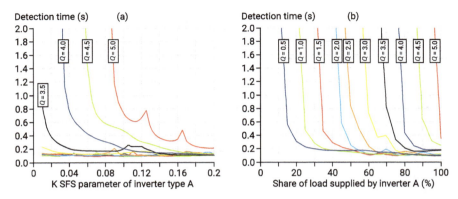

Figure 13.7 Detection time for a two-inverter system and different Q factors: (a) inverter type A (varying K) with an inverter type B (K = 0.05) supplying each inverter the 50% of the load power and (b) inverter type A (K = 0.15) with an inverter without anti-islanding when varying the share of the load being supplied by each inverter

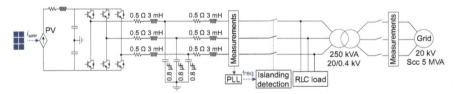

Figure 13.8 Scheme of the simulation model (in PSCAD) used for the validation

anti-islanding protection) does not only depend on the inverter itself but also on other inverters that could be connected to the same PCC, or even in other PCC.

The above results have been validated through a simulation model including a solar panel, a 6-pulse three-phase inverter controlled by current, an LCL filter, a phase lock loop (PLL) for frequency measurement, an RLC load, an LV-HV transformer and a grid model (Figure 13.8). This model permitted the evaluation of the influence of other system parameters, like the filter L and C values or the PLL constants. As a result, the filter parameters and the proportional constant of the PLL showed a weak influence on the detection time. However, the integral constant K_i of the PLL has a significant influence, as shown in Table 13.1.

This simulation model was also used to study a multi-inverter scenario. If all the inverters have the same SFS anti-islanding system with an equal value of K, their behaviour faced with an islanding situation is similar to the expected for a single inverter. However, the presence of one inverter without an SFS system or with an inadequate value of K can cause all the inverters located in the same or even in different PCCs to fail when faced with an islanding situation.

Another interesting conclusion is obtained when substituting a part of the load connected to the photovoltaic plant by an asynchronous motor. A commercial 11 kW

Table 13.1 Influence of the K_i constant of the PLL in the detection time, K_p is fixed to 1

K_i	120	400	640	720	800	880	960	1200
Time in s	> 3	1.09	0.75	0.69	0.63	0.57	0.53	0.43

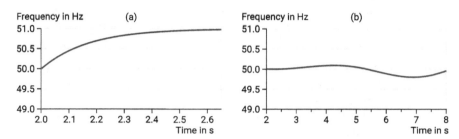

Figure 13.9 Frequency variation during an islanding situation ($K = 0.01$ and $Q = 2.5$): (a) with RLC load and (b) with a combined RLC and motor load

squirrel-cage motor has been used for this simulation. An RLC load is also connected in parallel to keep the same quality factor and a correct balance of active and reactive power (they have been adjusted to a maximum unbalance of 2%). Figure 13.9 shows the frequency variation for the same inverter with SFS constant $K = 0.01$ and the same quality factor of 2.5. The frequency limits match up with the established performance interval. Figure 13.9(a) corresponds to a pure RLC load and Figure 13.9(b) includes a motor. It can be observed that the presence of the motor causes the anti-islanding system to fail while the system performs well with RLC loads. It can also be observed that the way the frequency varies after the beginning of the islanding situation is different in both situations, as it moves forward in the same direction with RLC loads (following the positive-feedback concept of SFS) while it fluctuates around the reference value when a motor is connected.

The obtained result confirms and illustrates the affirmation found in the literature about the influence of the presence of motors among loads in the failure of anti-islanding systems [6]. Further studies must be done to determine how to avoid this interaction that could exist when active loads (motors) are present in the proximity of inverters.

13.7 Modelling of a real network

This section includes a simulation of the interaction among inverters with different anti-islanding configurations and power control parametrisations [7]. The performance of SFS has been studied and assessed in the presence of active and reactive power balance between generation and loads in the network. The static PV generator model used comprises the following elements:

- An ideal DC source represents a generic static generation system (e.g. PV system).
- A voltage source inverter (VSI) links the static DC stage with the AC system.
- An AC low pass filter to limit high-frequency harmonics generated by the inverter.
- The inverter control system affects its dynamic performance.

Figure 13.10 shows the inverter control scheme that follows the typical power control with an external PI control loop to regulate PQ power and an inner decoupled anti-windup PI current loop in the dq synchronous frame. (Information about control and behaviour of power systems can be found in Chapters 3 and 4.) The space angle and frequency of the voltage vector at the PCC are detected with a PLL.

The anti-islanding active frequency drifting method SFS was originally developed for a single-phase inverter and has been extended to three-phase inverters that utilise PLL. To implement the SFS positive feedback anti-islanding active method, the frequency deviation is used as a feedback signal to compute the inverter current phase angle Θ_{SFS} or chopping fraction according to the following expression (see Section 7.3.7):

$$\Theta = \frac{\pi}{2} cf \tag{13.4}$$

where cf is the chopping fraction given by (13.1).

In three-phase systems, a phase angle transformation is applied in the PQ controller to compute the modified current references to force frequency shift and thus frequency instability (see Section 7.3.7):

$$\begin{pmatrix} i'_{d,ref} \\ i'_{q,ref} \end{pmatrix} = \begin{bmatrix} \cos \Theta_{SFS} & -\sin \Theta_{SFS} \\ \sin \Theta_{SFS} & \cos \Theta_{SFS} \end{bmatrix} \cdot \begin{pmatrix} i_{d,ref} \\ i_{q,ref} \end{pmatrix} \tag{13.5}$$

During the simulations, the inverters were operated at unity power factor and the disconnection from the main grid occurs at the instant of 1 s. The results presented refer to a particular identified scenario in which the islanding detection becomes

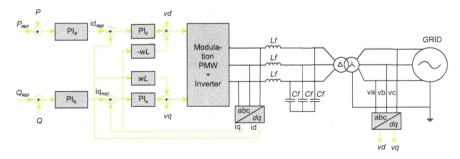

Figure 13.10   Inverter control scheme

more difficult by the active SFS method. Several simulations were run, varying different factors that may affect the reliability of the islanding detection, described below.

High gains on the outer power loop, especially in the Q control loop to inject no reactive power ($Q_{REF} = 0$) degrade the performance of the SFS method, as shown in Figure 13.11(a) (P control loop gains remain fixed). This effect is due to the cancellation of the reactive power injection on which active frequency drifting anti-islanding methods are based [3]. As a consequence, larger SFS gains would be required for faster and more successful islanding detection. This is shown in Figure 13.11(b), where a strong PQ power control loop was acting. Loads with dynamic voltage dependence seem to have the effect of accelerating or decelerating the frequency destabilisation depending on the percentages and the type of dependence.

Up to a certain percentage (10–15%), constant PQ power loads have an adverse effect, as previously stated, although a larger increment in constant power loads positively affects islanding detection [5]. This result is confirmed when passive loads are replaced directly by asynchronous motors with the same rate of power and the same PQ control (Figure 13.12(a)).

Under certain conditions, there is no simultaneous disconnection of generators; the disconnection of the first generator due to its anti-islanding system delays the disconnection of the second one. To analyse this effect, the SFS method is implemented with different frequency relay trigger delays (Figure 13.12(b)). Longer trigger delays cause a new power mismatch between generation and load in the islanded network after the disconnection of the first generator, changing the frequency drifting direction and, therefore, delaying the actuation of the anti-islanding system of the second generator.

The presence of more generators with different trigger delays could even lead to a better match between generation and load, causing the system to be within a non-detection zone (NDZ), making the islanding situation more likely.

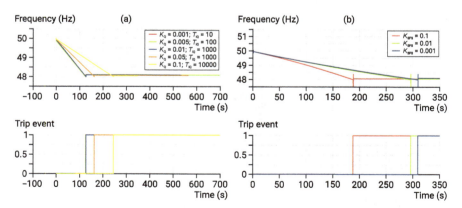

Figure 13.11 (a) Frequency deviation with different PQ control loop gains and (b) frequency deviation with different SFS gains

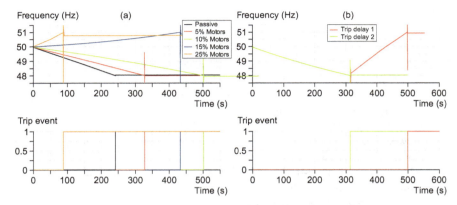

Figure 13.12 (a) Frequency deviation with different percentages of asynchronous motors and (b) frequency deviation depending on trigger event delays

13.8 Final remarks

Several simulations have been run to determine the causes of islanding situations detected in MV networks. In this analysis, different factors that may affect the reliability of the islanding detection have been studied, concluding that the detection time and the operation or not of the anti-islanding protection depend on several factors:

- The inverter itself, which may have active detection or not, and the way to implement the detection method.
- Other inverters that could be connected to the same point of the network, or even in other points.
- The presence of active loads, as motors, analysing the failures of anti-islanding systems that perform well with RLC loads.

Even having the same detection method, for instance SFS, there can be important differences between inverters. First, because the constant K of the SFS method has a significant influence. Another aspect that may degrade the performance of the SFS method is the reactive power control loop.

When several inverters are connected to the same network and all of them have the same SFS anti-islanding system with an equal value of K, their behaviour faced with an islanding situation is similar to the expected for a single inverter. However, the presence of inverters without SFS system or with an inadequate value of K may cause all the inverters located in the same network to fail when faced with an islanding situation. Moreover, the presence of generators with different trigger delays might increase the probability of islanding methods failing.

Finally, the way the frequency varies after the beginning of the islanding situation with motors is different from the behaviour with RLC, standardised for laboratory tests, since the dynamic voltage dependence seems to have the effect of accelerating or decelerating the frequency destabilization.

References

[1]   Pazos FJ. Power frequency overvoltages generated by solar plants. In: *CIRED 2009 – The 20th International Conference and Exhibition on Electricity Distribution – Part 2*; 2009. p. 1–4.

[2]   Pazos FJ. Operational experience and field tests on islanding events caused by large photovoltaic plants. In: *CIRED 2011 – The 21th International Conference and Exhibition on Electricity Distribution*; 2011. p. 1–4.

[3]   Bower W and Ropp M. Evaluation of Islanding Detection Methods for Utility-Interactive Inverters in Photovoltaic Systems (SAND2002-3591) [Report]. Albuquerque, New Mexico 87185 and Livermore, California 94: Sandia National Laboratories; 2002.

[4]   Sanchis P, Marroyo L, and Coloma J. Design methodology for the frequency shift method of islanding prevention and analysis of its detection capability. *Progress in Photovoltaics: Research and Applications*. 2005;13(5):409–428. Available from: https://onlinelibrary.wiley.com/doi/abs/10.1002/pip.613.

[5]   IEEE. IEEE Recommended Practice for Utility Interface of Photovoltaic (PV) Systems. IEEE Std 929-2000; 2000. p. 1–26.

[6]   Igarashi H, Sato T, Miyamoto K, *et al.* Power generation confirmation of induction motors and influence on islanding detection devices. *Electrical Engineering in Japan*. 2010;171(4):8–18. Available from: https://onlinelibrary.wiley.com/doi/abs/10.1002/eej.20962.

[7]   Pazos FJ, Romero-Cadaval E, González E, *et al.* Failure analysis of inverter based anti-islanding systems in photovoltaic islanding events. In: *22nd International Conference and Exhibition on Electricity Distribution (CIRED 2013)*; 2013. p. 1–4.

Chapter 14

Summary and outlook

Michael Finkel[1]

After presenting various practical examples of intended and unintended islanded grids in Chapters 10–13, this final chapter briefly summarises key aspects for the stable operation of islanded grids in the distribution grid and provides an outlook on current developments.

14.1 Concluding remarks

The practical examples of intended islanded grids presented were limited to real islands where it is not economically viable to establish a connection to the electricity grid (off-grid electrical islands), and independent grid-connected electrical islands which are used in emergency situations.

The basic supply of an off-grid electrical island is often provided by diesel generators or conventional power plants, which have grid-forming properties and are dimensioned in such a way that they can also compensate load fluctuations. The integration of RES into islanded grids can result in (temporary) energy surpluses that cannot be absorbed by the diesel generator. For this reason, it is very difficult, in terms of control, to replace the proportion of energy generation from conventional generation modules in favour of renewable energy from wind turbines and PV.

As presented in Chapter 10, storage systems (BESS, pumped storage, etc.) can increase the proportion of energy generated from renewable sources to 100%, as the storage system can react much more quickly to demand changes and readjust its output than a conventional generation unit and can also absorb energy in the event of power surpluses. Storage systems are, therefore, often essential in islanded grid operation, as they make a significant contribution to the stability of the islanded grid.

---

[1]Department of High Voltage Engineering and Electric Power Supply Systems, Augsburg Technical University of Applied Sciences, Germany

370 *Intended and unintended islanding of distribution grids*

In addition to the electrical islands on real islands, there are also electrical islands in remote areas. Therefore, some further examples should also be mentioned here: Electrifying villages in remote areas in India [1,2], Siberia, Northern Canada [3] and Africa [4] or on insular territories [5–7]. Remote industrial operations also possess a self-sufficient electrical system. Mines, in particular, require large and robust electrical installations [8,9].

One reason for operating a grid-connected electrical island is often to avoid power outages. In particular, the continued supply of critical infrastructure (e.g. hospitals, drinking water supply, servers) or sensitive consumers (e.g. airports, electricity-intensive industry, military bases) must be guaranteed in the event of a failure of the superordinate grid. Another reason that motivates grid-connected facilities to invest in an islanded grid is cost: With the help of PV systems, wind turbines, backup or main power generators and battery energy storage systems, it is possible to increase energy self-sufficiency and manage power outages.

The grids described so far are not part of the public electricity supply. In addition, it is also possible to activate and operate islanded grids within existing distribution grids and thus reduce the duration of the interruption in the event of a power outage and, in the event of a blackout, to support the restoration of the grid (cf. Chapters 11 and 12).

As presented in Chapter 13, the expansion of DGs also poses grid operators with the challenge of unintended islanding of grid areas. In order to ensure operational safety and reliability, these unintended islanded grids must be quickly recognised and switched off using suitable detection methods (Chapter 7).

To ensure stable islanded grid operation, the basic ancillary services of frequency and voltage control must be provided in the islanded grid as well as in the interconnected system. This includes the constant adjustment of generation capacity to consumption and maintaining the permissible bands for the system variables of voltage and frequency. This task is performed either by a centralised, dominant grid-forming generation unit, which can cover the entire existing load on its own, or by several DGs or electrical storage systems involved in grid formation. The load can be distributed to the individual generation units either via droop control or a centralised island grid controller.

A continuous full supply of the islanded grid can only be ensured if controllable generation capacity is available at the maximum load level. If this is not the case, active and reactive power must also be adapted to the existing generation level on the load side. Therefore, island grid operation concepts must always be adapted to the local conditions. In any case, however, island grid capability requires existing infrastructure to be upgraded and is accompanied by a number of organisational and technical challenges.

Figure 14.1 summarises the various aspects of stable islanded grid operation in the distribution grid covered in this book. This figure once again illustrates the complexity of the topic.

Summary and outlook 371

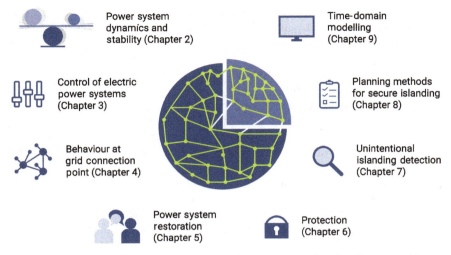

Figure 14.1 Aspects of stable islanded grid operation in the distribution grid covered in this book

14.2 Outlook and open questions

Current developments such as the decarbonisation of the (electrical) energy supply and the resulting phaseout of large fossil-fuelled power plants with synchronous generators, as well as the coupling of the electricity, heat and transport sectors, will have a massive impact on the way electricity grids are operated in the future. These developments not only have an impact on the large interconnected systems but often arise first in the context of insular power systems because islands are often the first power systems to reach very high instantaneous levels of inverter-based resources (IBR). Some of these current developments and resulting questions are therefore listed here without any claim to completeness (based on [10–12]):

- Frequency stability: Due to the decreasing system inertia, the decreasing – previously inherent – frequency dependency of consumers, and the simultaneous increase in demand (sector coupling), all available sources must be considered for the provision of the instantaneous reserve. The contribution of IBR, in particular, will play a decisive role here.
  - What is the minimum inertia required for frequency control?
  - Which technical properties must inertia from non-rotating generation have?
  - What should be the ratio of voltage-controlled resources (conventional generators, grid-forming inverters and synchronous condensers) to current-controlled resources (grid-following inverters) in a system for ensuring stability, and what are the driving factors?
  - If grid-forming inverters will be connected to the distribution system, have protection systems to be adapted and can unintended islanding be avoided?

372 *Intended and unintended islanding of distribution grids*

- Voltage stability: The static and dynamic reactive power requirements of the grids are increasing significantly due to the increased load on the lines. In addition, the share of conventional power plants for the provision of dynamic reactive power will decrease. Coordinated reactive power control is, therefore, becoming increasingly important. Static and dynamic reactive power sources must be available at all times.
- Grid codes: Do the requirements for generation units need to be adapted in terms of grid-forming characteristics and the ability to be capable of riding through a wide range of voltage and frequency events, including very fast rates of frequency change, jumps in the phase angle, transient overvoltages, and other events?
- Resonance stability: When is resonance stability guaranteed, or is there sufficient stability reserve in the grid? What is a suitable parameter for quantifying resonance stability? How is the need determined in practice?
- Short-circuit current: What are the short-circuit current requirements for grid protection at grid nodes at different grid levels, and what are the resulting requirements for power converters and protective devices? Do the protection concepts need to be adapted?
- Simulation models: When can traditional positive sequence transient analysis models be used, and when do we need to transition to EMT simulations?
- Grid restoration: Due to the fluctuating power supply of RES, defined requirements are necessary for their suitable integration into grid restoration concepts. Which generation modules must have black-start capability and the capability to take part in isolated network operations? The roles and tasks of the TSOs and DSOs in the grid and supply restoration process must take into account the increasing decentralisation of generation and be developed further.
- Flexibilisation and sector coupling: The use of grid-supporting flexibility can support grid operation and optimise the use of grid capacities in the distribution grid. To this end, planning and regulatory conditions must be created, and conflicts of interest with the energy market must be resolved. For the utilisation of consumers such as electric vehicles and power-to-X systems, further technical specifications are required for the provision of ancillary services.

The team of authors hopes to include the answers to some of these questions in the next edition of the book.

References

[1] Suryad VA, Doolla S, and Chandorkar M. Microgrids in India: Possibilities and Challenges. *IEEE Electrification Magazine*. 2017;5(2):47–55.
[2] IEEE Spectrum [Homepage on the Internet]. IEEE; 05 Oct 2021 [cited 2024 Feb 09]. How 14 Microgrids Set Off a Chain Reaction in a Himalayan Village. Available from: https://spectrum.ieee.org/how-14-microgrids-set-off-a-chain-reaction-in-a-himalayan-village.

Summary and outlook 373

[3] Nassif AB, Yazdanpanahi H, Wright M, *et al.* Planning and Optimization of a Small-Scale Hybrid Microgrid in a Northern Community. In: *2020 47th IEEE Photovoltaic Specialists Conference (PVSC)*; 2020: pp. 2554–2556.

[4] Motjoadi V, Bokoro PN, and Onibonoje MO. A Review of Microgrid-Based Approach to Rural Electrification in South Africa: Architecture and Policy Framework. *Energies*. 2020;13(9). Available from: https://www.mdpi.com/1996-1073/13/9/2193.

[5] Guerrero-Lemus R, Shepard LE, Graldez J, *et al.* Best practices for high penetration PV in insular power systems IEA, PVPS, Task 14 [Report IEA-PVPS T14-13:2021]. International Energy Agency (IEA), Photovoltaic Power System Programme; 2021. Available from: https://iea-pvps.org/wp-content/uploads/2021/09/IEA-PVPS-T13-24_2021_Best_practices_PV_island_report.pdf.

[6] Eales A, Smith K, and Galloway S. Sustainability Evaluation of Scottish Islanded Mini-grids – Towards Asset Management Strategies for Remote Off-grid Energy Infrastructure [Report]. University of Strathclyde and Energy Mutual; April 2020. Available from: https://strathprints.strath.ac.uk/id/eprint/82217.

[7] Hatziargyriou N, Dimeas A, Vasilakis N, *et al.* The Kythnos Microgrid: A 20-Year History. *IEEE Electrification Magazine*. 2020;8(4):46–54.

[8] Ellabban O and Alassi A. Optimal Hybrid Microgrid Sizing Framework for the Mining Industry with Three Case Studies from Australia. *IET Renewable Power Generation*. 2021 01;15.

[9] Althaus T. From Concept to Reality – The World's Largest Off-Grid Mining Hybrid Power System at Fekola Gold Mine. In: *5th International Hybrid Power Systems Workshop*, Online; 2021: pp. 1–27. Available from: https://hybridpowersystems.org/virtual2021/wp-content/uploads/sites/19/2023/03/1_4_HYB21_051_presentation_Althaus_Thorsten.pdf.

[10] BMWK. Roadmap Systemstabilität [Report]. 11019 Berlin: Bundesministerium für Wirtschaft und Klimaschutz (BMWK); 2023.

[11] ENTSO-E. High Penetration of Power Electronic Interfaced Power Sources and the Potential Contribution of Grid Forming Converters [Technical Report]. Brussels, Belgium: ENTSO-E Technical Group on High Penetration of Power Electronic Interfaced Power Sources; January 2020.

[12] Hoke A, Gevorgian V, Shah S, *et al.* Island Power Systems With High Levels of Inverter-Based Resources: Stability and Reliability Challenges. *IEEE Electrification Magazine*. 2021;9(1):74–91.

Glossary

The glossary provides terms and definitions used in the book.

Angle stability
Rotor angle stability is concerned with the ability of the interconnected synchronous machines in a power system to remain in synchronism under normal operating conditions and to regain synchronism after being subjected to a small or large disturbance [1].

Automatic voltage regulator
Automatic voltage regulator or AVR means the continuously acting automatic equipment controlling the terminal voltage of a synchronous power-generating module by comparing the actual terminal voltage with a reference value and controlling the output of an excitation control system [2].

Blackout
State within a given area of an electric power network, or of the complete electric power network, characterized by the complete loss of electric power [3].

Black-start capability
Black-start capability means the capability of recovery of a power-generating module from a total shutdown through a dedicated auxiliary power source without any electrical energy supply external to the power-generating facility [2].

Brownout
During a brownout, there is no complete power outage, but rather a temporary, slight voltage drop in the power network. Electricity continues to flow to the end user, but the voltage drops noticeably (flickering and dimming of light bulbs). Normally, short-term voltage drops do not cause any serious damage. Usually, a brownout occurs together with a network overload caused by an unexpectedly high demand for electricity. This temporary network instability can be caused by excessive electricity demand, severe weather events or technical malfunctions.

Capability of isolated (network) operation
Isolated network operation is the operation of asynchronous subsystems which may arise from network disturbances. In isolated network operation, a subsystem is supplied by at least one generating unit [4].

376   *Intended and unintended islanding of distribution grids*

Decoupling unit
Protective device that ensures that a distributed generation unit is safely disconnected from the mains in the event of faults or problems. A decoupling unit can, e.g., separate the generation unit in the event of voltage or frequency threshold violations.

Electric power system
An electric power system is a functional unit which can be delimited by technical, economic or other criteria within the electricity industry [4].

EMT simulation
The electromagnetic-transient (EMT) simulation method is used to simulate high-frequency (short time-frame) dynamics. Thus, the differential equations of the electrical network are modelled in detail for the sinusoidal voltages and currents. The simulation methods use small integration time steps of the order of 50 µs or less. This allows to model of power system components with high accuracy and makes EMT simulations valuable when studying the effects of power-electronic devices.

Energy cell
An energy cell consists of the infrastructure for different forms of energy, in which an energy cell management system in possible coordination with neighbouring cells organizes the balancing of generation and consumption across all existing forms of energy [5].

Distributed generation unit (DG)
DG can be defined as electric power generation within distribution networks or on the customer side of the network [6].

Droop
Droop means the ratio of a steady-state change of frequency to the resulting steady-state change in active power output, expressed in percentage terms. The change in frequency is expressed as a ratio to nominal frequency and the change in active power expressed as a ratio to maximum capacity or actual active power at the moment the relevant threshold is reached [2]; cf. [7].

Fault ride through (FRT)
Fault ride through means the capability of electrical devices to be able to remain connected to the network and operate through periods of low voltage at the connection point caused by secured faults [2].

Fast frequency response (FFR)
Fast frequency response is the ability of a grid-connected converter to provide an energy response in front of a frequency event slower than the rotating machine's inertia response but faster than the synchronous machine primary controller [8].

Glossary 377

Frequency

Frequency means the electric frequency of the system expressed in Hertz that can be measured in all parts of the synchronous area under the assumption of a consistent value for the system in the time frame of seconds, with only minor differences between different measurement locations. Its nominal value is 50/60 Hz [2].

Frequency stability

Frequency stability describes the controlling of frequency deviations as a result of imbalances between injection and withdrawal (active power control) and is achieved by primary and secondary control and by the use of minute reserves in the power stations [4].

Generating unit

A generating unit for electrical energy is a power station installation which can be delimited according to certain criteria. The generating unit may, e.g., be a power unit, a common-header power station, a combined cycle plant, the machine set of a hydro-electric power plant, a fuel cell stack or a solar module [4].

Grid code

The grid code is a catch-all term that encompasses a wide set of rules by which assets connected to a power system and market must abide, the goal of which is to support the cost-effective and reliable operation of the latter. It consists of four major parts: connection codes (discussed in this document), operation codes, planning codes and power market codes [9].

Grid connection point

The grid connection point is the point at which the supply connection of a connection user is connected to the network [4].

Grid forming converters (GFCs)

Grid forming converters (GFCs) shall be capable of supporting the operation of the AC power system (from EHV to LV) under normal, alerted, emerging, blackout and restoration states without having to rely on capabilities from SGs [8].

Grid restoration

Restoration of the power supply back to normal operation after a blackout. 'Top-down restoration' means a strategy that requires the assistance of other TSOs to re-energise parts of the system. 'Bottom-up restoration' means a strategy where part of the system of a system can be re-energised with its own resources (black-start capable units) [10].

Inertia

Inertia means the property of a rotating rigid body, such as the rotor of an alternator, such that it maintains its state of uniform rotational motion and angular momentum unless an external torque is applied [2].

378   *Intended and unintended islanding of distribution grids*

Inertia constant H
Ratio of the stored kinetic energy at nominal frequency and the nominal power [7].

Island detection method (IDM)
These procedures aim to detect and in some cases also directly shut down occurring electrical islands. Their goal is to be able to detect all possible islands under real network conditions (different network topologies, types of generation units, etc.) with little or no impact on voltage quality and low costs.

IDMs are divided into two different categories. While passive methods are only evaluating measured values and compare them to thresholds, active methods directly apply changes or small perturbations to the grid and measure the effect for detection.

Island operation
Island operation means the independent operation of a whole network or part of a network that is isolated after being disconnected from the interconnected system, having at least one power-generating module or HVDC system supplying power to this network and controlling the frequency and voltage [2].

Large signal stability, transient stability, of a power system
Power system stability in which disturbances may have large rates of change or large magnitudes [7].

Microgrid
A group of interconnected loads and distributed energy resources within clearly defined electrical boundaries that acts as a single controllable entity with respect to the grid. A microgrid can connect and disconnect from the grid to enable it to operate in both grid-connected or island-mode [11].

Non-detection zone (NDZ)
Totality of all cases in which an occurring electrical island cannot be successfully detected or switched off. The NDZ is often determined for different combinations of active and reactive power generation and consumption.

Point of connection (POC)
See grid connection point.

Point of common coupling (PCC)
The point of common coupling is the point in an electric power system, electrically nearest to a particular load (or generating device), at which other loads are, or may be, connected [3]. Often, the PCC is equivalent to the grid connection point.

Power frequency control
Power frequency control describes a control process with which TSOs maintain the mutually agreed electrical values at the boundaries of their control areas under normal operation and in particular under fault conditions. In this process, each TSO endeavours, by means of an appropriate contribution from his own control area, to

maintain both the interchange with other control areas within the agreed boundaries and the system frequency close to the setpoint value [4].

Power system strength
Power system strength in general is defined as the capability of the grid to preserve its voltage and frequency stability during and after the presence of a disturbance.

Primary control
Primary control is the stabilizing control, operating automatically in the seconds range, of the active power of the complete, coupled, synchronously operated three-phase interconnected network. It is produced from the active contribution of the power stations to changes in system frequency and is supported by the passive contribution of the loads which depend upon the system frequency (self-regulating effect) [4].

Rate of change of frequency (RoCoF)
A measure of the speed with which the frequency of the power system changes. RoCoF increases with low system inertia [9].

RMS simulation
The root-mean-square (RMS) simulation method is based on RMS positive sequence phasor equations to represent the electrical network and is used when only the fundamental frequency behaviour is of interest.

Rolling blackouts
Rolling blackouts are a deliberate shutdown of electricity supply for non-overlapping periods of time in different parts of the distribution region. They are used as a last-resort measure by electric utility companies to avoid a total blackout of the power system. Causes of rolling blackouts include generating capacity being below demand, power station outages, loss of renewable capacity, natural disasters, lack of fuel and conflict.

Secondary control
Secondary control is the influencing, in relation to a specific area, of generating units within a supply system for the purpose of maintaining the desired energy exchange of the control area with the rest of the interconnected system whilst at the same time providing integral frequency back-up control [4].

Self-regulation of loads
Inherent feature of loads to react under certain conditions autonomously on variations of frequency or voltage or both with a change in the power exchange with the electric power network [7].

Short-circuit power or fault level
The short-circuit power (or fault current) is the maximum power (or current) that flows to a given point of the power system in case of fault. Usually, the short-circuit

380 *Intended and unintended islanding of distribution grids*

power is indicated as S_k. More information about the short circuit power calculation can be found in IEC6909-0 [8].

Short-circuit ratio (SCR)
The SCR is the ratio of the short-circuit power of the system at the point of connection (PoC) at which an installation is connected to the system, and the rated power of the installation. In a conventional power system, it indicates the 'strength' of the system in relation to the electrical 'size' of an installation [8].

Stability
Power system stability is the ability of an electric power system, for a given initial operating condition, to regain a state of operating equilibrium after being subjected to a physical disturbance, with most system variables bounded so that practically the entire system remains intact [12].

Steady-state stability
'Steady-state stability' means the ability of a network or a synchronous power-generating module to revert and maintain stable operation following a minor disturbance [2].

Synthetic inertia
Synthetic inertia is the ability of a grid-connected power converter to exchange energy when a frequency event occurs. There is not a standard synthetic inertia implementation, although the responses typically involve measuring the frequency and commanding the converter to temporarily increase its active power output [8].

Transient stability
Should an electric power system which has suffered a 'major' failure progress through decaying transient phenomena to its original steady state, it demonstrates transient stability with regard to the nature, location and duration of this fault. The steady state following a fault may be identical to that before the fault or may differ from it. The non-linear formulae for synchronous machines must be used for analysis of the transient stability. The term 'overall stability' is commonly used in control technology [4].

Voltage control
The purpose of voltage control is to maintain an acceptable voltage profile throughout the network. This is achieved by balancing of the respective reactive power requirements of the network and the customers.

Voltage stability
Voltage stability refers to the ability of a power system to maintain steady voltages at all buses in the system after being subjected to a disturbance [1]. This is strongly connected to the provision of reactive power.

Glossary 381

References

[1] Hatziargyriou N, Milanović J, Rahmann C, *et al.* Stability Definitions and Characterization of Dynamic Behavior in Systems with High Penetration of Power Electronic Interfaced Technologies [Technical Report PES-TR77]. IEEE Power & Energy Society; 2020. Available from: https://resourcecenter.ieee-pes.org/publications/technical-reports/pes_tp_tr77_psdp_stability_051320.

[2] European Commission. Commission Regulation (EU) 2016/631 – Establishing a network code on requirements for grid connection of generators [Commission Regulation]. Brussels, Belgium; 14 April 2016.

[3] IEC. Electropedia: The World's Online Electrotechnical Vocabulary [International Electrotechnical Vocabulary]. Geneva 20, Switzerland: International Electrotechnical Commission (IEC). Available from: https://www.electropedia.org/.

[4] VDN. Transmission Code 2007 – Network and System Rules of the German Transmission System Operators [Transmission Code]. Robert-Koch-Platz 4, 10115 Berlin: Verband der Netzbetreiber VDN e.V. beim VDEW; 2007.

[5] Bayer J, Benz T, Erdmann N, *et al.* Zellulares Energiesystem [Study]. VDE Verband der Elektrotechnik Elektronik Informationstechnik e.V., Energietechnische Gesellschaft (ETG), Frankfurt am Main: VDE; 2019.

[6] Ackermann T, Andersson G, and Söder L. Distributed generation: a definition. *Electric Power Systems Research* 2001;57:195–204.

[7] DKE. 8B/105/CD:2021-11 – IEC TS 62898-3-3 ED1 – Microgrids – Part 3-3: Technical requirements? Self-regulation of dispatchable loads [standard]. Offenbach am Main: DKE Deutsche Kommission Elektrotechnik Elektronik Informationstechnik; 2021.

[8] ENTSO-E. High Penetration of Power Electronic Interfaced Power Sources and the Potential Contribution of Grid Forming Converters [Technical Report]. Brussels, Belgium: ENTSO-E, Wind Europe, SolarPower Europe, T&D Europe; 2020.

[9] IEA. System Integration of Renewables – An update on Best Practice, License: CC BY 4.0 [Report]. Paris: International Energy Agency (IEA); 2018. Available from: https://www.iea.org/reports/system-integration-of-renewables.

[10] European Commission. Commission Regulation (EU) 2017/2196 – Establishing a Network Code on Electricity Emergency and Restoration [Commission Regulation]. Brussels, Belgium; 24 November 2017.

[11] Ton DT and Smith MA. The U.S. Department of Energy's Microgrid Initiative. *The Electricity Journal* 2012;25(8):84–94.

[12] Kundur P, Paserba J, Ajjarapu V, *et al.* Definition and Classification of Power System Stability – IEEE/CIGRE Joint Task Force on Stability Terms and Definitions. *IEEE Transactions on Power Systems* 2004;19(3):1387–1401.

Index

active anti-islanding methods 359
active islanding detection methods
    (AIDM) 201, 359
  fault throwers 208
    earth fault 208
    short circuit 208–9
  frequency shift 205–6
  impedance insertion 208
  impedance measurement 205
  modulation of cos φ/ sin φ 207
  phase shift 207
  Q(f) control 207
active power
  balance 49, 69, 70, 79, 197–8,
    207, 269, 347
  control 16, 23, 267, 322, 341, 377
  reduction 18, 25, 84–5, 95, 126–7,
    192–3, 197, 277, 328
agglomerative hierarchical clustering
    (AHC) technique 226
anti-islanding detection method 361
artificial intelligence 168
asynchronous motor 346, 363,
    366–7
augmented admittance matrix 257–8
Austrian Power Grid (APG) 343
automatic frequency restoration
    reserves (aFRR) 74
automatic voltage controllers (AVCs)
    94
automatic voltage regulator (AVR)
    76, 91, 261, 262, 264
averaged models 38, 61
averaged switch model 39

battery energy storage systems
    (BESS) 139, 269, 295, 322,
    370

Bergeron model 258, 259
biogas plants 332–5
blackstart-capable unit (BSU) 152
blackstart phase 163–4
bottom-up approach 226
bottom-up grid restoration process
    17
  example of practical test of 343
    black-start and provision of
      stable initial configuration
      345–6
    grid configuration during test
      344–5
    grid protection during islanding
      tests 350–1
    start of storage pump in PP1
      346–8
    synchronising island and
      fall-back with power
      imbalance 348–9
    synchronising with second
      island 349–50
    test schedule 343–4
  key issues for 340
    black-start capability
      340
    communication 342
    consideration of distributed
      generation 342
    control requirements 340
    exit procedure 342
    interference with other grids
      342
    maximum load steps 341
    minimum load 341
    protection 341
    reactive power compensation
      341

bottom-up grid restoration process (*continued*)
    resonance and Ferro resonance 342
    resynchronisation 342
    strategy for energising grid 340–1
    training 342–3
    rules for grid restoration in Austria 343
bottom-up restoration 152–3
branch flow model (BFM) 221

Calheta power plant 300
Caribbean island St. Eustatius 288
    droop-based grid-forming control 288–9
    fast fault clearing and voltage stability after short-circuit faults 292–3
    fault current contribution 293
    frequency stability and uninterrupted power supply at sudden genset outage 291–2
    frequency stability at normal operation with large solar irradiance perturbations 290–1
    power quality and operation 293–4
    system operation without genset inertia 289–90
Centre of Inertia (CoI) 309
changes in the constraints (CIGs) 225–6
changing load behaviour 166
Clarke transformation 255–6
classical synchronous machine model 260
cold load pick-up (CLPU) 41, 87, 341
combined cycle gas turbines (CCGT) 127
combined heat and power (CHP) systems 9, 167

composite short-circuit ratio (CSCR) 56
composite ZIP model 259
Continuity equations 264
converter-interfaced generation (CIG) 31
Converter Interfaced Renewable Energy Sources (CI-RES) 310
coordinated reactive power control 372

Danish power system 137
data clustering techniques 226
DC–DC converter 268, 276
decarbonisation 8
decentralisation 8–9, 18, 166–7, 191, 372
detection methods 357
    purpose and principle of 199
    additional detection methods 201
    islanding protection with voltage and frequency thresholds 200–1
diesel engines 321
diesel generator 104, 139, 288, 319, 322–3, 336, 369
differential-algebraic equations (DAEs) 60, 252
differential equations 252
digitalisation 9
DIgSILENT PowerFactory 261
Dijkstra algorithms 336
discrete switch model 39
distributed energy resources (DERs) 215–16, 354
distributed generations (DGs) 3, 8–9, 15, 18, 95–7, 117, 119, 171, 191–2, 319–20, 323, 333, 342, 353
    fault-ride-through 193
    frequency-dependent active power reduction 192–3

distribution grid analysis
  model initialisation and numerical
      integration 253–5
  modelling of electromagnetic
      transient 251–2
  modelling of electromechanical
      interactions 253
  modelling of power system
      dynamics 250–1
  power system modelling 257
    electric grid model 257–8
    hydroelectric power plant
        264–5
    hydrogen storage power plant
        267–9
    load modelling 259–60
    modelling of conventional
        generation 260–1
    modelling of decentralised
        (renewable) generation
        266–7
    thermal power plant 261–4
    transmission and distribution
        line models 258–9
  provision of ancillary services in
      island network with high
      DER infeed 269
    example island network 270–1
    frequency regulation and HSPP
        performance evaluation
        271–6
    frequency regulation during
        high DER infeed 276–8
    frequency regulation without
        HSPPs 278–80
  transformation methods 255–7
distribution grids 11, 247, 250
  static voltage support through
      provision of reactive power
      from 97
    short-circuit current and
        dynamic voltage support
        98–101
  voltage control in 92
    allocation of voltage bandwidth
        95–7
    fundamentals 92–5

distribution networks (DNs) 215,
    295–6, 360
distribution system operators (DSOs)
    12, 87, 165, 247, 322, 339
doubly fed induction generator
    (DFIG) 100
dq coordinate system 256
droop-based grid-forming control
    288–9
dynamic generator models 261
dynamic grid simulators 342
dynamic limits 340
dynamic optimization problem 228
dynamic simulations 237, 250
dynamic voltage support 69,
    98–101

EirGrid 73, 137–8
electrical energy storage systems
    (EESS) 121
electrical grid 17, 75, 83, 92, 132,
    172, 191–3, 195, 201, 257,
    288
electrical islands 3, 12–15, 19, 83,
    102, 191–2, 194, 196–7,
    199–200, 203, 205–6, 208–9,
    212, 223, 357, 369–70
  in DIN VDE V 0126-1-1 15
  in IEEE 1547 14
  intentional electrical island 15
  in Mrugowsky 14–15
  regular electrical island 15
  unintentional electrical island 15
electrical loads 192, 194–5
electrical power systems 12, 32, 140,
    281
electrical storage systems 370
electric grid model 257–8
electricity grids 11, 31, 291, 371
electricity market 11
electric power systems 3–4, 7, 49,
    62, 69, 115, 149, 228
  control of reactive power and
      voltage 89
    generator control 91–2
    relationship between active and
        reactive powers and voltage
        89–91

electric power systems (*Continued*)
  requirements for generation
    units to support operation of
    the power system 97–101
  voltage control in distribution
    grids 92–7
  frequency characteristics of loads
    86
    rolling blackout 88–9
    under frequency-load shedding
    87–8
  frequency control and adjustment
    of generation to consumption
    70
  characteristics and control of
    synchronous generators
    75–83
  control behaviour of
    inverter-based resources
    83–5
  inertia management 71–3
  system frequency 70–1
  types of control reserve and
    provision 73–5
  for stable islanded grid operation
    101
  operation of islanded grid
    106–7
  resynchronising 104–6
  transition to islanded grid
    operation 101–4
Electric Reliability Council of Texas
  (ERCOT) 137
electric vehicles (EVs) 313
electrolysers 10, 268, 277
electromagnetic transient (EMT) 38,
  196, 250–2, 259, 261, 266
electromechanical interactions
  253
Empresa de Eletricidade da Madeira
  SA (EEM) 295
e-mobility 120
EMT-type models 60–1
equivalent circuit-based short-circuit
  ratio (ESCR) 56
Euler equations 264

Expected Energy not Supplied
  (EENS) 301
extra-high-voltage grids 174, 182–3,
  185, 249

fast fault current injection 131, 172
fast frequency response (FFR) 72,
  137
fault current contribution 293
fault ride-through (FRT) 49, 101,
  116, 128, 173, 187, 193, 228,
  229, 249, 294, 307, 359–60
Ferranti effect 155, 164
fibre optic cables (FOC) 204
field test 357–9
first-order approximation 237
flexibilisation 372
flow equation 265
frequency containment reserve (FCR)
  71
frequency control 23, 32–3, 36, 42,
  50–2, 69–70, 80–1, 91–2,
  101, 140–1, 157–9, 287, 290,
  340–1, 344, 346–8, 350, 371
frequency curve 157, 165, 349
frequency-dependent model 258,
  259
frequency reduction 49, 70, 82, 349
frequency response model 230–3
frequency-sensitive mode (FSM)
  122
frequency shift 205, 207, 211, 359,
  365
frequency stability 3, 20, 34, 39, 42,
  44–6, 49–52, 71, 86, 101,
  122, 127, 135–8, 141, 247,
  250, 253, 270, 278, 290–1,
  307, 309, 319, 329–31, 333,
  340, 342, 344, 371
fuel cells 10, 268–9, 276

gas engines 321
gas turbines 74, 127, 153
geothermal plant 302, 304, 305
German Federal Network Agency
  335

German grid codes 130, 320, 322, 327, 332
grid codes (GCs) 12, 22, 72, 115–19, 129–30, 133–5, 137–9, 142, 154, 249, 310, 319–21, 325, 327–8, 332–3, 342, 372
  comparison of selected grid codes 134
    black-start capability 139–40
    frequency stability 135–8
    robustness 138–9
  in electricity system regulation 118–19
  EU network code on requirements for generators 120
    limitation of active power reduction at under-frequency 127
    limited frequency-sensitive mode 124–6
    operating frequency ranges 123
    RoCoF withstand capability 123–4
    system restoration 132–4
  purpose of 119
  tailoring grid connection code requirements to system context 117–18
  variable renewable energy impacts the way power systems operated 116–17
grid-connected electrical island 370, 379
grid-connected inverters 56–8, 266
grid-following (GFL) inverters 38, 56–9, 266, 320–1
grid-following mode 42, 56, 266
grid-forming converter 42, 57, 309
grid-forming inverters 38, 42, 59, 266, 321, 371
grid-forming power converters 309
grid operator 73–4, 101–2, 106, 130, 152, 154, 161–2, 164–5, 174, 176, 183–5, 205, 208, 266, 287, 341–3, 353, 370
grid preparation phase 161–3

grid restoration 4, 17, 19, 24, 39, 42, 50, 54, 102, 165, 189, 320, 322, 339–40, 342–4, 351, 372

Heffron–Phillips model 36
high-voltage direct-current (HVDC) transmission 42–3, 167
high-voltage grids 174, 183, 187, 189
high voltage ride-through (HVRT) 61–2, 128, 311
Hill's model 41
Horizon 2020 SMILE project 314
hydroelectric power plants (HPP) 187, 250, 264–5, 321
hydrogen storage power plant (HSPP) 267–9
  frequency regulation without 278–80
  long time frame response 276
  medium time frame response 276
  short time frame response 274–6

impedance measurement (IDM) 186, 188, 202, 205, 359
impedance reduction 94–5
information and communications technology (ICT) 151, 160–1, 250
Initial-Value Problem of Differential-Algebraic Equations (IVP DAE) 228
insular power systems
  Caribbean island St. Eustatius 288
    droop-based grid-forming control 288–9
    fast fault clearing and voltage stability after short-circuit faults 292–3
    fault current contribution 293
    frequency stability and uninterrupted power supply at sudden genset outage 291–2
    frequency stability at normal operation with large solar irradiance perturbations 290–1

insular power systems (*Continued*)
   power quality and operation
      293–4
   system operation without genset
      inertia 289–90
Madeira Island 294
   generation expansion plan for
      Madeira electric power
      system 297–8
   grid expansion plan for the
      Madeira electric power
      system 298–9
Madeira Archipelago 294–6
Madeira Grid Code 310–13
security of supply through
   reliability and generation
   adequacy assessment
   299–306
sustainable energy action plan
   296–7
system dynamic performance
   306–10
using storage at secondary
   substation for voltage control
   313–16
intended islanding 23–5, 370–1
interconnected power system (IPS)
   3, 86, 91, 116, 152, 251, 253,
   257, 260–1
intertripping 204
inverter-based generation 32–3,
   38–9, 44, 49, 51, 55–6, 167,
   188, 289, 293, 310, 359
inverter-based resources (IBRs) 7–8,
   69, 83, 117, 247, 251, 371
   control behaviour of 83
   control hierarchies 83–4
   frequency-dependent active
      power reduction 84–5
   power sharing in islanded grids
      85–6
inverter control scheme 365
inverter-coupled systems 101, 153–5
inverters
   integrate control 356
   large-signal stability of 58–60

sensibility analysis 361–4
small-signal stability of 58
investment planning with static
   secure islanding constraints
   224
   changes in constraints 225–6
   changes in objective 224–5
   representative days 226–7
islanded grids 3, 31, 69, 116
   energy transition changes power
      grid 7–12
   formation of 16–19
   historic development 4–7
   operation 102
      operation of 106–7
      resynchronising 104–6
      transition to 102–4
   opportunities and threats of 23
      intended islanding 23–5
      unintended islanding 25–6
   power quality aspects in 140
      frequency 141–2
      harmonics 141
      voltage dip 141
      voltage fluctuation 141
      voltage unbalance 141
   power sharing in 85–6
   stable operation of 20
Islanded System Operation 250
island grids 187–9
Islanding Detection Method 354
isochronous control 75

Kärnten Netz GmbH (KNG) 343
k-means clustering technique 233
Kreskas algorithms 336

large-scale grid-forming inverters
   287
leading power plant (LPP) 320–4,
   333
limited frequency-sensitive mode
   (LFSM) 123–4
   LFSM-O 124–6
   LFSM-U 126
LINDA 2.0 333–4

LINDA 4 H$_2$O 334–6
LINDA projects 319
  basic requirements 320–1
  behaviour of DGs according to
    droop concept 326
    background 326–7
    estimation of
      frequency-dependent
      behaviour of PV systems 327
    measured behaviour of a mixed
      PV system population 327–9
  first LINDA project 332–3
  interaction between load and
    frequency 329
  increase in frequency stability
    330–1
  influence of load behaviour on
    frequency stability 329–30
  RoCoF measurement 332
  leading power plant and load
    management 321–4
  LINDA 2.0 333–4
  LINDA 4 H$_2$O 334–6
  main objectives 320
  motivation 319–20
  practical approach for derivation
    of a frequency droop
    characteristic 324–6
linear programming (LP) 237
line-commutated converter (LCC)
  42
line drop compensator (LDC) 94
liquid organic hydrogen carrier
  (LOHC) system 268
load management 106, 321–4
load modelling 259–60
load restoration phase 165
long-term voltage stability 54
Loss of Load Expectation (LOLE)
  301
low-voltage (LV) grid 15, 58, 92,
  95–7, 166, 173–4, 179–80,
  185, 321, 327, 334, 342
low-voltage ride-through (LVRT)
  129, 174

Madeira Archipelago 294
  energy mix 296
  generation 295
  transmission and distribution
    networks 295–6
Madeira Grid code 310–13
Madeira Island 294
  generation expansion plan for
    Madeira electric power
    system 297–8
  grid expansion plan for Madeira
    electric power system 298–9
  Madeira Archipelago 294–6
  Madeira Grid Code 310–13
  power system 310
  security of supply through
    reliability and generation
    adequacy assessment
    299–306
  sustainable energy action plan
    296–7
  system dynamic performance
    306–10
  using storage at secondary
    substation for voltage control
    313–16
Manual Frequency Restoration
  Reserves (mFRR) 74
Marktstammdatenregister (MaStR)
  324, 335
maximum power point (MPP) 269
mechanical power equation 265
medium-voltage grids 58, 174,
  183–5, 188–9
microgrids (MGs) 7, 215
mixed-integer linear programming
  (MILP) 236
mixed-integer non-linear
  programming (MINLP) 234

Nadir constraint 236
network power frequency
  characteristic 21, 81–2
neutral voltage displacement (NVD)
  208

non-detection zone (NDZ) 195–6,
366
calculation of simple arrangement
196–8
Non-dominated Sorting Genetic
Algorithm-II (NSGA-II)
algorithm 303
non-linear programming (NLP)
techniques 221
Nordic synchronous system 137
Norton equivalents 258

observability 10, 63
off-grid electrical islands 369
offshore grids 176
Ohm's law 257
on-load tap changer (OLTC) 41, 92
onshore grids 179–81
operational management 21, 69, 101,
107, 266, 351
operational planning with static
secure islanding constraints
217
day-ahead optimal planning model
with static islanding
constraints 218
AC power flow model 219
objective 218–19
static operational/technical
constraints 219–21
network modelling for power flow
constraints 221
adapted DistFlow relaxation
222
augmented DistFlow with line
shunts 223–4
extended DistFlow relaxation
with line shunts 223
modified Lin-DistFlow
relaxation 222

Park transformation 255
passive islanding detection methods
(PIDM) 201–5
Pelton turbines 340–1, 345

phase-locked loop (PLL) 32, 203,
255, 266, 363
phase shift 207
phase-to-phase short-circuit current
180
phasor measurement units (PMUs)
43
photovoltaic (PV) systems 7, 11, 88,
100, 106, 155, 166, 250, 269,
288, 296–7, 313, 315, 319,
322–8, 333, 353–7, 363, 365,
370
planning phase 161–2, 340
point of common coupling (PCC) 14,
97, 154, 216, 323, 354, 378
point of connection (PoC) 21, 55,
207, 349
positive- and negative-sequence
mode (PNSM) 172, 185
positive sequence mode (PSM) 172,
174
power electronic inverters 115
power-flow models 60, 62–3, 219
power line carrier communications
(PLCC) 205
power park modules (PPM) 120
power plant operators 12, 152, 162,
332, 339, 341–3
power quality and operation
293–4
power system dynamics 31–4, 60,
63–4, 250–1, 254, 261
power system modelling 257
electric grid model 257–8
hydroelectric power plant 264–5
hydrogen storage power plant
267–9
load modelling 259–60
modelling of conventional
generation 260–1
modelling of decentralised
(renewable) generation
266–7
thermal power plant 261–4
transmission and distribution line
models 258–9

power system restoration
  ancillary services and secondary
    technology during 153
    blackstart 153–4
    frequency control 157–9
    information and communication
      technology 160–1
    protection 160
    voltage control 154–7
  different phases of 161
    blackstart phase 163–4
    grid preparation phase 163
    load restoration phase 165
    planning phase 162
    resume normal operation 165
    system and network restoration
      phase 164–5
  emergency backup supply 152
  external impact due to renewable
    energy transition 166
    artificial intelligence in power
      system operation 168
    changing load behaviour 166
    decentralisation 166
    decreasing distinction between
      system and load restoration
      167
    HVDC 167
    inverter-based generation and
      load 167
    potential for distribution system
      islands 167
    start-up times 166
    weather dependency 166
  general 149
  historical blackouts 149–51
  restoration strategies 151
    bottom-up 152–3
    top-down 152
  techno-economic trade-off 151–2
power systems 31
  dynamics 32
    dynamic model basics 32–3
    high-voltage direct-current
      transmission 42–3

    inverter-based generation 38–9
    load 39–42
    main components involved in
      power systems dynamics
      33–4
    protection 43
    synchronous generators 34–7
    time/frequency range of concern
      33
    wide area monitoring and
      stability assessment 43–4
  frequency stability 49–52
  rotor angle stability 45
    steady-state rotor angle stability
      46–7
    transient rotor angle stability
      47–9
  simulation and analysis models
    and method 60
    comparison of models 63–5
    EMT-type models 60–1
    modal analysis 63
    power-flow models 62–3
    RMS-type dynamic models
      61–2
  stability 5, 44
    general definition of 44–5
    impact of high penetration of
      inverter-based generation on
      56–60
    relevance of stability
      phenomena 45
  system strength 55
  voltage stability 52–5
power system stabilizer (PSS) 261
power to heat (P2H) 10, 269
pressure head equation 265
primary control 12, 20–1, 71, 73–5,
  78, 82, 86, 118, 136, 173,
  268, 293
PROINVER project 354
protection 171
  behaviour of inverters in different
    grid constellations 176
  onshore grids 179–81

protection (*Continued*)
  wind power plant–offshore grids
    176–9
  development of short-circuit
    currents 181–3
  grid protection concepts and
    short-circuit contributions of
    183
    extra-high-voltage grids and
      high-voltage grids 183
    medium-voltage grids 183–5
  of inverter-based generators
    185–7
  for island grids 187–9
  short-circuit behaviour of inverters
    172
    feed-in of reactive current in
      positive and negative
      sequence 175–6
    feed-in of reactive current in
      positive sequence 174–5
    switch-off mode in event of
      fault 173
    zero-power mode 173–4
protection of inverters 356

Rain hydropower plant 333
rate of change of frequency (RoCoF)
    31, 123–4, 142, 203–4, 232,
    241, 247, 292, 307, 332, 340,
    358
reactive power capability 34, 84,
    121, 129–31, 154–6
reactive power control 23, 95, 97–8,
    131, 155–6, 314–15, 372
reactive power mode 154–6
reference values 38, 42, 56–7, 194,
    289, 349, 361
remote industrial operations 370
renewable energy generation systems
    189
renewable energy sources 7, 116,
    266, 269, 280–1, 287, 297,
    298, 309, 316, 320, 322
resonance stability 44, 372

resume normal operation 162,
    165
resynchronisation 24, 102, 104–5,
    150, 327–8, 342–4, 349
Ringhals nuclear power plant 151
RMS-type dynamic models 61–2
robustness 58, 127–32, 138–9, 140,
    164–5, 233
robust transmission 298
rolling blackout 87–9
root mean square (RMS) models 38,
    250

Sandia Frequency Shift (SFS) 354
secondary control 20–1, 73–5, 82–3,
    157, 159, 165, 192, 267–8,
    273–4, 277
secondary controllable generators 82
sector coupling 10, 269, 278, 320,
    371–2
selected loads
  frequency dependence of 40
  load self-regulation effect of 40
self-regulation effect 20, 39, 40, 52,
    71, 86, 330
semi-adaptive scheme 87
Sequential Monte Carlo Simulation
    (SMCS) 302
short-circuit capacity 21, 31, 142,
    184, 248, 341–2, 351
short-circuit current 20–1, 34, 55,
    61–2, 69, 98–101, 160, 167,
    171, 173, 175, 177, 179–81,
    183–9, 208, 372
  development of 181–3
short-circuit ratio (SCR) 3, 20–1,
    55–6, 100, 248, 320–1
short-term voltage stability 54–5,
    61–2, 98
simple source model 39
simplified/reduced model 229
simulation models 248–51, 253–4,
    266, 331–2, 363, 372
solar farms 297
Spanish power system 353
sparse matrix 257

stability analysis 39, 43, 45–6, 48, 53, 62
stable frequency 20
stable islanded grid 25, 101–107, 370–371
stable operation 18, 20–23, 59, 83, 97, 102, 140, 255, 270–271, 291, 320–321, 369
stable voltage 42, 140, 153, 339
stand-alone generator 75–7
start-up grids 17, 19, 152–3, 155
start-up times 166
STATCOM function 154
Statia Utility Company (STUCO) 288
static reference system 256
static reserve probability distribution 301
static security 62, 215, 218, 227, 234, 238
storage systems 9, 12, 95, 120, 126, 139, 167, 250, 269, 281, 295, 298, 321–2, 325, 330, 369–70
sub-grids 3, 12–13, 15–20, 24, 101–4, 116, 153, 164
  definition of electrical islands in literature 14–15
  degrees of decoupling sub-grids 13–14
  disconnection of 12–13
  formation of 16–19
  stable operation of 20
    grid-forming and system-supporting properties of generation units 22–3
    power system strength, rotating inertia, and short-circuit ratio 20–2
sub-synchronous control interactions (SSCI) 64
sub-synchronous resonance (SSR) 64
supervisory control and data acquisition (SCADA) systems 43
swing equation 35–6, 260
switch-off mode (SOM) 172–3

switch-off threshold 328
synchronous generators 34–7, 56–8, 69, 171, 261, 341, 371
  characteristics and control of 75
  droop control characteristic and generator power frequency characteristic 77–9
  parallel operation of two and more generators 79–80
  resulting droop of parallel generators 80–2
  secondary control 82–3
  stand-alone generator 75–7
synchronous power-generating modules (SPGMs) 120
system frequency 61, 70–1, 73, 81, 103, 123–4, 138, 258, 272–4, 278–80, 291, 346, 379
system restoration 17, 87, 101, 103, 132, 134, 149, 152–3, 159–62, 166–8, 339

Taylor expansion 63, 236
techno-economic trade-off 151–2
thermal power plant (TPP) model 261–4
three-phase systems 255, 269, 365
time-domain models 60
top-down approaches 339
top-down restoration 152
Torricelli equation 265
total harmonic distortion (THD) 202–3
transfer function 230–1, 267
transformation methods 255–7
transient islanding constraints in planning problems 227
  case study results 238
    dynamic performance 240–242
    planning costs 239–240
  decomposition algorithm for solving MG planning problem with dynamic constraints 233–238
  frequency response model 230–233

transmission and distribution line
models 258–9
transmission system operators
(TSOs) 12, 151, 339
transparency 10, 117, 119
type 1 units 171
type 2 units 171

under frequency load-shedding
(UFLS) schemes 86–8
uniform system frequency dynamics
model 231
unintended islanded grids
field test 357–9
inverter sensibility analysis 361–4
modelling of real network 364–7
operational experience on
islanding events caused by
large photovoltaic plants 354
damaged inverters 356
dangerous work conditions 355
failure of network automation
356
impossibility of network
operation or maintenance
355–6
potential solutions 359–60
technical reasons 356
behaviour of islanding network
357
behaviour of photovoltaic
inverters 356–7
unintended islanding 25–6, 186, 189,
370–1
unintentional islanding detection
description of selected islanding
detection methods 202
communication 204–205
detection of voltage harmonics
202–3
frequency shift 205–206
impedance insertion 208–209
impedance measurement 205
modulation of cos φ/sin φ
207–208
phase jump detection 204

phase shift 207
Q(f) control 207
rate of change of frequency
203–4
voltage and frequency
thresholds 202
occurrence of unintentional
electrical islands 191
approach for simple islanding
scenarios 195
behaviour of distributed
generation units 192–3
behaviour of electrical loads
194–5
definition of non-detection zone
195–6
influence of P(f) and real load
model 198–9
NDZ calculation of simple
arrangement 196–8
phases of electrical islands 192
Union for the Co-ordination of
Production and Transmission
of Electricity (UCPTE) 5

Verbund Hydro Power (VHP) 343
Virtual Synchronous Machine (VSM)
control schemes 217
voltage control 22–3, 54, 69, 92, 95,
97, 101–3, 116, 129, 132,
154–7, 215, 221, 306, 313–14,
340, 370, 380
voltage measurement criterion 186
voltage range 41, 92, 127–9, 134
voltage-regulated distribution
transformers (VRDTs) 94,
96
voltage source converters (VSC) 42
voltage source inverter (VSI) 365
voltage stability 52–5, 127–32

weather dependency 166
weighted short-circuit ratio (WSCR)
56
Western Electricity Coordinating
Council (WECC) 266

wide area measurement system
(WAMS) 43
wind power plants (WPP) 9, 121,
138, 176, 187, 250
wind turbines 72, 138–9, 159, 166,
176, 188, 266, 297, 307,
369–70

zero-crossing algorithm 332
zero-power mode (ZPM) 172–4
zero-sequence current 173
ZIP model 40–1, 195, 259

Printed in the USA
CPSIA information can be obtained
at www.ICGtesting.com
JSHW051919241024
72254JS00002BA/4